Riemannian Geometry

Oxford University Press, Walton Street, Oxford OX2 6DP
Oxford   New York
Athens   Auckland   Bangkok   Bombay
Calcutta   Cape Town   Dar es Salaam   Delhi
Florence   Hong Kong   Istanbul   Karachi
Kuala Lumpur   Madras   Madrid   Melbourne
Mexico City   Nairobi   Paris   Singapore
Taipei   Tokyo   Toronto
and associated companies in
Berlin   Ibadan

Oxford is a trade mark of Oxford University Press

Published in the United States
by Oxford University Press Inc., New York

© T. J. Willmore, 1993
First published 1993
First published in paperback 1996 (with corrections)

All rights reserved. No part of this publication may be
reproduced, stored in a retrieval system, or transmitted, in any
form or by any means, without the prior permission in writing of Oxford
University Press. Within the UK, exceptions are allowed in respect of any
fair dealing for the purpose of research or private study, or criticism or
review, as permitted under the Copyright, Designs and Patents Act, 1988, or
in the case of reprographic reproduction in accordance with the terms of
licences issued by the Copyright Licensing Agency. Enquiries concerning
reproduction outside those terms and in other countries should be sent to
the Rights Department, Oxford University Press, at the address above.

This book is sold subject to the condition that it shall not, by way of trade
or otherwise, be lent, re-sold, hired out, or otherwise circulated without
the publisher's prior consent in any form of binding or cover other than
that in which it is published and without a similar condition including
this condition being imposed on the subsequent purchaser.

A catalogue record for this book is available from the British Library

Library of Congress Cataloging in Publication Data
Willmore, T. (Thomas), 1919–
Riemannian geometry/T. J. Willmore.
1. Geometry, Riemannian.   I. Title.
QA649.W53   1993   516.3'73–dc20   92-45861

ISBN 0 19 851492 1

Printed in Great Britain on acid-free paper by
Bookcraft Ltd., Midsomer Norton, Avon

Riemannian Geometry

T. J. WILLMORE
Emeritus Professor of Pure Mathematics
University of Durham

CLARENDON PRESS · OXFORD

To my wife, Joyce, who has encouraged me constantly through 52 years of married life.

Preface

It is assumed that the reader has some knowledge of linear algebra and advanced calculus such as is contained in the early chapters of Boothby (1986). The necessary results are carefully stated so that a reader who accepts them should be able to understand almost the entire book with minimal prerequisites.

Some selection of material is inevitable in treating such a vast subject — an obvious omission is the treatment of holomony groups but this is deliberate because of the comprehensive study of this topic in the book by Salamon (1989). Our last chapter gives the first account in book form of the theory of Willmore surfaces, a subject which is proving of widespread interest in current research as can be seen from the bibliography.

It is impossible to write a book of this nature without receiving help from numerous friends, colleagues, and students. In particular I owe a debt to A. G. Walker, A. Lichnerowicz, S. S. Chern, A. Gray, and L. Vanhecke. I thank D. Wilkins for his assistance in dealing with twistor spaces, and D. Ferus, U. Pinkall, and R. Bryant for their interest and distinguished research activities in developing ideas associated with Willmore surfaces. In particular I thank U. Pinkall for providing the material for Figs 7.1 to 7.13 and A. Bobenko for Figs 7.14 and 7.15.

Finally I thank the staff of Oxford University Press for guiding the work from my manuscript to the finished product.

Durham T. J. W.
May 1993

Contents

1 **Differentiable manifolds** — 1
  - 1.1 Preliminaries — 1
  - 1.2 Differentiable manifolds — 2
  - 1.3 Tangent vectors — 5
  - 1.4 The differential of a map — 10
  - 1.5 The tangent and cotangent bundles — 14
  - 1.6 Submanifolds; the inverse and implicit function theorems — 14
  - 1.7 Vector fields — 16
  - 1.8 Distributions — 18
  - 1.9 Exercises — 20

2 **Tensors and differential forms** — 21
  - 2.1 Tensor product — 21
  - 2.2 Tensor fields and differential forms — 27
  - 2.3 Exterior derivation — 28
  - 2.4 Differential ideals and the theorem of Frobenius — 32
  - 2.5 Orientation of manifolds — 34
  - 2.6 Covariant differentiation — 38
  - 2.7 Identities satisfied by the curvature and torsion tensors — 42
  - 2.8 The Koszul connexion — 43
  - 2.9 Connexions and moving frames — 48
  - 2.10 Tensorial forms — 54
  - 2.11 Exercises — 55

3 **Riemannian manifolds** — 56
  - 3.1 Riemannian metrics — 56
  - 3.2 Identities satisfied by the curvature tensor of a Riemannian manifold — 58
  - 3.3 Sectional curvature — 60
  - 3.4 Geodesics — 65
  - 3.5 Normal coordinates — 72
  - 3.6 Volume form — 79
  - 3.7 The Lie derivative — 82

	3.8	The Hodge star operator	84
	3.9	Maxwell's equations	93
	3.10	Consequences of the theorems of Stokes and of Hodge	95
	3.11	The space of curvature tensors	99
	3.12	Conformal metrics and conformal connexions	105
	3.13	Exercises	113

4 Submanifold theory 115

	4.1	Submanifolds	115
	4.2	The equations of Gauss, Codazzi, and Ricci	121
	4.3	The method of moving frames	125
	4.4	The theory of curves	129
	4.5	The theory of surfaces	132
	4.6	Surface theory in classical notation	139
	4.7	The Gauss–Bonnet theorem	142
	4.8	Exercises	151

5 Complex and almost-complex manifolds 152

	5.1	Complex structures on vector spaces	152
	5.2	Hermitian metrics	155
	5.3	Almost-complex manifolds	157
	5.4	Walker derivation	164
	5.5	Hermitian metrics and Kähler metrics	166
	5.6	Complex manifolds	168
	5.7	Kähler metrics in local coordinate systems	170
	5.8	Holomorphic sectional curvature	174
	5.9	Curvature and Chern forms of Kähler manifolds	177
	5.10	Review of complex structures on vector spaces	178
	5.11	The twistor space of a Riemannian manifold	180
	5.12	The twistor space of S^4	181
	5.13	Twistor spaces of spheres	184
	5.14	Riemannian immersions and Codazzi's equation	185
	5.15	Isotropic immersions	187
	5.16	The bundle of oriented orthonormal frames	189
	5.17	The relationship between the frame bundle and the twistor space	190
	5.18	Integrability of the almost-complex structures on the twistor space	192
	5.19	Interpretation of the integrability condition	195
	5.20	Exercises	200

6	**Special Riemannian manifolds**	**201**
6.1	Harmonic and related metrics	201
6.2	Normal coordinates	202
6.3	The distance function Ω	206
6.4	The discriminant function	208
6.5	Geodesic polar coordinates	210
6.6	Jacobi fields	213
6.7	Definition of harmonic metrics	223
6.8	Curvature conditions for harmonic manifolds	228
6.9	Consequences of the curvature conditions	234
6.10	Harmonic manifolds of dimension 4	240
6.11	Mean-value properties	246
6.12	Commutative metrics	253
6.13	The curvature of Einstein symmetric spaces	255
6.14	D'Atri spaces	257
6.15	Exercises	267
7	**Special Riemannian submanifolds**	**269**
7.1	Introduction	269
7.2	Surfaces in E^3	270
7.3	Conformal invariance	276
7.4	The Euler equation	279
7.5	Willmore surfaces and minimal submanifolds of S^3	283
7.6	Pinkall's inequality	284
7.7	Related work by Robert L. Bryant	287
7.8	The conformal invariants of Li and Yau, and of Gromov	291
7.9	Contributions from the German school and their colleagues	293
7.10	Other recent contributions to the subject	300
7.11	Exercises	304
Appendix: Partitions of unity		305
Bibliography		311
Index		317

1
Differentiable manifolds

1.1 PRELIMINARIES

Differentiable manifolds and their geometry appear naturally in the study of diverse areas of mathematics, including Lie group theory, homogeneous spaces, probability theory, differential equations, the theory of functions of a single complex variable and of several complex variables, algebraic geometry, classical mechanics, relativity theory, and the theory of elementary particles of physics. Riemannian geometry is a special geometry associated with differentiable manifolds and has many applications to physics. It is also a natural meeting place for several branches of mathematics.

Because of the vast domain of the subject we shall be selective and in particular we shall confine our studies to elementary topics, though we shall indicate where the reader should go for further information about the deeper and more sophisticated items. There are several different techniques available for tackling problems in differential geometry, including classical tensor calculus, the so-called suffix-free calculus, the calculus of exterior forms, and the method of moving frames. Each technique has its own devotees and sometimes discourteous noises are made about devotees of rival techniques. We shall adopt a more catholic outlook and use whatever technique seems more appropriate to each particular problem. We do believe, however, that the calculus of exterior forms has been undervalued in applications to physics and, in particular, to engineering. At times we sympathize with the view of the late Sir Arthur Eddington when he remarked that apparently the Creator of the universe preferred antisymmetry to symmetry!

Classically differential geometry consisted of the study of curves and surfaces embedded in three-dimensional Euclidean space by means of the differential and integral calculus. Intensive study of surfaces showed that some of the properties were intrinsic, that is, they depended in some sense upon the surface itself and not on the way in which it was embedded in the containing Euclidean space. Other properties were extrinsic and definitely depended on the way that the surface was embedded. Indeed, some curvatures initially defined extrinsically were later shown to be intrinsic invariants. This led to a need to define a surface 'in its own right', and the problems arising from embedding the surface in Euclidean space were another matter. The development of classical mechanics led to the need to consider analogues to surfaces but with dimensions greater than two. The concept of a differentiable manifold finally evolved and this clarified matters which were previously confusing. In a similar way the concept of a fibre bundle which arose during the late 1940s made 'natural' ideas which previously had appeared rather strange. Similarly the modern definition of a connexion clarifies situations which previously appeared mysterious. However the student who meets the subject for the first time in the concise form of definition, theorem, proof may find equally mysterious why the definitions and concepts were chosen in that particular form. We shall try to motivate our definitions and concepts and this will inevitably lead to a less concise treatment. We hope, however, that it will lead to a better understanding of the subject matter.

1.2 DIFFERENTIABLE MANIFOLDS

Nearly all the geometric objects studied in this book live on differentiable manifolds, so we begin our studies by examining the concept of manifold and differentiable structures associated with it.

Let d be an integer satisfying the condition $d \geq 1$. Let $\mathbb{R}^d$ denote the d-dimensional vector space over the real numbers, that is, $\mathbb{R}^d$ denotes the set of real numbers $(a_1, \ldots, a_d)$. In particular when $d = 1$ we denote $\mathbb{R}^1$ by $\mathbb{R}$, the real line. The point $(0, \ldots, 0)$ is the origin of $\mathbb{R}^d$ and will be denoted by O. The function $r_i: \mathbb{R}^d \to \mathbb{R}$ given by $(a_1, \ldots, a_d) \to a_i$ is called the ith *coordinate function* on $\mathbb{R}^d$.

Let X be some set and let $f: X \to \mathbb{R}^d$ be some map. Then $f_i = r_i \circ f$ is called the ith *component* of f.

We shall assume that the reader is familiar with the differential calculus of real-valued functions of several real variables. In terms of the notation above, if $f: \mathbb{R}^n \to \mathbb{R}: (t_1, \ldots, t_n) \to f(t_1, \ldots, t_n)$ is a differentiable function, then the partial derivative of f with respect to r_i is given by

$$\frac{\partial}{\partial r_i}\bigg|_t f = \frac{\partial f}{\partial r_i}\bigg|_t = \lim_{h \to 0} \frac{f(t_1, \ldots, t_{i-1}, t_i + h, t_{i+1}, \ldots, t_n) - f(t)}{h}.$$

By a cubic neighbourhood $C(r)$ of the origin in $\mathbb{R}^d$ we mean the set $C(r) = \{(a_1, \ldots, a_d): |a_i| < r \text{ for } i = 1, \ldots, d\}$.

If $\alpha = (\alpha_1, \ldots, \alpha_d)$ is a set of d non-negative integers, we set $[\alpha] = \Sigma \alpha_i$, and $\partial^\alpha/\partial r^\alpha = \partial^{[\alpha]}/\partial r_1^{\alpha_1}, \ldots, \partial r_d^{\alpha_d}$. For consistency if $\alpha = (0, \ldots, 0)$ we set $\partial^\alpha f/\partial r^\alpha = f$. With this notation settled we proceed with the definition of a differentiable manifold. We wish to define a manifold in such a way that it retains many of the properties of

1.2 Differentiable manifolds

Euclidean space. Moreover we wish to talk about the continuity of objects which live on the manifold, so clearly we must start with a topological space and add further restrictions. We shall assume that the reader is familiar with the idea that a topological space is determined by its collection of open sets, and that such a space is called Hausdorff if it has the property that given any two points of the space one can find open neighbourhoods of each which have empty intersection. The main reason why we ask for the Hausdorff condition is to ensure that a convergent sequence of points can have no more than one limit point.

The essential point about a manifold is that each point has a neighbourhood which 'looks like' Euclidean space. We already know what we mean by a continuous map from $\mathbb{R}^d$ to $\mathbb{R}^q$ where d and $q \in \mathbb{Z}$. We say that a space M is *locally Euclidean of dimension d* if M is a Hausdorff topological space for which each point has a neighbourhood homeomorphic to an open set of $\mathbb{R}^d$. Let ϕ be a homeomorphism of a connected open set U of M onto an open subset of $\mathbb{R}^d$. Then the pair (U, ϕ) is called a *coordinate system* or *local chart*, and the functions $x_i = r_i \circ \phi$ are called *coordinate functions*. If m is a point of U and if $\phi(m) = 0$, we say that the coordinate system is *centred* at m.

A differentiable structure $\mathcal{F}$ of class $C^k (1 \leq k \leq \infty)$ on a locally Euclidean space M is a collection of coordinate systems $\{U_\alpha, \phi_\alpha): \alpha \in I\}$ such that

(i) $\bigcup_{\alpha \in I} U_\alpha = M$;
(ii) $\phi_\alpha \circ \phi_\beta^{-1}$ is C^k for all $\alpha, \beta \in I$;
(iii) the collection is *maximal* with respect to (i) and (ii); that is, if (U, ϕ) is a coordinate system such that $\phi_\alpha \circ \phi^{-1}$ and $\phi \circ \phi_\alpha^{-1}$ are C^k for all $\alpha \in I$, then $(U, \phi) \in \mathcal{F}$.

In practice it is sufficient to consider coordinate systems $\mathcal{F}_0$ which satisfy just (i) and (ii) because then there exists a uniquely determined maximal system $\mathcal{F}$ satisfying (i), (ii), and (iii) such that $\mathcal{F} \supset \mathcal{F}_0$. This follows by defining $\mathcal{F}$ as follows:

$$\mathcal{F} = \{(U, \phi): \phi \circ \phi_\alpha^{-1} \text{ and } \phi_\alpha \circ \phi^{-1} \text{ are } C^k \text{ for all } \phi_\alpha \in \mathcal{F}_0\}.$$

We can now give a precise definition of the type of space which forms the background for our deliberations.

Definition (1.2.1) A d-dimensional differentiable (or differential) manifold of class C^k is a pair $(M, \mathcal{F})$ where M is a d-dimensional, second countable, locally Euclidean space and $\mathcal{F}$ is a differentiable structure of class C^k.

We remind the reader that 'second countable' means that M as a topological space admits a countable basis of open sets. We shall see later that this condition implies that M is paracompact and hence admits partitions of unity—a valuable property as we later discover.

An analytic structure is one for which the maps $\phi_\alpha \circ \phi_\beta^{-1}$, for all α, β, are analytic in the sense of being expressed locally as a convergent power series. An analytic manifold $(M, \mathcal{F}^\omega)$ is similarly defined except that $\mathcal{F}^\omega$ is now an analytic structure.

Examples of differentiable manifolds

1. $\mathbb{R}^d$ is seen to be a differentiable manifold by taking for $\mathcal{F}$ the maximal collection containing $(\mathbb{R}^d, i)$ where i is the identity map.
2. A finite d-dimensional vector space V is a d-dimensional differentiable manifold. To see this, let $\{e_i\}$ be a basis of V and let $\{r_i\}$ be the dual basis. Then $\{r_i\}$ are the coordinate functions of a global coordinate system on V, which canonically determines a differentiable structure $\mathcal{F}$ on V. Moreover this structure is independent of the choice of basis, since different bases yield C^ω-coordinate systems.
3. An open set U of a differentiable manifold $(M, \mathcal{F})$ is itself a differentiable manifold with differentiable structure

$$\mathcal{F}_U = \{(U_\alpha \cap U), \phi | (U_\alpha \cap U): (U, \phi_\alpha) \in \mathcal{F}\}.$$

4. The general linear group $\mathrm{Gl}(n, \mathbb{R})$ is a differentiable manifold. This is the set of all non-singular $n \times n$ matrices each of which can be represented by a point in $\mathbb{R}^{n^2}$. Since the determinant is a continuous function on $\mathbb{R}^{n^2}$, $\mathrm{Gl}(n, \mathbb{R})$ becomes a differentiable manifold as the open set of $\mathbb{R}^{n^2}$ where the determinant function is non-zero.

These examples are all analytic manifolds and, a fortiori, C^∞-manifolds. However we prefer to consider mainly C^∞-manifolds, as we shall see later that several powerful techniques are available for use with C^∞-manifolds but not within the context of analytic manifolds. Moreover the study of C^k-manifolds for finite k introduces problems of analysis associated with loss of differentiability. Since we are primarily interested in geometry and do not wish to be sidetracked with analytical difficulties, we again prefer to consider mainly C^∞-manifolds.

Product manifolds

Suppose we are given two differentiable manifolds $(M, \mathcal{F})$ and $(M', \mathcal{F}')$ of dimensions d and d', respectively. Then $M \times M'$ becomes a differentiable manifold of dimension $(d + d')$ by assigning to it the differentiable structure determined as the maximal collection containing

$$\{(U_\alpha \times V_\beta, \phi_\alpha \times \psi_\beta): U_\alpha \times V_\beta \to \mathbb{R}^d \times \mathbb{R}^{d'}: (U_\alpha, \phi_\alpha) \in \mathfrak{F}, (V_\beta, \psi_\beta) \in \mathfrak{F}'\}.$$

Differentiable maps

Let U be an open subset of $\mathbb{R}^d$ and let f be a real-valued function defined on U. We say that f is differentiable of class C^k on U, where k is an integer, if the partial derivatives of all orders less than or equal to k exist and are continuous on U. Thus we require $\partial^\alpha f / \partial r_\alpha$ to exist and be continuous for $[\alpha] \leqslant k$. In particular f is C^0 if f is continuous.

Note that the function $f: \mathbb{R}^2 \to \mathbb{R}$ defined by

$$f(x, y) = xy/(x^2 + y^2), \quad x^2 + y^2 \neq 0; \ f(0, 0) = 0,$$

admits both partial derivatives $\partial f/\partial x$, $\partial f/\partial y$ at $(0, 0)$ although f is not continuous there. Mere existence of partial derivatives is too weak a property for most geometric considerations.

We say that a mapping $f: U \to \mathbb{R}^n$ is differentiable of class C^k if each component $f_i = r_i \circ f$ is of class C^k. We say that f is of class C^∞ if it is of class C^k for all k.

In order to define differentiable maps from one differentiable manifold M to another M' we use local charts to reduce the problem to consideration of maps of $\mathbb{R}^d$ to $\mathbb{R}^n$.

Let U be an open subset of M and let $f: U \to \mathbb{R}$ be a map. We say that f is a C^∞-function on U (denoted by $f \in C^\infty(U)$) if f composed with ϕ^{-1} is C^∞ for each coordinate map on M.

More generally, if $(M, \mathfrak{F})$, $(M', \mathfrak{F}')$ are two differentiable manifolds, we say that a map $\psi: M \to M'$ is C^∞ if and only if $\phi \circ \psi \circ \tau^{-1}$ is C^∞ for each coordinate map τ on M and ϕ on M'.

In the next section we define tangent vectors and cotangent vectors at points on differentiable manifolds and investigate their behaviour under differentiable maps.

1.3 TANGENT VECTORS

Let v be a vector with components $v_1, \ldots, v_d$ at a point p of Euclidean space $\mathbb{R}^d$. Let f be a real-valued function which is differentiable in some neighbourhood of p. Then we may associate with v and f the real number

$$v(f) = v_1 \left.\frac{\partial f}{\partial r_1}\right|_p + \cdots + v_d \left.\frac{\partial f}{\partial r_d}\right|_p,$$

so that v can be regarded as an operator on differentiable functions. The number $v(f)$ is called the *directional derivative* of f in the direction v.

Let f and g be two differentiable functions defined in some neighbourhood of p and let λ be a real number. Then the directional derivative satisfies

$$v(f + \lambda g) = v(f) + \lambda v(g),$$
$$v(fg) \quad = f(p)v(g) + g(p)v(f).$$

These properties show that v is a linear derivation. This motivates our definition of a tangent vector at a point p on a manifold M, that is, we think of a tangent vector as a derivation acting on functions differentiable in some neighbourhood of p.

Germs

Let real-valued functions f and g be defined on open sets of M containing the point m. We say that f and g have the same *germ* at m if they agree on some neighbourhood of m. This gives rise to an equivalence relation on C^∞-functions defined on neighbourhoods of m, two functions being equivalent if and only if they have the same germ. The equivalence classes are called *germs*. Denote the set of germs at m by $\tilde{F}_m$. If f is a C^∞-function defined in a neighbourhood of m we denote the corresponding germ by $\mathbf{f}$. The value of the germ $\mathbf{f}(p)$ is determined by the value of any representative function f at p. The set of germs becomes an algebra over $\mathbb{R}$ with respect to addition, scalar multiplication, and multiplication of functions.

We denote by $F_m \subset \tilde{F}_m$ the subset of germs which vanish at m. Clearly F_m is an ideal in $\tilde{F}_m$ and we denote by F_m^k the ideal of $\tilde{F}_m$ consisting of all finite linear combinations of k-fold products of F_m. We have

$$\tilde{F}_m \supset F_m \supset F_m^2 \supset F_m^3 \cdots.$$

Definition (1.3.1) *A tangent vector v at a point $m \in M$ is a linear derivation of the algebra $\tilde{F}_m$.*

Thus for all $\mathbf{f}, \mathbf{g} \in \tilde{F}_m$ we have

$$v(\mathbf{f} + \lambda \mathbf{g}) = v(\mathbf{f}) + \lambda v(\mathbf{g})$$
$$v(\mathbf{fg}) \quad = \mathbf{f}(m)v(\mathbf{g}) + \mathbf{g}(m)v(\mathbf{f}).$$

We denote by M_m the set of all tangent vectors to M at m, and call M_m the *tangent space* to M at m. The set M_m becomes a real vector space by defining addition and scalar multiplication in the obvious way:

1.3 Tangent vectors

$$(v + w)(\mathbf{f}) = v(\mathbf{f}) + w(\mathbf{f}),$$
$$(\lambda v)(\mathbf{f}) = \lambda v(\mathbf{f}),$$

for $v, w \in M_m$ and $\lambda \in \mathbb{R}$.

If our definition of tangent vector is sensible, we want the dimension of M_m as a vector space to be equal to the dimension of M. Moreover, in the particular case when M is $\mathbb{R}^d$, we want this definition to be consistent with what we normally mean by a vector in $\mathbb{R}^d$. We now verify that our definition does in fact satisfy these requirements.

We first prove that if v is a tangent vector and $\mathbf{c}$ is the germ of a function which has constant value c in a neighbourhood of m, then $v(\mathbf{c}) = 0$. This follows quickly since

$$v(\mathbf{c}) = cv(\mathbf{1}),$$
$$v(\mathbf{1}) = v(\mathbf{1}.\mathbf{1}) = 1v(\mathbf{1}) + 1v(\mathbf{1}) = 2v(\mathbf{1})$$

giving

$$v(\mathbf{c}) = c.0 = 0.$$

To obtain the result on dimensionality, we first prove that M_m is isomorphic to the dual of the space (F_m/F_m^2) which we denote by $(F_m/F_m^2)^*$.

Let $v \in M_m$. Since germs belonging to F_m vanish at m, it follows from the derivation property that v vanishes on F_m^2. This v determines a linear mapping of F_m/F_m^2 into the reals, that is, v determines uniquely an element of $(F_m/F_m^2)^*$.

Conversely, let $l \in (F_m/F_m^2)^*$. Then the corresponding tangent vector v_l is defined by

$$v_l(\mathbf{f}) = l(\mathbf{f} - \mathbf{f}(\mathbf{m})) \qquad \text{for } f \in \tilde{F}_m.$$

We must now check that v_l so defined is indeed a tangent vector. The linearity property follows immediately. To prove that it is a derivation we have

$$v_l(\mathbf{f} \cdot \mathbf{g}) = l(\{\mathbf{f} \cdot \mathbf{g} - \mathbf{f}(\mathbf{m})\mathbf{g}(\mathbf{m})\})$$
$$= l(\{(\mathbf{f} - \mathbf{f}(\mathbf{m}))(\mathbf{g} - \mathbf{g}(\mathbf{m})) + \mathbf{f}(\mathbf{m})(\mathbf{g} - \mathbf{g}(\mathbf{m}))$$
$$+ (\mathbf{f} - \mathbf{f}(\mathbf{m}))\mathbf{g}(\mathbf{m})$$
$$= l(\{(\mathbf{f} - \mathbf{f}(\mathbf{m}))(\mathbf{g} - \mathbf{g}(\mathbf{m}))\}) + \mathbf{f}(\mathbf{m})l(\mathbf{g} - \mathbf{gm}))$$
$$+ \mathbf{g}(\mathbf{m})l(\mathbf{f} - \mathbf{f}(\mathbf{m}))$$
$$= \mathbf{f}(\mathbf{m})v_l(\mathbf{g}) + \mathbf{g}(\mathbf{m})v_l(\mathbf{f})$$

where we use the notation $\mathbf{f}(\mathbf{m})$ to denote the germ with the constant value $\mathbf{f}(m)$. Thus we have obtained maps of M_m into $(F_m/F_m^2)^*$ and

8   1 Differentiable manifolds

conversely. Moreover these maps are checked to be inverses of each other. Thus we establish an isomorphism between M_m and $(F_m/F_m^2)^*$. We now prove that $\dim (F_m/F_m^2) = \dim M$.

We first prove that if g is of class $C^k (k \geq 2)$ on a star-shaped open set U about the point p in $\mathbb{R}^d$, then for each $q \in U$ we have

$$g(q) = g(p) + \sum_{i=1}^{d} \left.\frac{\partial g}{\partial r_i}\right|_p (r_i(q) - r_i(p))$$

$$+ \sum_{i,j} (r_i(q) - r_i(p))(r_j(q) - r_j(p)) \int_0^1 (1-t) \left.\frac{\partial^2 g}{\partial r_i \partial r_j}\right|_{p+t(q-p)} dt.$$

In particular when $g \in C^\infty$, then the second summation determines an element of F_p^2.

Let $H: t \to H(t)$ be a C^k real-valued function for $0 \leq t \leq 1$ and $k \geq 2$. Integration by parts gives

$$\int_0^1 (1-t)H''(t)\,dt = [(1-t)H'(t)]_0^1 + \int_0^1 H'(t)\,dt$$

$$= -H'(0) + H(1) - H(0).$$

Thus

$$H(1) = H(0) + H'(0) + \int_0^1 (1-t)H''(t)\,dt.$$

Write

$$H(t) = g(p + t(q-p)).$$

Then

$$H(0) = g(p), \quad H(1) = g(q),$$

$$H'(0) = \sum_{i=1}^{d} \left.\frac{\partial g}{\partial x_i}\right|_p (x_i(q) - x_i(p)),$$

$$H''(t) = \sum_{i,j} (x_i(q) - x_i(p))(x_j(q) - x_j(p)) \left.\frac{\partial^2 g}{\partial x_i \partial x_j}\right|_{p+t(q-p)}.$$

Thus our formula for the expansion of $g(q)$ is justified.

Let (U, ϕ) be a coordinate system of M around m with coordinate functions $x_1, \ldots, x_d$. Let $f \in F_m$. We consider the expansion of $f \circ \phi^{-1}$, and compose with ϕ to get

$$f = \sum_{i=1}^{d} \left.\frac{\partial}{\partial r_i}(f \circ \phi^{-1})\right|_{\phi(m)} (x_i - x_i(m)) + \sum_{i,j} (x_i - x_i(m))(x_j - x_j(m))h$$

1.3 Tangent vectors

on a neighbourhood of m where $h \in C^\infty$. Hence

$$\mathbf{f} = \sum_{i=1}^{d} \frac{\partial}{\partial r_i}(f \circ \phi^{-1})\bigg|_{\phi(m)} (\mathbf{x}_i - \mathbf{x}_i(\mathbf{m})), \qquad \mathrm{mod}\, F_m^2.$$

It follows that $\{\mathbf{x}_i - \mathbf{x}_i(\mathbf{m})\}: i = 1, \ldots, d$ spans F_m/F_m^2. Hence dim $F_m/F_m^2 \leq d$. We now claim that these elements are linearly independent. For suppose

$$\sum a_i(\mathbf{x}_i - \mathbf{x}_i(\mathbf{m})) \in F_m^2.$$

Since

$$\sum_i a_i(x_i - x_i(m)) \circ \phi^{-1} = \sum_i a_i(r_i - r_i(\phi(m)))$$

it follows that $\Sigma_i^d(\mathbf{r}_i - \mathbf{r}_i(\phi(\mathbf{m}))) \in F_{\phi(m)}^2$. This implies that

$$\frac{\partial}{\partial r_j}\bigg|_{\phi(m)} \left[\sum a_i(r_i - r_i(\phi(m)))\right] = 0$$

for $j = 1, \ldots, d$ and hence $a_i = 0$ for all i. Thus the elements $\{\mathbf{x}_i - \mathbf{x}_i(\mathbf{m})\}$ form a basis for F_m/F_m^2 which is therefore d-dimensional. Hence $(F_m/F_m^2)^*$ is d-dimensional and hence we have proved that dim $M_m = \dim M$.

In practice we treat tangent vectors as operating on functions rather than on germs of functions. If f is a differentiable function defined on a neighbourhood of m and $v \in M_m$, we define

$$v(f) = v(\mathbf{f}).$$

Clearly for such functions f and g we have

$$v(f + \lambda g) = v(f) + \lambda v(g),$$
$$v(f \cdot g) = f(m)v(g) + g(m)v(f),$$

where $f + \lambda g$ and $f \cdot g$ are defined on the intersection of the domains of f and g.

Let (U, ϕ) be a coordinate system with coordinate functions $x_1, \ldots, x_d$ and let $m \in U$. For each $i \in (1, \ldots, d)$ we define a tangent vector $(\partial/\partial x_i)|_m \in M_m$ by setting

$$\left(\frac{\partial}{\partial x_i}\bigg|_m\right) f = \frac{\partial}{\partial r_i}(f \circ \phi^{-1})\bigg|_{\phi(m)}$$

for each function f which is C^∞ on a neighbourhood of m. We interpret $(\partial/\partial x_i|_m)f$ as the direction derivative of f at m in the x_i coordinate direction. This is helped by using the notation

$$\left.\frac{\partial f}{\partial x_i}\right|_m = \left(\left.\frac{\partial}{\partial x_i}\right|_m\right) f.$$

Clearly $(\partial/\partial x_i|_m)(f)$ depends only on the germ of f at m, and satisfies the linearity and derivation properties. Thus $(\partial/\partial x_i)|_m$: $i = 1, \ldots, d$ is a tangent vector at m. Moreover we claim that it is a basis of M_m. Indeed it is the basis of M_m dual to the basis $\{\{(\mathbf{x}_i - \mathbf{x}_i(\mathbf{m}))\}: i = 1, \ldots, d\}$. This follows because

$$\left.\frac{\partial}{\partial x_i}\right|_m (x_j - x_j(m)) = \delta_{ij}.$$

Moreover we have

$$v = \sum_{i=1}^{d} v(x_i) \left.\frac{\partial}{\partial x_i}\right|_m$$

as can be verified by applying both sides to the function $(x_j - x_j(m))$. Thus *a tangent vector is determined by its action on the coordinate functions.*

We note that in the particular case when $M = \mathbb{R}^d$ referred to the canonical coordinate system $r_1, \ldots, r_d$, the tangent vectors associated with these coordinates are precisely the usual partial derivative operators, and the tangent vectors act as directional derivatives.

1.4 THE DIFFERENTIAL OF A MAP

Let M, N be C^∞-manifolds and let $\varphi: M \to N$ be a C^∞-map. Let $m \in M$. Then the differential $d\psi$ of ψ at m is the linear map

$$d\psi: M_m \to N_{\psi(m)}$$

defined as follows. If $v \in M_m$, its image $d\psi(v)$ is determined by its action on differentiable functions of N at $\psi(m)$. We define this action by

$$d\psi(v) \cdot g = v(g \circ \psi)$$

where g is a C^∞-function on a neighbourhood of $\psi(m)$. Strictly we should denote this map by $d\psi|M_m$ but when there is no fear of confusion we delete the restriction symbol.

The map ψ is called *non-singular* when the kernel of $d\psi$ consists of the zero vector. The *dual map*

$$\delta\psi: N^*_{\psi(m)} \to M^*_m$$

is defined in the usual way by requiring

1.4 The differential of a map

$$\delta\psi(\omega)v = \omega(d\psi(v))$$

for $\omega \in N^*_{\psi(m)}$ and $v \in M_m$. In the special case when $N = \mathbb{R}^1$ so that ψ is a real-valued function $f: M \to N$ and $r_0 = f(m)$ we have

$$df(v) = v(f)\frac{d}{dr}\bigg|_{r_0}.$$

It is convenient to regard df as the element of M^*_m given by $df(v) = v(f)$.

Technically df is identified with $\delta f(\omega)$ where ω is the dual basis of the one-dimensional vector space $\mathbb{R}_{r_0}$ with basis $(d/dr)|_{r_0}$.

Consequences of our definition of differential

1. Let $(U, x_1, \ldots, x_d)$, $(V, y_1, \ldots, y_{d'})$ be coordinate systems about $m \in M$ and $\psi(m) \in N$. Then we have

$$d\psi\left(\frac{\partial}{\partial x_j}\bigg|_m\right) = \sum_{i=1}^{d'} \frac{\partial(y_i \circ \psi)}{\partial x_j}\bigg|_m \frac{\partial}{\partial y_i}\bigg|_{\psi(m)}.$$

The matrix $\{\partial(y_i \circ \psi)/\partial x_j\}$ is called the *Jacobian* of ψ with respect to the coordinate systems. For maps of Euclidean spaces $\mathbb{R}^d \to \mathbb{R}^{d'}$, and for canonical coordinate systems this becomes the usual Jacobian matrix of multivariable calculus.

2. Let $(U, x_1, \ldots, x_d)$ be a coordinate system of M valid in a neighbourhood of m. We have seen that $\{\partial/\partial x_i|_m\}$ is a basis for M_m. It follows that $\{dx_i|_m\}$ is a basis for the dual space M^*_m. Moreover if f is a real-valued C^∞-function, then

$$df|_m = \sum_{i=1}^{d} \frac{\partial f}{\partial x_i}\bigg|_m dx_i|_m.$$

3. Let $\psi: M \to N$ and $\phi: N \to P$ be C^∞-maps. Then

$$d(\phi \circ \psi)|_m = d\phi|_{\psi(m)} \circ d\psi_m$$

which we can write simply as

$$d(\phi \circ \psi) = d\phi \circ d\psi.$$

This is none other than the well-known chain rule when M, N and P are Euclidean spaces.

4. A *smooth curve* in M is a C^∞-mapping

$$\sigma: (a, b) \to M \quad \text{for } a, b \in \mathbb{R}.$$

The tangent vector to the curve at $t \in (a, b)$ is the vector

$$d\sigma\left(\frac{d}{dr}\bigg|_t\right) \in M_{\sigma(t)}.$$

We denote this tangent vector by $\dot{\sigma}(t)$.

We now prove that for a given non-zero tangent vector v at m there is a curve through m whose tangent vector coincides with v. To do this we choose a coordinate system (U, ϕ) centred at m for which

$$v = d\phi^{-1}\left(\frac{\partial}{\partial r_1}\bigg|_0\right).$$

Then v is the tangent vector at O to the curve $t \to \phi^{-1}(t, 0, \ldots, 0)$. In classical terminology we can choose a coordinate system centred at m so that v has components δ_1^i and is the tangent vector to the curve expressed locally as $x_i = \delta_1^i t$. Clearly there are many different curves through m which all have v as tangent vector.

If $M = \mathbb{R}^d$ so that σ is a curve in Euclidean space $\mathbb{R}^d$, then

$$\dot{\sigma}(t) = \frac{d\sigma_1}{dr}\bigg|_t \frac{\partial}{\partial r_1}\bigg|_{\sigma(t)} + \cdots + \frac{d\sigma_d}{dr}\bigg|_t \frac{\partial}{\partial r_d}\bigg|_{\sigma(t)}.$$

If we identify this tangent vector with the element

$$\left(\frac{d\sigma_1}{dr}, \ldots, \frac{d\sigma_d}{dr}\right)$$

of $\mathbb{R}^d$ we have

$$\dot{\sigma}(t) = \lim_{h \to 0} \frac{\sigma(t+h) - \sigma(t)}{h}.$$

This agrees with the geometric notion of tangent vector to a curve in $\mathbb{R}^d$. Thus we have confirmed that our abstract definition of tangent vector is sensible in that it is consistent with what we intuitively mean by tangent vector in the well-known Euclidean case.

In this section we have followed closely the treatment of tangent vector given in the book by Warner (1983).

A question arises naturally: What is the analogue of the elementary calculus result that $dy/dx = 0$ implies that y is a constant function? The answer is given by the following theorem.

Theorem (1.4.1) Let ψ be a C^∞-mapping of the connected manifold M into the manifold N, and let $d\psi_m = 0$ for each $m \in M$. Then ψ is a constant map, that is, $\psi(m) = q$ for all $m \in M$.

We break the proof of the theorem into two parts.

1.4 The differential of a map

Lemma (1.4.2) A locally constant function f on a connected space is globally constant.

To prove the lemma, let $m \in M$ and let $f(m) = c$ in some neighbourhood U of m. Consider now the subset of M given by $f^{-1}(c)$. Since M is Hausdorff and f is continuous, it follows that $f^{-1}(c)$ is closed. We now prove that $f^{-1}(c)$ is also open. However this follows immediately since if $f(p) = c$, the local constancy guarantees that there is a neighbourhood of p for which every point has this property. Since M is connected it follows that $f^{-1}(c) = M$ and the lemma is proved.

Lemma (1.4.3) If $d\psi_m = 0$ for all $m \in M$, then ψ is locally constant.

To prove this lemma let $(U, x^1, \ldots, x^d)$ be a coordinate system around m, and let $(V, y^1, \ldots, y^n)$ be a coordinate system around $q = \psi(m) \in N$. Then locally the map ψ is given by

$$y^\alpha = \psi^\alpha(x^1, \ldots, x^d).$$

We have

$$d\psi\left(\frac{\partial}{\partial x^i}\right) = \sum_{\alpha=1}^{n} \frac{\partial}{\partial x^i}(y^\alpha \circ \psi) \frac{\partial}{\partial y^\alpha}, \quad i = 1, \ldots, d.$$

Since $d\psi_m = 0$ and the vectors $\partial/\partial y^\alpha$ form a basis for $N_{\psi(m)}$, it follows that $\partial(y^\alpha \circ \psi)/\partial x^i = 0$. Thus in some neighbourhood of m, the functions $y^\alpha \circ \psi$ are constant, and the lemma is proved.

The proof of the theorem follows immediately from the two previous lemmas.

Orientation

Definition (1.4.4) A differentiable manifold M is said to be orientable provided that there is a collection of coordinate systems in M whose domains cover M and such that for each pair of overlapping charts (U, x), (V, y) the Jacobian determinant function det $(\partial y^i/\partial x^j)$ is positive.

As an example of an orientable manifold we have $\mathbb{R}^n$ since this can be covered by a single coordinate system. A further example is the sphere S^2, covered by two charts whose domains are S^2 punctured by the north and south poles, respectively, the maps of the charts being given by stereographic projection. In the next chapter we shall return to this topic when we have suitable tools at our disposal.

1.5 THE TANGENT AND COTANGENT BUNDLES

Let M be a C^∞-manifold with differentiable structure $\mathfrak{F}$. Let $T(M) = \bigcup_{m \in M} M_m$, $T^*(M) = \bigcup_{m \in M} M_m^*$. We shall indicate how $T(M)$ and $T^*(M)$ can be given a topological and a differentiable structure so that they themselves become differentiable manifolds. We deal first with $T(M)$. We have a natural projection $\pi: T(M) \to M$ given by $\pi(v) = m$ if $v \in M_m$. Now let (U, ϕ) be a coordinate system of M, that is, $(U, \phi) \in \mathfrak{F}$, with coordinates $x^1, \ldots, x^d$. Corresponding to $\phi: U \to \mathbb{R}^d$ we have the map $\tilde{\phi}: \pi^{-1}(U) \to \mathbb{R}^{2d}$ given by

$$\tilde{\phi}(v) = (x^1(\pi(v)), \ldots, x^d(\pi(v)), dx^1(v), \ldots, dx^d(v)).$$

First check that if (U, ϕ), (V, ψ) are overlapping coordinate systems for M, then $\tilde{\psi} \circ \tilde{\phi}^{-1}$ is a C^∞-map: $\mathbb{R}^{2d} \to \mathbb{R}^{2d}$.

Secondly in order to give $T(M)$ a topology we assert that the collection $\{\tilde{\phi}^{-1}(W): W \text{ open in } \mathbb{R}^{2d}, (U, \phi) \in \mathfrak{F}\}$ forms a basis for a topology on $T(M)$ which makes it into a $2d$-dimensional, second countable, locally Euclidean space. Finally, we give $T(M)$ a differentiable structure by defining $\tilde{\mathfrak{F}}$ to be the maximal collection containing

$$\{(\pi^{-1}(U), \tilde{\phi}): (U, \phi) \in \mathfrak{F}\}.$$

Then $\tilde{\mathfrak{F}}$ satisfies the requirements for a C^∞-structure on $T(M)$. In this way $T(M)$ becomes a differentiable manifold called the tangent bundle.

In a similar way we can show that the cotangent bundle $T^*(M)$ becomes a differentiable manifold. These are important manifolds as their properties will motivate the definition of a more general object called a vector bundle which will play an important role in later chapters.

1.6 SUBMANIFOLDS; THE INVERSE AND IMPLICIT FUNCTION THEOREMS

We now consider special types of maps of manifolds. Let $\psi: M \to N$ be a differentiable map. We say that ψ is an immersion if $d\psi_m$ is injective; alternatively, if $d\psi_m$ has maximal rank considered as a linear transformation. If, in addition, we assume that ψ is injective, then the pair (M, ψ) is called a submanifold of N. If, in addition, we assume that ψ is a homeomorphism of M into $\psi(M)$ with its relative topology, then we call ψ an imbedding. The term diffeomorphism is used to describe a map $\psi: M \to N$ which is bijective and such that both ψ and ψ^{-1} are C^∞-maps.

The problem of deciding whether a given manifold can support non-diffeomorphic differentiable structures is very difficult. An example

1.6 Submanifolds; the inverse and implicit function theorems

was given by Kervaire (1960) of a manifold which admits no differentiable structure. In a pioneer paper J. Milnor (1956) showed that the seven-dimensional sphere admits 28 different differentiable structures. Recently it was shown that $\mathbb{R}^4$ admits an exotic differentiable structure, that is, one different from the usual one. These are problems of differential topology, an important subject which lies outside the domain of the present book. However we are primarily concerned with properties of objects which live on a given differentiable manifold.

The inverse function theorem of advanced calculus can be stated as follows.

Theorem (1.6.1) Let $U \subset \mathbb{R}^d$ be open and let $f: U \to \mathbb{R}^d$ be a C^∞-map. If the Jacobian matrix

$$\left(\frac{\partial r^i \circ f}{\partial r^j} \right)_{i,j=1,\ldots,d}$$

is non-singular at $r_0 \in U$, then there exists an open set V with $r_0 \in V \subset U$ such that the restriction of f to V, written $f|V$, maps V bijectively into the open set $f(V)$, and moreover $(f|V)^{-1}$ is a C^∞-map.

A second important theorem of advanced calculus is the implicit function theorem. This generalizes to manifolds to give the following result.

Theorem (1.6.2) Assume that $\psi: M^d \to N^n$ is C^∞, that q is a point of N, that $P = \psi^{-1}(q)$ is non-empty, and that $d\psi: M_m \to N_{\psi(m)}$ is surjective for all $m \in P$. Then P has a unique manifold structure such that (P, i) is a submanifold of M, where i is the inclusion map. Moreover, $i: P \to M$ is an imbedding, and the dimension of P is $d - n$.

These theorems have the following consequences.

1. Let M, N be differentiable manifolds and let $\psi: M \to N$ be a C^∞-map. Let $m \in M$ such that $d\psi: M_m \to N_{\psi(m)}$ is an isomorphism. Then there is a neighbourhood U of m such that $\psi: U \to \psi(U)$ is a diffeomorphism into the open set $\psi(U)$ in N.

2. Suppose that $\dim M = d$ and that $y^1, \ldots, y^d$ is a set of independent functions at m_0 in the sense that the differentials $dy^1, \ldots, dy^d$ form an independent set in M_m^*. Then these functions form a coordinate system in a neighbourhood of m_0.

3. Suppose that we are given less than d independent functions at m; in fact, suppose that $y^1, \ldots, y^l$ are independent with $l < d$. Then these functions form part of a coordinate system at m in the sense that $(d - l)$ other functions $y^{l+1}, \ldots, y^d$ can be chosen so that

the complete set form a coordinate system on a neighbourhood of m.

4 Let $\psi: M \to N$ be C^∞ and assume that $d\psi: M_m \to N_{\psi(m)}$ is surjective. Let $x^1, \ldots, x^n$ form a coordinate system for some neighbourhood of $\psi(m)$. Then $x^1 \circ \psi, \ldots, x^n \circ \psi$ form part of a coordinate system on some neighbourhood of m.

5 Suppose that $y^1, \ldots, y^k$, $k > d$, is a set of functions defined in a neighbourhood of m such that their differentials span M_m^*. Then a subset of the y's form a coordinate system in a neighbourhood of m.

6 Let $\psi: M \to N$ be such that $d\psi: M_m \to N_{\psi(m)}$ is injective. Let $x^1, \ldots, x^n$ form a coordinate system in a neighbourhood of $\psi(m)$. Then a subset of the functions $\{x^i \circ \psi\}$ forms a coordinate system on a neighbourhood of m. In particular ψ is injective on a neighbourhood of m.

In our treatment of these topics we have been influenced by the book of Chevalley (1946) and also by the book by Warner (1983). In particular we refer the reader to Warner's book for proof of the above results.

1.7 VECTOR FIELDS

A vector field X over M is a section of the tangent bundle $T(M)$, that is, a C^∞-mapping $s: M \to T(M)$ such that $\pi \circ s$ is the identity.

Let (U, ϕ) be a coordinate system with coordinates $x^1, \ldots, x^d$ over an open set U of M. Then $X = s(U)$ can be represented by

$$X = \sum_{i=1}^d X^i \frac{\partial}{\partial x^i}$$

where the X^i are C^∞-functions over U. Alternatively we could have defined a vector field X over M as a law which associates to each point $m \in M$ an element $X(m)$ of M_m in such a way that when X is represented locally in a system of coordinates around m, the corresponding components are C^∞-functions. The vector field X, considered as a differential operator, maps smooth functions defined in a neighbourhood U to smooth functions on U by

$$X(af + bg) = aX(f) + bX(g),$$
$$X(fg) \quad\quad = f \cdot X(g) + g \cdot X(f), \quad \text{where } a, b \in \mathbb{R}.$$

We have considered a tangent vector at a point m by examining curves through m, and it is natural to consider whether there are curves

1.7 Vector fields

through m which are specially related to a given vector field X. We are looking for curves through m which have the property that at each point p on the curve, the tangent vector to the curve is precisely $X(p)$. Intuitively we can ask whether the vectors $X(p)$ fit together or integrate to form an integral curve. We define an integral curve of the vector field X, passing through m, as a C^∞-mapping γ of a real open interval (a, b) containing 0 such that $\gamma(0) = m$, and $\dot\gamma(t) = X(\gamma(t))$, $t \in (a, b)$. Then relative to a system of coordinates $x^1, \ldots, x^d$ around m, the conditions satisfied by an integral curve become

$$\dot\gamma^i = X^i(\gamma^1(t), \ldots, \gamma^d(t)), \qquad i = 1, \ldots, d.$$

However there are fundamental existence and uniqueness theorems for solutions of such a system of first-order differential equations (see, for example, Hurewicz (1958)) and since everything in our case is smooth the theorems apply. It follows that there is an integral curve of X passing through m. In fact, there is a maximal integral curve through m in the sense that any integral curve must be a subset of this maximal integral curve. We refer the reader to Warner (1983) for detailed proofs.

We now show how two given vector fields determine a third vector field. Let X, Y be given vector fields. We define the commutator $[X, Y]$ by

$$[X, Y]_m = (X_m Y - Y_m X).$$

We now prove that $[X, Y]$ is indeed a vector field. We first observe that

$$[X, Y](af + bg) = a[X, Y]f + b[X, Y]g.$$

Moreover, we have

$$\begin{aligned}[][X, Y](f \cdot g) &= X\{Y(f \cdot g)\} - Y\{X(f \cdot g)\} \\ &= X\{f(Yg) + g \cdot (Yf)\} - Y\{f \cdot (Xg) + g \cdot (Xf)\} \\ &= (Xf) \cdot (Yg) + f \cdot (XYg) + (Xg) \cdot (Yf) + g \cdot (XYf) \\ &\quad - (Yf) \cdot (Xg) - f \cdot (YXg) - (Yg) \cdot (Xf) - g \cdot (YXf) \\ &= f \cdot ([X, Y]g) + g \cdot ([X, Y]f),\end{aligned}$$

so that $[X, Y]$ is a derivation. Moreover, if $[X, Y] = [X, Y]^i \partial/\partial x^i$ computation gives

$$[X, Y]^i = \sum_{j=1}^{d} \left(X^j \frac{\partial Y^i}{\partial x^j} - Y^j \frac{\partial X^i}{\partial x^j} \right),$$

so $[X, Y]$ is a smooth vector field. Sometimes $[X, Y]$ is called the Lie bracket of X and Y, after the mathematicians Sophus Lie, who used it extensively. It is easy to check that

(i) $[X, Y] = -[Y, X]$;
(ii) $[fX, gY] = f \cdot g[X, Y] + f \cdot (Xg)Y - g \cdot (Yf)X$;
(iii) $[[X, Y], Z] + [[Y, Z], X] + [[Z, X], Y] = 0$;

for all vector fields X, Y, Z on M. Relation (iii) is called the Jacobi identity. A vector space with a bilinear operation satisfying (i) and (iii) is called a Lie algebra. We see that the set of all vector fields on M forms a natural (infinite-dimensional) Lie algebra.

We now see how the Lie bracket operation behaves under maps. Let M, N be smooth manifolds and let $\psi: M \to N$ be a smooth map. Let X, Y be vector fields on M, N, respectively. We say that X and Y are ψ-related if for every point $m \in M$ we have

$$d\psi(X_m) = Y_{\psi(m)}.$$

Proposition (1.7.1) *Let X_1, X_2 be vector fields on M which are ψ-related respectively to vector fields Y_1, Y_2 on N. Then $[X_1, X_2]$ and $[Y_1, Y_2]$ are ψ-related.*

To prove this we remark that the fact that X_i, Y_i are ψ-related ($i = 1, 2$) may be expressed by

$$(X_i(g \circ \psi))_p = (Y_i g)_{\psi(p)},$$

or

$$(X_i(g \circ \psi))_p = (Y_i g \circ \psi)_p$$

holds for any point p in some neighbourhood of m in M. Hence

$$(Y_1 Y_2 g)_{\psi(p)} = (X_1(Y_2 g \circ \psi))_p = (X_1 X_2(g \circ \psi))_p.$$

We get a similar formula by interchanging the suffixes 1 and 2. On subtracting we get

$$([Y_1, Y_2]g)_{\psi(p)} = ([X_1, X_2](g \circ \psi))_p$$

from which

$$[Y_1, Y_2]_{\psi(p)} = d\psi([X_1, X_2]_p),$$

and the proposition is proved.

1.8 DISTRIBUTIONS

A k-dimensional vector subspace of the d-dimensional tangent space M_m is called a contact element of dimension k at m. A law which

1.8 Distributions

assigns to each point $m \in M$ a k-dimensional contact element is called a *k-dimensional distribution*.

We denote by $\mathcal{K}$ a distribution of dimension k on M, and by $\mathcal{K}_m$ the k-dimensional subspace which is assigned to the point m. We shall assume that $\mathcal{K}$ is a C^∞-distribution — by this we mean that there exists a neighbourhood U of the point m and a system of C^∞-vector fields $X_1, \ldots, X_k$ on U such that, for each point $q \in U$, the vectors $(X_1)_q, \ldots, (X_k)_q$ form a basis for the space $\mathcal{K}_q$.

The reader should be warned not to confuse a one-dimensional distribution with a vector field. We have seen that there are special one-dimensional submanifolds (the integral curves) associated with a vector field. The question arises whether there are k-dimensional submanifolds associated with a k-dimensional distribution, that is, whether the contact elements fit together or integrate to yield a k-dimensional submanifold. More precisely, we call a submanifold N of M an integral manifold of $\mathcal{K}$ if, for every point $p \in N$, $\mathcal{K}_p$ coincides with the tangent space to N at p. The distribution is then said to be *integrable*.

It is not difficult to see that, unlike the corresponding result for vector fields, certain conditions must be satisfied if a distribution has an integral submanifold through p. To prove this, suppose that such a submanifold exists, and let X be a vector field on M such that $X(p) \in \mathcal{K}_p$ for every point $p \in N$. The differential of the inclusion map $I: N \to M$ is injective and hence there is one and only one vector field Y on N defined in a neighbourhood of p such that $X(p) = \mathrm{d}I(Y(p))$. Let X_1, X_2 be two such vector fields on M and let Y_1, Y_2 be the corresponding vector fields on N. Using the property of $[\,,\,]$ under maps, we see that since $[Y_1, Y_2]$ is a vector field on N, it follows that $[X_1, X_2]_p \in \mathcal{K}_p$.

We say that a distribution $\mathcal{D}$ is *involutive* if $X_1, X_2 \in \mathcal{D}$ implies that $[X_1, X_2] \in \mathcal{D}$. From the previous argument we see that a necessary condition for a distribution to be integrable is that it is involutive. An important result, known as the theorem of Frobenius, says that the converse is true, that is, an involutive distribution is integrable. We refer the reader to Warner (1983) for the proof of this theorem. We shall see later that the set of all contact elements of order k on a manifold forms a fibre bundle, a more sophisticated analogue of the tangent bundle, and an alternative definition of a distribution would be a cross-section of this bundle. Global cross-sections of fibre bundles do not usually exist because of topological obstructions. It follows that a manifold does not always admit globally a k-dimensional distribution. For example it is known that a three-dimensional distribution cannot exist globally on the 4-sphere.

In the next chapter we shall obtain an alternative description of the Frobenius theorem in terms of differential forms.

1.9 EXERCISES

1. Use the implicit function theorem to prove that S^2 given by
$$S^2 = [(x, y, z) : x^2 + y^2 + z^2 = 1]$$
is a smooth manifold.

2. Show that S^2 is a smooth manifold by considering two coordinate charts obtained by stereographic projection from the north and from the south poles.

3. If $\psi: M \to N$ is a smooth map and X is a vector field on M, construct an example to show that $d\psi(X)$ is not necessarily a vector field on N.

4. The torus $S^1 \times S^1$ is represented as a submanifold of $\mathbb{R}^4$ by the map
$$\begin{aligned} y^1 &= \cos 2\pi u, & 0 \leqslant u \leqslant 1 \\ y^2 &= \sin 2\pi u, & 0 \leqslant u \leqslant 1 \\ y^3 &= \cos 2\pi v, & 0 \leqslant v \leqslant 1 \\ y^4 &= \sin 2\pi v, & 0 \leqslant v \leqslant 1. \end{aligned}$$

Consider the subset of $S^1 \times S^1$ given by
$$\begin{aligned} y^1 &= \cos 2\pi t, \\ y^2 &= \sin 2\pi t, \\ y^3 &= \cos 2\pi \alpha t, \\ y^4 &= \sin 2\pi \alpha t, \end{aligned}$$

where α is an irrational number. Show that this subset is a submanifold of $S^1 \times S^1$ which is dense. Does this submanifold have as its topology that induced from that of $S^1 \times S^1$?

2
Tensors and differential forms

2.1 TENSOR PRODUCT

In this chapter we associate new vector spaces to given vector spaces. We have already seen that corresponding to a given vector space V we have another vector space V^*, namely the dual vector space consisting of linear maps of V into the field $\mathbb{R}$. Moreover, the reader will have met already the sum $V \oplus W$ of two vector spaces V and W, consisting of pairs (v, w) of $V \times W$ with the usual operations acting separately on each component. Let U be a vector subspace of V, and define an equivalence relation $\sim$ on V by asserting $v_1 \sim v_2$ if and only if $v_1 - v_2 \in U$. The set of equivalence classes itself forms a vector space, called the *quotient space*, denoted by V/U.

Another vector space derived from V and W is the free vector space denoted by $F(V, W)$. This consists of all finite linear combinations of pairs (v, w) with $v \in V$ and $w \in W$. This space is too large for our purpose, so we introduce an equivalence relation $\sim$ by considering the subspace $R(V, W)$ of $F(V, W)$ generated by elements of the form

$$(v_1 + v_2, w) - (v_1, w) - (v_2, w),$$
$$(v, w_1 + w_2) - (v, w_1) - (v, w_2),$$
$$(av, w) - a(v, w),$$
$$(v, aw) - a(v, w),$$

where $a \in \mathbb{R}$; $v, v_1, v_2 \in V$; $w, w_1, w_2 \in W$. The corresponding quotient space $F(V, W)/R(V, W)$ is called the *tensor product* of V and W and is denoted by $V \otimes W$.

An important property of the tensor product is the so-called universal mapping property, described as follows. Let ϕ denote the bilinear map $(v, w) \to v \otimes w$ of $V \times W$ into $V \otimes W$. Let U be another vector space, and let $l: V \times W \to U$ be a bilinear map. Then there exists a unique linear map $\tilde{l}: V \times W \to U$ such that the following diagram commutes:

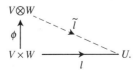

It is not difficult to prove that the tensor product also has the following properties:

(1) $V \otimes W$ is canonically isomorphic to $W \otimes V$.
(2) $V \otimes (W \otimes U)$ is canonically isomorphic to $(V \otimes W) \otimes U$.
(3) Let $\text{Hom}(V, W)$ denote the vector space of linear transformations from V to W, and consider the bilinear map of $V^* \times W$ into $\text{Hom}(V, W)$ given by

$$(f, w)(v) = f(v)w$$

for $f \in V^*$, $v \in V$, and $w \in W$. By the universal mapping property, this determines uniquely a linear map $V^* \otimes W \to \text{Hom}(V, W)$ which is an isomorphism.

(4) Let $\{e_i : i = 1, \ldots, m\}$, $\{f_j : j = 1, \ldots, n\}$ be bases for V and W. Then $\{e_i \otimes f_j : i = 1, \ldots, m; j = 1, \ldots, n\}$ is a basis for $V \otimes W$. It follows that

$$\dim(V \otimes W) = (\dim V)(\dim W).$$

Contravariance and covariance

The tensor space $V_{r,s}$ of type (r, s) associated with V is the space

$$V \otimes \cdots \otimes V \otimes V^* \otimes \cdots \otimes V^*,$$

where there are r copies of V and s copies of V^*. This is known in the literature as the space of tensors r times *contravariant* and s times *covariant*. This is an unfortunate historical accident, as we shall see later that contravariant tensors are really 'covariant' in nature — however we do not have the courage to correct this terminology which is almost universally used.

The direct sum

$$T(V) = \sum V_{r,s} \quad (r, s \geq 0),$$

where $V_{0,0} = \mathbb{R}$, is called the tensor algebra of V. Elements of $T(V)$ are called *tensors*.

Let $u = u_1 \otimes \cdots \otimes u_{r_1} \otimes u_1^* \otimes \cdots \otimes u_{s_1}^*$ belong to V_{r_1, s_1} and $v = v_1 \otimes \cdots \otimes v_{r_2} \otimes v_1^* \otimes \cdots \otimes v_{s_2}^*$ belong to V_{r_2, s_2}. Then their product $u \otimes v$ is defined by

$$u \otimes v = u_1 \otimes \cdots u_{r_1} \otimes v_1 \otimes \cdots \otimes v_{r_2} \otimes u_1^* \otimes \cdots \otimes u_{s_1}^* \otimes v_1^* \otimes \cdots \otimes v_{s_2}^*$$

which belongs to $V_{r_1 + r_2, s_1 + s_2}$.

If a tensor belongs to the tensor space $V_{r,s}$ it is called *homogeneous*

2.1 Tensor product

of degree (r, s). Such a tensor is called *decomposable* if it can be written in the form

$$v_1 \otimes \cdots \otimes r_r \otimes v_1^* \otimes \cdots \otimes v_s^*,$$

where $v_i \in V$ and $v_j^* \in V^*$, $(i = 1, \ldots, r; j = 1, \ldots, s)$. A general tensor of V is a linear combination of decomposable tensors.

Symmetric and skew-symmetric tensors

We shall find in our application of tensors to Riemannian geometry that an important role is played by symmetric and skew-symmetric tensors. We now define precisely what we mean by these concepts.

Let $C(V)$ denote the subalgebra of tensors given by

$$C(V) = \sum_{k=0}^{\infty} V_{k,0}.$$

Consider the subset $I(V)$ of $C(V)$ generated by elements of the form $v \otimes v$ for $v \in V$. It is readily verified that $I(V)$ is both a left and right ideal of $C(V)$. We set

$$I_k(V) = I(V) \cap V_{k,0}$$

so that

$$I(V) = \sum_{k=0}^{\infty} I_k(V).$$

We form the quotient algebra $\Lambda(V) = C(V)/I(V)$, from the equivalence relation asserting that two elements of $C(V)$ are equivalent if their difference lies in $I(V)$. The quotient $\Lambda(V)$ is called the exterior algebra of V. If we write

$$\Lambda_k(V) = V_{k,0}/I_k(V), \quad k \geq 2,$$
$$\Lambda_1(V) = V,$$
$$\Lambda_0(V) = \mathbb{R},$$

then $\Lambda(V) = \Sigma_{k=0}^{\infty} \Lambda_k(V)$.

Multiplication in the algebra $\Lambda(V)$ is called the *wedge* or *exterior product*. The equivalence class containing $v_1 \otimes \cdots \otimes v_k$ is denoted by $v_1 \wedge v_2 \wedge \cdots \wedge v_k$.

The multilinear map $f: V \times V \times \cdots \times V \to W$, where there are k copies of V, is called alternating if

$$f(v_{\pi(1)}, \ldots, v_{\pi(k)}) = (\text{sgn } \pi) f(v_1, \ldots, v_k), \quad v_1, \ldots, v_k \in V$$

for all permutations π in the permutation group S on k letters, and

2 Tensors and differential forms

sgn $\pi = \pm 1$ according as π is an even or odd permutation. The set of all alternating multilinear maps

$$V \times V \times \cdots \times V \to \mathbb{R}$$

will be denoted by $A_k(V)$. We set $A_0(V) = \mathbb{R}$. It is not difficult to establish the following properties of the exterior algebra.

(1) Let $u \in \Lambda_k(V)$, $v \in \Lambda_l(V)$; then $u \wedge v \in \Lambda_{k+l}(V)$ and $u \wedge v = (-1)^{kl} v \wedge u$. In particular if either k or l is even, then $u \wedge v = v \wedge u$: also if $k = l = 1$, we have $u_1 \wedge u_1 = 0$.

(2) If $e_1, \ldots, e_n$ is a basis of V, then

$\{e_i\}$, $i = 1, \ldots, n$, is a basis of $\Lambda_1(V)$,
$\{e_{i_1} \wedge e_{i_2}\}$, $i_1 < i_2$, is a basis for $\Lambda_2(V)$,
$\{e_{i_1} \wedge e_{i_2} \wedge e_{i_3}\}$, $i_1 < i_2 < i_3$, is a basis for $\Lambda_3(V)$, etc.
Moreover $\Lambda_{n+j}(V) = \{0\}$ when $j > 0$. We see that

$$\dim \Lambda_k(V) = \binom{n}{k} = \frac{n!}{(k!)(n-k)!}, \quad 0 \leqslant k \leqslant n,$$

and

$$\dim \Lambda(V) = \sum_{k=0}^{n} \binom{n}{k} = 2^n.$$

(3) We have a universal mapping theorem for multilinear maps. More precisely, let ϕ denote the mapping $(v_1, \ldots, v_k) \to v_1 \wedge \cdots \wedge v_k$ so that ϕ is an alternating multilinear map of $V \times \cdots \times V \to \Lambda_k(V)$. To each alternating multilinear map f of $V \times \cdots \times V$ (k copies) into a vector space W, there corresponds a unique linear map $\tilde{f}: \Lambda_k(V) \to W$ (unique up to isomorphism) such that $\tilde{f} \circ \phi = f$; that is, the following diagram commutes:

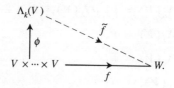

By taking $W = \mathbb{R}$ we see immediately that there is a natural isomorphism between $\Lambda_k(V)^*$ and $A_k(V)$.

It will be clear by now that we could have defined a tensor of type (r, s) as a multilinear mapping of $V^* \times \cdots \times V^* \times V \times \cdots \times V$ into $\mathbb{R}$ where we have taken r copies of V^* and s copies of V. This indeed is done in many books. We have preferred our approach because many properties of tensor products are easier to prove with this definition.

2.1 Tensor product

Change of basis

Let $\{e_i\}$ be a basis for V and let $\{e_{i'}\}$ be a new basis related to the previous one by

$$e_{i'} = \sum_{i=1}^{n} p_{i'}^i e_i, \qquad e_i = \sum_{i=1}^{n} p_i^{i'} e_{i'}$$

so that $(p_{i'}^i)$ and $(p_i^{i'})$ are reciprocal matrices. The dual bases $\{\omega^i\}$, $\{\omega^{i'}\}$ will be related by

$$\omega^{i'} = \sum p_i^{i'} \omega^i, \qquad \omega^i = \sum p_{i'}^i \omega^{i'}.$$

We often use the Einstein summation convention which deletes the summation sign and sums over repeated indices.

A tensor T of type (r, s) can be expressed relative to a basis $\{e_i\}$ of V as

$$T = T^{i_1 \cdots i_r}_{j_1 \cdots j_s} \quad e_{i_1} \otimes \cdots \otimes e_{i_r} \otimes \omega^{j_1} \otimes \cdots \otimes \omega^{j_s}.$$

The same tensor referred to the basis $\{e_{i'}\}$ of V will be expressed as

$$T = T^{i'_1 \cdots i'_r}_{j'_1 \cdots j'_s} \quad e_{i'_1} \otimes \cdots \otimes e_{i'_r} \otimes \omega^{j'_1} \otimes \cdots \otimes \omega^{j'_s}.$$

Using the formula for change of basis we have

$$T^{i'_1 \cdots i'_r}_{j'_1 \cdots j'_s} = p^{i'_1}_{i_1} \cdots p^{i'_r}_{i_r} p^{j_1}_{j'_1} \cdots p^{j_s}_{j'_s} T^{i_1 \cdots i_r}_{j_1 \cdots j_s},$$

together with a similar relation with the dashed and undashed indices interchanged.

Historically a tensor was considered as a set of components which transformed in the above manner when a change of basis was made. Conversely, if the numbers $T^{i_1 \cdots i_r}_{j_1 \cdots j_s}$ transform like the components of a tensor, then they are the components of a tensor. We have given an intrinsic definition independent of basis, and have deduced the necessary change of components associated with a change of basis as a consequence. However in practice it often happens that a set of indexed symbols is thrown up naturally in the course of calculations, and it is useful to examine how these symbols transform under a change of basis in order to discover whether or not the symbols are components of a tensor. However the reader should be warned that not all objects of interest in geometry are tensors, though many are.

Contraction

This is an operation which converts a tensor of type (r, s), $r \geq 1, s \geq 1$ to one of type $(r-1, s-1)$. We define this operation first for decomposable tensors and then extend to a general tensor by linearity.

Let $T = v_1 \otimes \cdots \otimes v_r \otimes v_1^* \otimes \cdots \otimes v_s^*$, $v_i \in V$, $v_j^* \in V^*$. We select one of the v's, say v_i, and one of the v^*'s, say v_j^*. Then the corresponding contract tensor T is given by

$$T = v_j^*(v_i) v_1 \otimes \cdots \otimes v_{i-1} \otimes v_{i+1} \otimes \cdots \otimes v_r \otimes v_1^* \otimes \cdots$$
$$\otimes v_{j-1}^* \otimes v_{j+1}^* \otimes \cdots \otimes v_s^*.$$

Suppose we choose a basis $\{e_i\}$ for V. Then any tensor of type (r, s) is represented by components $T^{i_1 \cdots i_r}_{j_1 \cdots j_s}$. We select the ith upper index and the jth lower index, equate these indices, and sum over the common value. This gives a new set

$$T^{i_1 \cdots i_r}_{j_1 \cdots j_s} = \sum_{i=1}^{n} T^{i_1 \cdots i \cdots i_r}_{j_1 \cdots i \cdots j_s},$$

where the upper index i and the lower index j have been omitted from the left-hand symbol. We claim that the new symbols are components of the contracted tensor. This follows from the transformation law when we use the fact that $p^i_{i'} p^{j'}_i = \delta^{j'}_{i'}$.

It will be seen that although in our invariant definition we restricted ourselves in the first instance to decomposable tensors, the advantage of the index definition is that it applies to any tensor. Moreover, it is very useful in practical calculations.

Interior multiplication

There is another operation, analogous to contraction, which is particularly useful when applied to tensors of the algebra $\Lambda(V)$. Let u, v belong to $\Lambda(V)$. We define an endomorphism $\varepsilon(u)$ of $\Lambda(V)$ by left multiplication by u as follows:

$$\varepsilon(u)v = u \wedge v, \quad v \in \Lambda(V).$$

The transpose of this endomorphism, say $i(u)$, is an endomorphism of $\Lambda(V)^*$. Since $\Lambda(V)^*$ and $\Lambda(V^*)$ are isomorphic, $i(u)$ can be considered as an endomorphism of $\Lambda(V^*)$. Then $i(u)$ is called *interior multiplication by u*. This is often denoted by $\lrcorner u$ in the literature. We shall usually restrict ourselves to the case when $u \in V$ and $v^* \in \Lambda_r(V)$. Then

$$(i(u)v^*)(v_2, v_3, \ldots, v_r) = rv^*(u, v_2, \ldots, v_r),$$

so that $i(u)$ sends an element of $\Lambda_r(V)$ to an element of $\Lambda_{r-1}(V)$. In particular if $v^* \in \Lambda_1(V)$, then $i(u)v^* = v^*(u)$.

An endomorphism l of $\Lambda(V)$ is called

(i) a derivation if $l(u \wedge v) = l(u) \wedge v + u \wedge l(v)$;

(ii) an anti-derivation if $l(u) \wedge v) = l(u) \wedge v + (-1)^p u \wedge l(v)$, where $u \in \Lambda_p(V)$, $v \in \Lambda(V)$.

2.2 Tensor fields and differential forms

We note that when $u \in V$, then $i(u)$ is an anti-derivation, since

$$i(u)(v_1 \wedge \cdots \wedge v_j) = \sum_{i=1}^{j} (-1)^{i+1} v_1 \wedge \cdots \wedge i(u)v_i \wedge \cdots \wedge v_j.$$

2.2 TENSOR FIELDS AND DIFFERENTIAL FORMS

We have already seen that each point m of a differentiable manifold M possesses a tangent vector space M_m of dimension n, so it is natural to apply the previous algebraic constructions by identifying V with this tangent space. However, we wish to do this in such a way that the resulting tensors vary smoothly over M. We make these ideas more precise as follows.

We define:

$$T_{r,s}(M) = \bigcup_{m \in M} (M_m)_{r,s}, \quad \text{the tensor bundle of type } (r,s) \text{ over } M;$$

$$\Lambda_k^*(M) = \bigcup_{m \in M} \Lambda_k(M_m^*), \quad \text{the exterior } k\text{-form bundle over } M;$$

$$\Lambda^*(M) = \bigcup_{m \in M} \Lambda(M_m^*), \quad \text{the exterior algebra bundle over } M.$$

We now indicate how these bundles can be given natural differentiable structures so that they become differentiable manifolds in such a way that the projection maps to M are smooth. It will be sufficient to consider $T_{r,s}(M)$ as the other two cases are treated similarly. Let (U, ϕ) be a coordinate system on M with coordinates $y_1, y_2, \ldots, y_n$. Then coordinate bases for M and M^* are given respectively by $\{\partial/\partial y^i\}$ and $\{dy^i\}$, for $m \in U$, and these determine a basis for $T_{r,s}(M_m)$, $m \in U$. These give rise to a mapping of the inverse image of U in $T_{r,s}(M)$ under the projection map to $\phi(U) \times \mathbb{R}^n \times \cdots \times \mathbb{R}^n$. These maps give coordinate systems which themselves determine a differentiable manifold structure.

We define a tensor field of type (r, s) on M as a C^∞-mapping of M into $T_{r,s}(M)$ such that this composed with the canonical projection $T_{r,s}(M) \to M$ yields the identity mapping. Briefly we can say that such a tensor field is a section of the tensor bundle. Alternatively we can say that $\alpha: M \to T_{r,s}(M)$ is a tensor field if and only if for each coordinate system $(U, y_1, \ldots, y_n)$ of M we have

$$\alpha | U = a_{j_1 \cdots j_s}^{i_1 \cdots i_r} \frac{\partial}{\partial y^{i_1}} \otimes \cdots \otimes \frac{\partial}{\partial y^{i_r}} \otimes dy^{j_1} \otimes \cdots \otimes dy^{j_s}$$

where $a_{j_1 \cdots j_s}^{i_1 \cdots i_r}$ are C^∞-functions over U.

Similarly we define a field of alternating tensors of type $(0, r)$, commonly called a differential r-form, as a C^∞-mapping $M \to \Lambda_r^*(M)$ such that this composed with the canonical projection yields the identity. Moreover, the map $\beta: M \to \Lambda_r^*(M)$ is a differential r-form if and only if we have

$$\beta|U = \sum_{i_1 < \cdots < i_r} b_{i_1 \ldots i_r} dy^{i_1} \wedge \cdots \wedge dy^{i_r}$$

where $b_{i_1 \ldots i_r}$ are C^∞-functions on U. The set of 0-forms is seen to be the ring of C^∞-functions $C^\infty(M)$ defined over M. We see that forms can be added, multiplied by C^∞-functions, and given a product $\wedge$ since we can define these operations pointwise. Let $E^r(M)$ denote the set of all smooth r-forms on M and let $E(M)$ denote the set of all differential forms over M. Then $E(M)$ is a module over the ring of C^∞-functions on M. Let $X_1, \ldots, X_r$ be vector fields on M and let $\omega \in \Lambda_r(M^*)$. Then $\omega(X_1, \ldots, X_r)$ makes sense — it is a function defined pointwise by

$$\omega(X_1, \ldots, X_r)(p) = \omega_p(X_1(p), \ldots, X_r(p)).$$

Let $\mathfrak{X}(M)$ denote the C^∞-module of vector fields on M. Then

$$\omega: \mathfrak{X}(M) \times \cdots \times \mathfrak{X}(M) \to C^\infty(M)$$

(k copies of $\mathfrak{X}(M)$) is an alternating map which is multilinear over the module $\mathfrak{X}(M)$ into $C^\infty(M)$. For example, if ω is a 2-form, then

$$\omega(X, fY + gZ) = f\omega(X, Y) + g\omega(X, Z),$$

where f and g are C^∞-functions.

Let $\omega_1, \ldots, \omega_r$ be 1-forms and let $X_1, \ldots, X_r \in \mathfrak{X}(M)$. Then $(\omega_1 \wedge \cdots \wedge \omega_r)(X_1, \ldots, X_r)$ is the determinant of the matrix whose element in the jth row and kth column is $\omega_j(X_k)$. In an analogous way a tensor field T of type (r, s) may be considered as a multilinear map with respect to $C^\infty(M)$ given by

$$T: E^1(M) \times \cdots \times E^1(M) \times \mathfrak{X}(M) \times \cdots \times \mathfrak{X}(M) \to C^\infty(M)$$

(r copies of $E^1(M)$ and s copies of $\mathfrak{X}(M)$).

2.3 EXTERIOR DERIVATION

We now define an operator $d: E(M) \to E(M)$ which is characterized by the following properties:

(1) d is an $\mathbb{R}$-linear mapping from $E(M)$ into itself such that

$$d(E^r) \subset E^{r+1};$$

2.3 Exterior derivation

(2) Let $f \in E^0(M)$ $(= C^\infty(M))$; then df is the 1-form given by
$$df(X) = Xf;$$

(3) Let $\omega \in E^r$ and $\eta \in E^s$. Then
$$d(\omega \wedge \eta) = d\omega \wedge \eta + (-1)^r \omega \wedge d\eta;$$

(4) $d^2 = 0$.

In terms of a local coordinate system, if
$$\omega = \sum f_{i_1 \ldots i_r} dy^{i_1} \wedge \cdots \wedge dy^{i_r} \quad \text{where } i_1 < i_2 < \cdots < i_r,$$

we define $d\omega$ by
$$d\omega = \sum df_{i_1 \ldots i_r} \wedge dy^{i_1} \wedge \cdots \wedge dy^{i_r}.$$

This definition is not yet satisfactory because we have to show that it is independent of the coordinate system chosen. We note that our definition satisfies properties (1), (2), (3), and (4). Certainly (1) and (2) are trivially satisfied. To check that (3) is satisfied it will be sufficient to consider
$$\omega = f dy^{i_1} \wedge \cdots \wedge dy^{i_r}; \quad \eta = g dy^{j_1} \wedge \cdots \wedge dy^{j_s}$$
where the non-trivial cases involve $r > 0$, $s > 0$. Indeed for $r = s = 0$ the equation is satisfied because
$$d(fg) = df \cdot g(p) + f(p) \cdot dg.$$
If ω and η involve the same differential dy^i, both sides are zero and the equation is trivially satisfied. Assume that
$$\omega \wedge \eta = \varepsilon fg \, dy^{l_1} \wedge \cdots \wedge dy^{l_{r+s}}$$
where $l_1 < l_2 < \cdots < l_{r+s}$ and $\varepsilon = \pm 1$ according to the sign of the permutation giving this ordering. Then

$$\begin{aligned}
d(\omega \wedge \eta) &= \varepsilon d(fg) \wedge dy^{l_1} \wedge \cdots \wedge dy^{l_{r+s}} \\
&= \varepsilon (df_p g(p) + f(p) dg_p) \wedge dy^{l_1} \wedge \cdots \wedge dy^{l_{r+s}} \\
&= (df_p \wedge dy^{i_1}|_p \wedge \cdots \wedge dy^{i_r}|_p) \wedge (g(p) dy_p^{j_1}|_p \wedge \cdots \wedge dy^{j_s}|_p) \\
&\quad + (-1)^r (f(p) dy^{i_1}|_p \wedge \cdots \wedge dy^{i_r}|_p) \wedge \\
&\qquad (dg_p \wedge dy_p^{j_1}|_p \wedge \cdots \wedge dy^{j_s}|_p). \\
&= d\omega \wedge \eta + (-1)^r \omega \wedge d\eta.
\end{aligned}$$

To show that (4) is satisfied, let f be defined on a coordinate neighbourhood $(U, y^1, \ldots, y^n)$. Then

2 Tensors and differential forms

$$df = \left(\frac{\partial f}{\partial y^i}\right) dy^i$$

and

$$d(df)_p = d\left(\frac{\partial f}{\partial y^i}\right)\bigg|_p \wedge dy^i\bigg|_p = \left(\frac{\partial^2 f}{\partial y^i \partial y^j}\right)_p dy^i|_p \wedge dy^j|_p.$$

This is clearly zero, since the coefficients are symmetric in j and i while the term $dy^i \wedge dy^j$ is skew-symmetric in j and i. Thus our definition satisfies the required conditions.

To show that our definition is independent of the particular coordinate system chosen, let d' be the corresponding operator relative to the coordinate system $(U, y', \ldots, y'^n)$. There is no loss of generality in taking the same U. Let

$$\omega = a_{i_1 \ldots i_r} dy^{i_1} \wedge \cdots \wedge dy^{i_r} = a_{i_1 \ldots i_r}' dy'^{i_1} \wedge \cdots \wedge dy'^{i_r}$$

Since d' has the same properties (1), (2), (3), and (4) we have

$$d'\omega|_p = d' \sum a_{i_1 \ldots i_r} dy^{i_1} \wedge \cdots \wedge dy^{i_r}|_p$$

$$= \sum d'(a_{i_1} \cdots a_{i_r} \wedge dy^{i_1} \wedge \cdots \wedge dy^{i_r})|_p$$

$$= \sum d' a_{i_1} \cdots a_{i_r} \wedge dy^{i_1} \wedge \cdots \wedge dy^{i_r}|_p$$

$$+ \sum (-1)^{k-1} a_{i_1} \cdots a_{i_r}|_p dy^{i_1} \wedge \cdots \wedge d'(dy^{i_k})|_p \wedge \cdots dy^{i_r}|_p$$

$$= \sum d(a_{i_1} \cdots a_{i_r})|_p dy^{i_1} \wedge \cdots dy^{i_r}|_p$$

$$= d\omega|_p,$$

thus showing that our definition is independent of the coordinate system.

We still have to show that there is a uniquely defined operator with properties (1), (2), (3), and (4). To do this we prove the following lemma.

Lemma (2.3.1) The operators d and d' coincide entirely if they coincide with their action on $C^\infty(M)$ and on $E^1(M)$.

Suppose that an r-form ω vanishes on an open set U. Then we prove that $d\omega$ vanishes on U. To prove this, for each $p \in U$ let f be a function such that $f(p) = 0$ and $f = 1$ outside U. Then $\omega = f\omega$ and hence $d\omega = df \wedge \omega + f d\omega$. Since ω and f vanish at p so does $d\omega$. Thus, if two forms ω and ω' coincide on an open set U, then $d\omega = d\omega'$ on U.

We now proceed with the proof of the lemma. Let $D = d - d'$. We

2.3 Exterior derivation

have to prove that if D annihilates $C^\infty(M)$ and $E^1(M)$, then it is zero on $\Lambda^*(M)$. Let p be an arbitrary point of M, and let $(V, y^1, \ldots, y^n)$ be a coordinate system valid in some neighbourhood of p. Let

$$\omega = \sum a_{i_1 \ldots i_r} dy^{i_1} \wedge \cdots \wedge dy^{i_r}, \qquad i_1 < i_2 < \cdots < i_r.$$

If we write $\omega^j = dy^j$ we may extend $a_{i_1 \ldots i_r}$, ω^i to the whole of M. Let us assume that $D = 0$ in some smaller neighbourhood U of V. We have to show that

$$D(a_{i_1 \ldots i_r} \omega^{i_1} \wedge \cdots \wedge \omega^{i_r}) = 0,$$

and this clearly follows from property (3). Thus we have $D\omega = 0$. Since we have now proved that the actions of d and d' coincide on $C^\infty(M)$ and $E^1(M)$, the uniqueness of the exterior derivative operator is now established.

The effect of mappings

Let M, N be two differentiable manifolds and let $\psi: M \to N$ be a differentiable mapping. We have seen in Chapter 1 that this map induces the differential $d\psi: W_m \to N_{\psi(m)}$ and its transpose $\delta\psi: N^*_{\psi(m)} \to M^*_m$. The latter induces an algebra homomorphism $\delta\psi: \Lambda(N^*_{\psi(m)}) \to \Lambda(M^*_m)$. Thus if ω is an r-form on N, we may 'pull it back' to give an r-form on M by setting

$$\delta\psi(\omega)|_m = \delta\psi(\omega|_{\psi(m)}).$$

This is one of the advantages of working with forms rather than vector fields, because the image of a vector field under $d\psi$ is not necessarily a vector field.

An important property of the pull-back operator is that it commutes with d. That is, if we start with a form on N, take its exterior derivative, and pull it back to M we get precisely the same result as if we had first pulled the form back to M and then taken the exterior derivative. To prove this we observe that

$$d(\delta\psi(w))|_m = d\left[\sum (a_{i_1 \ldots i_r} \circ \psi) d(y_{i_1} \circ \psi) \wedge \cdots \wedge d(y_{i_r} \circ \psi)\right]\bigg|_m$$

$$= \left[\sum d(a_{i_1 \ldots i_r} \circ \psi) \wedge d(y_{i_1} \circ \psi) \wedge \cdots \wedge d(y_{i_r} \circ \psi)\right]\bigg|_m$$

$$= \delta\psi\left[\sum da_{i_1 \ldots i_r} \wedge dy_{i_1} \wedge \cdots \wedge dy^{i_r}\right]\bigg|_m$$

$$= \delta\psi(d\omega)|_m$$

thus completing the proof.

2.4 DIFFERENTIAL IDEALS AND THE THEOREM OF FROBENIUS

Let D be a p-dimensional distribution on M. We say that the q-form ω annihilates D if for each $m \subset M$,

$$\omega_m(v_1, \ldots, v_q) = 0$$

whenever $v_1, \ldots, v_q \in D$.

A form $\omega \in E(M)$ is said to annihilate D if each of the homogeneous parts of ω annihilates D. Let

$$I(D) = \{\omega \in E(M): \omega \text{ annihilates } D\}.$$

We say that a collection of 1-forms $\omega_1, \ldots, \omega_k$ on M is independent if the restriction to each $m \in M$ gives an independent set of M_m^*. We now have the following result.

Proposition (2.4.1) Let D be a smooth p-dimensional distribution on M. Then

(i) $I(D)$ is an ideal in $E(M)$;
(ii) $I(D)$ is locally generated by $n - p$ independent 1-forms;
(iii) If $I \subset E(M)$ is an ideal locally generated by $n - p$ independent 1-forms, then there exists a unique C^∞-distribution D of dimension p on M such that $I = I(D)$.

The proof is straightforward and we leave the details to the reader.

Definition (2.4.2) The ideal I is called a differential ideal if it is closed under exterior differentiation, that is, $d(I) \subset I$.

The importance of this concept is shown by the following proposition.

Proposition (2.4.3) A distribution D on M is involutive if and only if $I(D)$ is a differential ideal.

To prove this let ω be an r-form in $I(D)$ and let $X_0, \ldots, X_r$ be smooth vector fields lying in D. From our formula for the exterior derivative of an r-form we have

$$d\omega(X_0, \ldots, X_r) = \sum_{i=0}^{r} (-1)^r X_i \{\omega(X_0, \ldots, \hat{X}_i, \ldots, X_r)\}$$
$$+ \sum_{i<j} (-1)^{i+j} \omega([X_i, X_j], X_0,$$
$$\ldots, \hat{X}_i, \ldots, \hat{X}_j, \ldots, X_r).$$

If we assume that D is involutive, the right-hand side vanishes and so $d\omega(X_0, \ldots, X_r) = 0$. Hence $d\omega \in I(D)$ and I is a differential ideal.

2.4 Differential ideals and the theorem of Frobenius

To prove the converse, suppose that $I(D)$ is a differential ideal. Let X_0 and X_1 be vector fields lying in D and let $m \in M$. From the previous proposition we know that there exist independent 1-forms $\omega_1, \ldots, \omega_{n-r}$ generating $I(D)$ on a neighbourhood U of m. By the usual partition of unity argument we may multiply the forms ω_i by a C^∞-function which assumes the value 1 on a neighbourhood of m and has compact support in U. For convenience we shall use the same notation to describe these forms which extend over the whole of M. Then, from the formula for exterior differentiation, we have

$$d\omega_i(X_0, X_1) = X_0 \omega_i(X_1) - X_1 \omega_i(X_0) - \omega_i[X_0, X_1].$$

Using the fact that $I(D)$ is a differential ideal and $\omega_i \in I(D)$ we have

$$\omega_i(\lfloor X_0, X_1 \rfloor)(m) = 0 \qquad \text{for } i = 1, \ldots, n - r.$$

It follows that $[X_0, X_1](m) \in D(m)$, and D is therefore involutive. We are now able to restate the theorem of Frobenius in terms of differential forms as follows.

Theorem (2.4.4) Let $I \subset E(M)$ be a differential ideal locally generated by $n - r$ independent 1-forms. Then for each $m \in M$ there exists a unique maximal, connected integral manifold of I through m which has dimension r.

We close this section by stating and proving an important result which has numerous applications as we shall see later.

Lemma (2.4.5) (Cartan's lemma) Let $r \leqslant n$ and let $\omega_1, \ldots, \omega_r$ be 1-forms on M which are linearly independent at each $m \in M$. Here $n = \dim M$. Let $\theta_1, \ldots, \theta_r$ be 1-forms on M with the property that

$$\sum_{i=1}^{r} \theta_i \wedge \omega_i = 0.$$

Then there exist C^∞-functions h_{ij} on M with $h_{ij} = h_{ji}$ such that

$$\theta_i = \sum_{j=1}^{r} h_{ij} \omega_j, \qquad (i = 1, \ldots, r).$$

Proof. Introduce $\{\omega_\alpha\}$, $\alpha = r + 1, \ldots, n$, so that $\{\omega_A\}$, $A = 1, \ldots, n$ forms a basis for M_m^*. Then

$$\theta_i = \sum_{A=1}^{n} a_{Ai} \omega_A$$

so that

$$\sum_{A_i=1}^{r} \theta_i \wedge \omega_i = 0 \Rightarrow \sum a_{Ai} \omega_A \wedge \omega_i = 0.$$

We must have $a_{ij} = a_{ji}$, $a_{\alpha i} = 0$. Hence, writing $h_{ij} = a_{ij}$, we get

$$\theta_i = \sum_{j=1}^{n} h_{ij}\omega_j.$$

It is convenient to consider here another lemma which will be very useful later on.

Lemma (2.4.6) *Let C_{ijk} be a set of numbers such that C_{ijk} is symmetric with respect to the first two indices and skew-symmetric with respect to the last two indices. Then $C_{ijk} = 0$.*

The proof is trivial because

$$C_{ijk} = -C_{ikj} = -C_{kij} = C_{kji} = C_{jki} = -C_{jik} = -C_{ijk}.$$

We shall see later that this trivial lemma has highly non-trivial consequences concerning the uniqueness of a set of 1-forms.

2.5 ORIENTATION OF MANIFOLDS

We are now in a position to give a fuller treatment of the topic of orientation of manifolds mentioned briefly in Chapter 1. We begin with the concept of an orientable vector space. Let V be a vector space over the real numbers and let $(e_1, \ldots, e_n)$, $(f_1, \ldots, f_n)$ be bases of V. Set

$$f_i = \sum_{j=1}^{n} \alpha_i^j e_j.$$

Then the bases are said to have the same orientation if $\det(\alpha_i^j) > 0$. If $(g_1, \ldots, g_n)$ is another basis such that

$$g_i = \sum_{j=1}^{n} \beta_i^j f_j$$

then

$$g_i = \sum_{j,k} \beta_i^j \alpha_j^k e_k$$

and since

$$\det\left(\sum \beta_i^j \alpha_j^k\right) = \det(\beta_i^j) \det(\alpha_j^k),$$

it follows that having the same orientation is an equivalence relation. There are, in fact, just two equivalence classes of bases. This leads to the following definition.

2.5 Orientation of manifolds

Definition (2.5.1) An oriented vector space is a vector space together with an equivalence class of bases, consisting of all those bases with the same orientation as a given basis. These bases are called positively oriented.

Theorem (2.5.2) Let Ω be a non-zero alternating covariant tensor on V of order $n = \dim V$. Let $e_1, \ldots, e_n$ be a basis of V. Then for any set of vectors $v_1, \ldots, v_n$ with $v_i = \Sigma \alpha_i^j e_j$ we have

$$\Omega(v_1, \ldots, v_n) = \det(\alpha_j^i)\Omega(e_1, \ldots, e_n).$$

Proof. $\Omega(v_1, \ldots, v_n) = \sum_{j_1 \cdots j_n} \alpha_1^{j_1} \alpha_2^{j_2} \cdots \alpha_n^{j_n} \Omega(e_1, \ldots, e_n)$

$$= \sum_{\sigma \in \sigma_n} (\text{sign } \sigma) \alpha_1^{\sigma(1)} \alpha_2^{\sigma(2)} \cdots \alpha_n^{\sigma(n)} \Omega(e_1, \ldots, e_n)$$

$$= \det(\alpha_j^i)\Omega(e_1, \ldots, e_n),$$

where σ_n is the symmetric group of n letters.

An immediate corollary to the theorem is that the non-vanishing tensor Ω has the same sign on two similarly oriented bases, but opposite signs on two bases which are differently oriented. Thus the choice of a non-zero tensor such as Ω determines an orientation.

In particular if V is a Euclidean space with a positive definite inner product, we may choose an orthonormal basis $e_1, e_2, \ldots, e_n$ to determine the orientation and an n-form Ω whose value on $(e_1, \ldots, e_n)$ is $+1$. Thus the value of Ω on any orthonormal basis is ± 1, and Ω is uniquely determined up to sign. When $n = 2$, $\Omega(v_1, v_2)$ is interpreted as the area of the parallelogram determined by v_1, v_2. When $n = 3$, $\Omega(v_1, v_2, v_3)$ is the volume of the parallelopiped determined by v_1, v_2, v_3. However Euclidean space is exceptional because one can identify the tangent space at a point with the Euclidean space itself. In order to extend these ideas to a differentiable manifold, it is necessary to orient each tangent space M_m in such a way that the orientation of neighbouring tangent space agree.

Definition (2.5.3) The differentiable manifold M is said to be orientable if it is possible to choose a $C^\infty - n$ form Ω on M which is non-zero at any point of M.

Clearly such an n-form orients each tangent space of M. Moreover, it is by no means unique, for any form $\Omega' = \lambda\Omega$ where $\lambda > 0$ on M would give the same orientation to M. Consider $\mathbb{R}^n$ with the form

$$\tilde{\Omega} = dx^1 \wedge \cdots \wedge dx^n$$

corresponding to the orientation of the frame $\left(\dfrac{\partial}{\partial x^1}, \ldots, \dfrac{\partial}{\partial x^n}\right)$.

This is called the natural orientation of $\mathbb{R}^n$ but it depends on the coordinate system chosen. Let U, V be open subsets of $\mathbb{R}^n$, and denote the restrictions of $\tilde{\Omega}$ to U and V by $\tilde{\Omega}_U, \tilde{\Omega}_V$, respectively. A diffeomorphism $F: U \to V$ is said to be orientation preserving if $F^*(\tilde{\Omega}_V) = \lambda \tilde{\Omega}_U$ for some strictly positive function λ on U.

We can generalize this notion by saying that a diffeomorphism $F: M_1 \to M_2$ of two manifolds oriented respectively by Ω_1, Ω_2 is orientation preserving if

$$F^*(\Omega_2) = \lambda \Omega_1$$

where λ is a strictly positive C^∞-function on M_1. We now show that our present definition is consistent with that given in Chapter 1.

Definition (2.5.4) *A differentiable manifold M is orientable if it can be covered with coherently oriented coordinate neighbourhoods $\{U_\alpha, \phi_\alpha\}$; that is, such that if $U_\alpha \cap U_\beta \neq \emptyset$, then $\phi_\alpha \circ \phi_\beta^{-1}$ is an orientation preserving diffeomorphism of subsets of $\mathbb{R}^n$.*

We now establish the consistency.

Theorem (2.5.5) *A differentiable manifold is orientable if and only if it has a covering $\{U_\alpha, \phi_\alpha\}$ of coherently oriented coordinate neighbourhoods.*

To prove that a manifold orientable with respect to Definition (2.5.3) is also orientable with respect to Definition (2.5.4) is straightforward. For suppose Ω is a nowhere vanishing n-form which determines the orientation. Choose any covering $\{U_\alpha, \phi_\alpha\}$ with local coordinates $x_\alpha^1, \ldots, x_\alpha^n$. If we denote by Ω_{U_α} the restriction of Ω to U_α we have

$$\phi_\alpha^{-1*} \Omega_{U_\alpha} = \lambda_\alpha(x) \, dx_\alpha^1 \wedge \cdots \wedge dx_\alpha^n.$$

We can easily arrange that $\lambda_\alpha(x)$ is positive because a change of coordinates from $(x^1, \ldots, x^n)$ to $(-x^1, \ldots, x^x)$ will change the sign of $\lambda_\alpha(x)$. For U_β we can choose coordinates $(x_\beta^1, \ldots, x_\beta^n)$ so that $\phi_\beta^{-1*} \Omega_{U_\beta} = \lambda_\beta(x) \, dx_\beta^1 \wedge \cdots \, dx_\beta^n$, with λ_β positive. If $U_\alpha \cap U_\beta \neq \emptyset$ we see that the formula for the change (of component) of the n-form is given by

$$\lambda_\alpha \det\left(\frac{\partial x_\alpha^i}{\partial x_\beta^j}\right) = \lambda_\beta.$$

Since λ_α and λ_β have been chosen positive, we see that the determinant of the Jacobian is positive and hence the coordinate neighbourhoods are coherently oriented.

To prove the converse is more difficult because we are given information locally about each point from which we have to construct a global

2.5 Orientation of manifolds

n-form. We make use of a partition of unity (see the appendix for details of this tool).

Definition (2.5.6) A C^∞-partiton of unity on M is a collection of C^∞-functions $\{f_\gamma\}$ defined on M such that

(i) $f_\gamma \geq 0$ on M;
(ii) $\{\text{support}(f_\gamma)\}$ form a locally finite covering of M;
(iii) $\Sigma_\gamma f_\gamma(x) = 1$.

A partition of unity is said to be *subordinate* to an open covering $\{A_\alpha\}$ of M if for each γ there is an A_α such that support $(f_\gamma) \subset A_\alpha$. In the appendix we prove that every open covering of M has a partition of unity subordinate to it. To proceed with the proof of the equivalence of the definitions, we suppose that M has a covering by a coherently oriented coordinate neighbourhood system $\{U_\alpha, \phi_\alpha\}$.

Let $\{f_i\}$ be a subordinate partition of unity, and for each $i = 1, 2, \ldots$, choose a coordinate neighbourhood $U_{\alpha_i}, \phi_{\alpha_i}$ such that support $(f_i) \subset U_{\alpha_i}$; relabel these neighbourhoods $\{U_i, \phi_i\}$. Then $\{U_i\}$ cover M. Moreover, if $U_i \cap U_j \neq \emptyset$, our hypothesis shows that the Jacobian matrix of $\phi_i \circ \phi_j^{-1}$ has positive determinant on $U_i \cap U_j$. We define Ω by

$$\Omega = \sum_i f_i \phi_i^*(dx_i^1 \wedge dx_i^2 \wedge \cdots \wedge dx_i^n),$$

and extend to the whole of M since each summand is zero outside the closed set support (f_i). We now have to show that $\Omega_p \neq 0$ at each point $p \in M$. Let (V, ψ) be a coordinate neighbourhood of p coherently oriented with the $\{U_i, \phi_i\}$ such that V intersects only a finite number of the U_i, say, $i = i_1, \ldots, i_k$. Let $y^1, \ldots, y^n$ be local coordinates in V. Then we have

$$\Omega_p = \sum_{j=1}^k f_{i_j}(p) \phi_{i_j}^*(dx_{i_j}^1 \wedge \cdots \wedge dx_{i_j}^n)$$

$$= \sum f_{i_j}(p) \det \left(\frac{\partial x_{i_j}^k}{\partial y^l}\right)_{\psi(p)} \psi^*(dy^1 \wedge \cdots \wedge dy^n).$$

Now each $f_{i_j} \geq 0$ on M and at least one is strictly positive at p. Moreover, by hypothesis, the Jacobian determinants are all positive. Hence $\Omega_p \neq 0$ and since p was arbitrary, the theorem is proved.

Example (2.5.7) The real projective spaces.
Consider first the real projective plane defined as follows. Let $X = \mathbb{R}^3 \setminus \{0\}$, that is, the space of triples of real numbers

$x = (x_1, x_2, x_3)$ excluding $(0, 0, 0)$. We define $x \sim y$ if there is a real non-zero number t such that $y = tx$, that is, $(y_1, y_2, y_3) = (tx_1, tx_2, tx_3)$. The equivalence classes $[x]$ can be considered as lines through the origin. The quotient space is denoted by $\mathbb{R}P^2$ and called the real projective plane. An equivalent picture is obtained by identifying antipodal points on the unit sphere $S^2 = \{(x_1, x_2, x_3): x_1^2 + x_2^2 + x_3^2 = 1\}$, that is, by imposing the condition that the points (x_1, x_2, x_3) and $(-x_1, -x_2, -x_3)$ are equivalent. In this way $\mathbb{R}P^2$ becomes the spaces of antipodal points of S^2. Since S^2 is Hausdorff the same property holds for $\mathbb{R}P^2$. To prove that $\mathbb{R}P^2$ has a countable basis, it is sufficient to show that $\mathbb{R}P^2$ is covered by three coordinate neighbourhoods defined as follows.

Let $\tilde{U}_i = \{x \in S^2 : x^i \neq 0\}$, and let $U_i = \pi(\tilde{U}_i)$ where π is the projection map sending x to the equivalence class to which it belongs. Then $\phi_i: U_i \to \mathbb{R}^2$ is defined by choosing any $x = (x^1, x^2, x^3)$ representing $[x] \in U_i$ and putting

$$\phi_1(x) = (x^2/x^1, x^2/x^1), \quad \phi_2(x) = (x^1/x^2, x^2/x^2),$$
$$\phi_3(x) = (x^1/x^3, x^2/x^3).$$

Then each ϕ_i is properly defined, continuous and bijective. If $z \in \mathbb{R}^2$, then

$$\phi_1^{-1}(z^2, z^3) = \pi(1, z^2, z^3)$$

and hence ϕ_1^{-1} is continuous. Similarly for ϕ_2^{-1} and ϕ_3^{-1}. Thus $\mathbb{R}P^2$ is a C^∞-manifold since $\phi_i \circ \phi_j^{-1}$ is C^∞ where it is defined.

Exercise (2.5.8) If $\mathbb{R}P^n$ is defined in an analogous way from $X = \mathbb{R}^{n+1} \setminus \{0\}$ by taking $x \sim y$ if there is a non-zero t such that $y = tx$, prove that $\mathbb{R}P^n$ is a C^∞-differentiable manifold.

Exercise (2.5.9) Show that the antipodal map $x \to -x$ on the n-sphere S^n is orientation preserving if and only if n is odd. Hence deduce that $\mathbb{R}P^n$ is orientable if n is odd and non-orientable if n is even.

2.6 COVARIANT DIFFERENTIATION

We have already seen that the operation of exterior derivation maps p-forms to $(p+1)$-forms, $p \geq 0$. A natural question arises whether the partial derivative operator maps tensor fields to tensor fields. This is certainly the case for tensors of type $(0, 0)$, since on making a change of local coordinates from (x^i) to $(x^{i'})$ we have

$$\frac{\partial f}{\partial x^{i'}} = \left(\frac{\partial x^i}{\partial x^{i'}}\right) \frac{\partial f}{\partial x^i}$$

2.6 Covariant differentiation

which is of the required tensorial type. However, if λ^i are the components of a tensor of type (1, 0), we have under a similar coordinate transformation

$$\lambda^{i'} = \left(\frac{\partial x^{i'}}{\partial x^i}\right) \lambda^i,$$

$$\frac{\partial \lambda^{i'}}{\partial x^{j'}} = \left(\frac{\partial x^{i'}}{\partial x^i}\right) \frac{\partial \lambda^i}{\partial x^j} \frac{\partial x^j}{\partial x^{j'}} + \lambda^i \left(\frac{\partial^2 x^{i'}}{\partial x^i \partial x^j}\right) \frac{\partial x^j}{\partial x^{j'}}.$$

The latter equation is not of the required type. If we wish the tensorial character to be preserved, clearly we must modify the operator $\partial/\partial x^i$. A reasonable suggestion is to write

$$\nabla_k \lambda^i = \partial_k \lambda^i + \Gamma^i_{kj} \lambda^j$$

for some suitable functions Γ^i_{kj}, and to determine if possible the law of transformation of the Γ's if the tensorial character of $\nabla_k \lambda^i$ is to be preserved. We have

$$\nabla_{k'} \lambda^{i'} = \partial_{k'} \lambda^{i'} + \Gamma^{i'}_{k'j'} \lambda^{j'} = \partial_{k'}\left(\lambda^i \frac{\partial x^{i'}}{\partial x^i}\right) + \Gamma^{i'}_{k'j'} \frac{\partial x^{j'}}{\partial x^j} \lambda^j$$

$$= \frac{\partial x^{i'}}{\partial x^i} \frac{\partial x^k}{\partial x^{k'}} \frac{\partial \lambda^i}{\partial x^k} + \lambda^i \frac{\partial^2 x^{i'}}{\partial x^k \partial x^i} \frac{\partial x^k}{\partial x^{k'}} + \Gamma^{i'}_{k'j'} \frac{\partial x^{j'}}{\partial x^j} \lambda^j.$$

We substitute in the first term for $\partial \lambda^i / \partial x^k$ in terms of $\nabla_k \lambda^i$ to get

$$\nabla_{k'} \lambda^{i'} = \frac{\partial x^{i'}}{\partial x^i} \frac{\partial x^k}{\partial x^{k'}} \left(\nabla_k \lambda^i - \Gamma^i_{kj} \lambda^j\right) + \lambda^i \frac{\partial^2 x^{i'}}{\partial x^k \partial x^i} \frac{\partial x^k}{\partial x^{k'}} + \Gamma^{i'}_{k'j'} \frac{\partial x^{j'}}{\partial x^j} \lambda^j.$$

We see that the tensorial character will be preserved if the Γ's satisfy

$$-\frac{\partial x^{i'}}{\partial x^i} \frac{\partial x^k}{\partial x^{k'}} \Gamma^i_{kj} + \frac{\partial^2 x^{i'}}{\partial x^k \partial x^j} \frac{\partial x^k}{\partial x^{k'}} + \Gamma^{i'}_{k'j'} \frac{\partial x^{j'}}{\partial x^j} = 0,$$

which may be written

$$\Gamma^{i'}_{k'h'} = \Gamma^i_{kh}\left(\frac{\partial x^{i'}}{\partial x^i}\right)\left(\frac{\partial x^h}{\partial x^{h'}}\right)\left(\frac{\partial x^k}{\partial x^{k'}}\right) - \left(\frac{\partial^2 x^{i'}}{\partial x^h \partial x^k}\right)\left(\frac{\partial x^h}{\partial x^{h'}}\right)\left(\frac{\partial x^k}{\partial x^{k'}}\right). \quad (2.1)$$

Thus, given an arbitrary set of n^3 differentiable functions Γ^i_{kj} in a coordinate system (x^i), the operator ∇_k will be tensorial provided that the components of Γ in the new coordinate system are as given above.

So we can now define a *connexion* to be a system of n^3 functions with respect to a coordinate system, whose law of transformation is given above. We can define the covariant derivative of a tensor of type (0, 0) to be the ordinary partial derivative. We see also that the covariant derivative of a tensor of type (0, 1) is determined by the formula

$$\nabla_k \omega_i = \partial \omega_i / \partial x^k - \Gamma^h_{ki} \omega_h$$

provided that we want the operator ∇_k to satisfy the Leibnitz rule for operating on a product and to commute with contraction. This follows because, for an arbitrary vector field λ^i,

$$\nabla_k(\lambda^i\omega_i) = (\partial_k\lambda^i)\omega_i + \lambda^i\partial_k\omega_i,$$

$$(\nabla_k\lambda^i)\omega_i + \lambda^i(\nabla_k\omega_i) = (\nabla_k\lambda^i)\omega_i - (\Gamma^i_{kj}\lambda^j)\omega_i + \lambda^i\partial_k\omega_i.$$

Thus

$$\nabla_k\omega_i = \partial_k\omega_i - \Gamma^h_{ki}\omega_h,$$

since we can 'cancel' λ^i after interchanging dummy suffixes i and j in the term involving Γ. In a similar way we can show that for any decomposable tensor of type (1, 1), say

$$T^i_j = \lambda^i\omega_j$$

we have

$$\nabla_k T^i_j = \partial_k T^i_j + \Gamma^i_{kh}T^h_j - \Gamma^h_{kj}T^i_h.$$

Since any tensor can be expressed as a linear combination of decomposable tensors, we see that the above equation holds for any tensor field of type (1, 1). Similarly we find

$$\nabla_k T^{ij} = \partial_k T^{ij} + \Gamma^i_{kh}T^{hj} + \Gamma^j_{kh}T^{ih},$$

$$\nabla_k T_{ij} = \partial_k T_{ij} - \Gamma^h_{ki}T_{hj} - \Gamma^h_{kj}T_{ih}.$$

In general

$$\nabla_k T^{r_1\cdots r_m}_{s_1\cdots s_p} = \partial_k T^{r_1\cdots r_m}_{s_1\cdots s_p}$$
$$+ \sum_{\alpha=1}^{m} T^{r_1\cdots r_{\alpha-1}jr_{\alpha+1}\cdots r_m}_{s_1\cdots s_p}\Gamma^{r_\alpha}_{kj} - \sum_{\beta=1}^{p} T^{r_1\cdots r_m}_{s_1\cdots s_{\beta-1}hs_{\beta+1}\cdots s_p}\Gamma^h_{ks_\beta}.$$

(2.2)

There seems little doubt that this is the way in which covariant differentiation with respect to a connexion was first considered. It has many disadvantages, the first obvious objection being that it depends upon coordinate systems, and the second because it seems to be an analytical device devoid of geometric meaning. However it is undoubtedly a powerful tool. Many mathematicians use it (perhaps in private!) and subsequently translate their discoveries into an invariant suffix-free notation which has more geometric significance. We shall feel free to use suffixes when convenient to do so.

Although the components Γ^k_{ij} do not transform like the components of a tensor, the quantities

$$T^k_{ij} = \Gamma^k_{ij} - \Gamma^k_{ji}$$

2.6 Covariant differentiation

are components of a tensor T, called the *torsion tensor* of the connexion. The connexion is said to be *symmetric* if $T = 0$. Then in every coordinate system we have $\Gamma^k_{ij} = \Gamma^k_{ji}$ and this motivates the word 'symmetric'. We shall find, however, that this terminology can be misleading so we prefer to use the description 'connexion with zero torsion' or 'torsion-free connexion'.

Proposition (2.6.1) The torsion tensor T satisfies $T(p) = 0$ if and only if there is a coordinate neighbourhood (U, x^i) around p such that $\Gamma^k_{ij}(p) = 0$ for all i, j, k.

Trivially $\Gamma^k_{ij}(p) = 0$ implies $T^k_{ij}(p) = 0$ so that $T(p) = 0$ in all systems of coordinates around p. Conversely, suppose $T^k_{ij}(p) = 0$ for all i, j, k relative to some coordinate system (x^k) valid in a neighbourhood of p. Define functions x'^k by

$$x'^k = [x^k(q) - x^k(p)] + \frac{1}{2}\Gamma^k_{ij}(p)[x^i(q) - x^i(p)][x^j(q) - x^j(p)].$$

Then $\partial x'^k/\partial x^l = \delta^k_l + \Gamma^k_{il}(p)[x^i(q) - x^i(p)]$ so that $\partial x'^k/\partial x^l|_p = \delta^k_l$ and hence (x'^k) is a coordinate system valid in some neighbourhood of p. Moreover we have

$$\frac{\partial^2 x'^k}{\partial x^l \partial x^i} = \Gamma^k_{il}(p).$$

Using the tranformation law for the Γ's gives $\Gamma^{k'}_{i'j'} = 0$, and the proof is complete.

We note that some authors use an alternative notation for covariant differentiation. For example, instead of $\nabla_j \lambda^i$ they write $\lambda^i_{;j}$ or $\lambda^i_{,j}$. Similarly $\nabla_k \nabla_j \lambda^i$ becomes $\lambda^i_{;jk}$ or $\lambda^i_{,jk}$ — note the different order of suffixes k and j in the alternative notation.

Let $f: M \to R$ be a smooth function. Then

$$\nabla_i f = \frac{\partial f}{\partial x^i},$$

$$\nabla_j \nabla_i f = \frac{\partial^2 f}{\partial x^i \partial x^j} - \Gamma^k_{ji} \frac{\partial f}{\partial x^k}.$$

Thus

$$(\nabla_j \nabla_i - \nabla_i \nabla_j) f = (\Gamma^k_{ij} - \Gamma^k_{ji}) \frac{\partial f}{\partial x^k} = T^k_{ij} \frac{\partial}{\partial x^k},$$

and only when $T = 0$ do the operators ∇_i and ∇_j commute. Even this is insufficient for commutation when the operators act on general tensor fields. For example we have

$$\nabla_k \nabla_j \lambda^i - \nabla_j \nabla_k \lambda^i = -R^i_{ljk} \lambda^l + T^l_{jk} \nabla_l \lambda^i,$$

$$\nabla_k \nabla_j \omega_i - \nabla_j \nabla_k \omega_i = R^l_{ijk} \omega_l + T^l_{jk} \nabla_l \omega_i,$$

where we have written

$$R^i_{jkl} = \frac{\partial \Gamma^i_{lj}}{\partial x^k} - \frac{\partial \Gamma^i_{kj}}{\partial x^l} + \Gamma^h_{lj}\Gamma^i_{kh} - \Gamma^h_{kj}\Gamma^i_{lh}.$$

The fact that R^i_{jkl} are the components of a tensor may be verified by checking that they transform appropriately under a change of coordinates. Later we shall see why this follows without such tedious computations. The tensor R is called the *curvature tensor* of the connexion — we will justify the use of the word 'curvature' later on.

2.7 IDENTITIES SATISFIED BY THE CURVATURE AND TORSION TENSORS

We claim that the following identities are valid:

$$R^i_{jkl} = -R^i_{jlk}. \tag{2.3}$$

$$R^i_{jkl} + R^i_{klj} + R^i_{ljk}$$
$$= \nabla_j T^i_{kl} + \nabla_k T^i_{jl} + \nabla_l T^i_{jk} + T^h_{jk}T^i_{hl} + T^h_{kl}T^i_{hj} + T^h_{lj}T^i_{hk}. \tag{2.4}$$

$$\nabla_l R^h_{ijk} + \nabla_j R^h_{ikl} + \nabla_k R^h_{ilj} + T^s_{jk}R^h_{ise} + T^s_{kl}R^h_{isj} + T^s_{lj}R^h_{isk} = 0. \tag{2.5}$$

When $T = 0$ identities (2.4) and (2.5) take the simpler form

$$R^i_{jkl} + R^i_{klj} + R^i_{ljk} = 0, \tag{2.4'}$$

$$\nabla_l R^h_{ijk} + \nabla_j R^h_{ikl} + \nabla_k R^h_{ilj} = 0. \tag{2.5'}$$

and these are known respectively as the first and second Bianchi identities. The proof of (2.3) is trivial. To prove (2.4) and (2.5) the reader is advised to take a large sheet of paper, write very small and compute! However we now give a short proof of identities (2.4') and (2.5').

Clearly it will be sufficient to prove that (2.4') and (2.5') hold at an arbitrary point p of M. Since we assume that $T = 0$ it follows from the previous proposition that we can choose a coordinate system so that the connexion coefficients Γ^k_{ij} are zero at p. Then

$$R^i_{jkl} = \frac{\partial \Gamma^i_{lj}}{\partial x^k} - \frac{\partial \Gamma^i_{kj}}{\partial x^l}$$

and identity (2.4') follows immediately. Moreover at p we have

$$\nabla_l R^h_{ijk} = \frac{\partial^2 \Gamma^h_{ki}}{\partial x^l \partial x^j} - \frac{\partial^2 \Gamma^h_{ji}}{\partial x^l \partial x^k},$$

from which identity (2.5') follows.

2.8 THE KOSZUL CONNEXION

Our previous treatment of connexions has been more or less on classical lines of the development of the subject. We now give a more modern treatment due to Koszul, known as the 'suffix-free tensor calculus', which is more satisfactory aesthetically and perhaps gives more insight.

Definition (2.8.1) A Koszul connexion on a C^∞-manifold M is a mapping $\nabla: \mathfrak{X}(M) \times \mathfrak{X}(M) \to \mathfrak{X}(M)$ which associates to any two C^∞-vector fields X, Y the vector field $\nabla_X Y$ such that:

(1) $\nabla_{X_1+X_2} Y = \nabla_{X_1} Y + \nabla_{X_2} Y$;
(2) $\nabla_X(Y_1 + Y_2) = \nabla_X Y_1 + \nabla_X Y_2$;
(3) $\nabla_{fX} Y = f\nabla_X Y$;
(4) $\nabla_X(fY) = f\nabla_X Y + (Xf)Y$.

Property (3) implies that at the point p, $\nabla_X Y(p)$ depends only on the vector $X(p)$. If we take $X = \partial/\partial x^i$, $Y = \partial/\partial x^j$, then we must have

$$\nabla_{\frac{\partial}{\partial x^i}} \left(\frac{\partial}{\partial x^j} \right) = L_{ij}^k \frac{\partial}{\partial x^k},$$

for some functions L_{ij}^k. If we choose a new coordinate system $x^{i'}$ then

$$\frac{\partial}{\partial x^i} = \frac{\partial x^{i'}}{\partial x^i} \frac{\partial}{\partial x^{i'}}$$

and we have

$$\nabla_{\left(\frac{\partial x^{i'}}{\partial x^i} \frac{\partial}{\partial x^{i'}}\right)} \left(\frac{\partial x^{j'}}{\partial x^j} \frac{\partial}{\partial x^{j'}} \right) = \left(\frac{\partial x^{i'}}{\partial x^i} \right) \left[\frac{\partial^2 x^{j'}}{\partial x^j \partial x^p} \cdot \frac{\partial x^p}{\partial x^{i'}} \frac{\partial}{\partial x^{j'}} + \frac{\partial x^{j'}}{\partial x^j} L_{i'j'}^{k'} \frac{\partial}{\partial x^{k'}} \right].$$

$$L_{ij}^k \frac{\partial}{\partial x^k} = \frac{\partial^2 x^{j'}}{\partial x^j \partial x^i} \frac{\partial}{\partial x^{j'}} + L_{i'j'}^{k'} \frac{\partial x^{i'}}{\partial x^i} \frac{\partial x^{j'}}{\partial x^j} \frac{\partial}{\partial x^{k'}}.$$

This gives, on interchanging dashed and undashed letters,

$$L_{i'j'}^{k'} \frac{\partial}{\partial x^{k'}} = \frac{\partial^2 x^j}{\partial x^{j'} \partial x^{i'}} \frac{\partial}{\partial x^j} + L_{ij}^k \frac{\partial x^i}{\partial x^{i'}} \frac{\partial x^j}{\partial x^{j'}} \frac{\partial}{\partial x^k}$$

so that the L's obey the same transformation law as the Γ's of a classical connexion. Conversely, given a set of n^3 functions this determines

$$\nabla_{\frac{\partial}{\partial x^i}} \left(\frac{\partial}{\partial x^j} \right)$$

and hence $\nabla_X Y$ for any vector fields X and Y. In fact, let $X = X^i \partial/\partial x^i$, $Y = Y^j \partial/\partial x^j$. Then

44   2   *Tensors and differential forms*

$$\nabla_X Y = X^i \left(\frac{\partial Y^j}{\partial x^i} \frac{\partial}{\partial x^j} + Y^j \Gamma^k_{ij} \frac{\partial}{\partial x^k} \right)$$

$$= X^i \left(\frac{\partial Y^j}{\partial x^i} + \Gamma^j_{ik} Y^k \right) \frac{\partial}{\partial x^j} = X^i (\nabla_i Y^j) \frac{\partial}{\partial x^j}.$$

We now define $\nabla_X f = Xf$, so that we have extended the domain of definition of ∇ to the set of C^∞-functions. We now show how the domain can be extended to include all tensor fields. Suppose ω is a tensor field of type $(0, 1)$, so that $\omega(Y)$ is a function. We want ∇_X to obey the Leibnitz law

$$\nabla_X (A \otimes B) = \nabla_X A \otimes B + A \otimes \nabla_X B$$

where A and B are tensor fields. Moreover we want ∇_X to commute with the contraction of tensors. This gives

$$\nabla_X (\omega(Y)) = X(\omega(Y)) = (\nabla_X \omega) Y + \omega(\nabla_X Y)$$

so we must have

$$(\nabla_X \omega)(Y) = X(\omega(Y)) - \omega(\nabla_X Y)$$

for all Y. This extends the operator to tensors of type $(0, 1)$. Since any tensor field is the sum of products of functions times tensor products of vector fields and tensors of type $(0, 1)$, we see that ∇_X extends in a uniquely determined way to operate on tensor fields of type (p, q). When expressed in terms of coordinates the resulting formula is precisely that given by (2.2).

If we write $X = X^i \partial/\partial x^i$, $Y = Y^j \partial/\partial x^j$, then

$$\nabla_X Y = X^i (\nabla_i Y^j) \frac{\partial}{\partial x^j}.$$

Thus for a given vector field Y we have for each $p \in M$ a transformation, linear over $\mathbb{R}$, $\nabla Y(p): M_p \to M_p$ given by

$$\nabla Y(p) X(p) = \nabla_{X_p} Y = (\nabla_X Y)(p).$$

In fact we have

$$\nabla Y = \nabla_i Y^k \, dx^i \otimes \frac{\partial}{\partial x^k}$$

showing that ∇Y is a tensor field of type $(1, 1)$, given in terms of its components by $\nabla_i Y^k$.

There is one great advantage of the Koszul approach, namely that the basis chosen for the vector fields need not be a coordinate basis. If $\{X_\alpha\}$ is an arbitrary basis for the vector fields defined over some neighbourhood U, then the equation

2.8 The Koszul connexion

$$\nabla_{X_\alpha} X_\beta = \gamma^\varepsilon_{\alpha\beta} X_\varepsilon$$

certainly determines n^3 functions but these are no longer to be identified with the Γ^k_{ij} which arise from coordinate bases.

Covariant differentiation along a curve

Let c be a parametrized curve in M, that is, c is a smooth map from the real numbers into M. A *vector field along* c is a map which assigns to each $t \in \mathbb{R}$ a tangent vector $V_t \in M_{c(t)}$. We want this map to be smooth in the sense that for any smooth function f on M the map $t \to V_t f$ must be a smooth function of t. As an example consider the velocity vector field dc/dt along c which is the vector field defined by

$$\frac{dc}{dt} = c_* \frac{d}{dt}$$

where d/dt is the standard vector field on the real numbers.

Let M have a connexion Γ. We now define an operation called *covariant differentiation along a curve* which associates to any vector field V_t along c a new vector field along c which we denote by DV/dt. We require this operator to satisfy the axioms:

(1) $\dfrac{D}{dt}(V + W) = \dfrac{DV}{dt} + \dfrac{DW}{dt}$;

(2) $\dfrac{D}{dt}(fV) = \dfrac{df}{dt}V + f\dfrac{DV}{dt}$,

where f is any smooth real-valued function on $\mathbb{R}$:

(3) if V is induced from a vector field Y on M, so that $V_t = Y(c(t))$ for each t, then DV/dt is equal to $\nabla_{\frac{dc}{dt}} Y$.

The problem is that previously we have defined covariant differentiation for vector fields on some neighbourhood of M, and we have to ensure that DV/dt depends only on the connexion and c, and is independent of the particular choice of the vector field Y. We now prove that there is only one operator satisfying (1), (2), and (3).

Choose a local system of coordinates and let the point $c(t)$ have coordinates $u^i(t)$, $i = 1, \ldots, n$. The given vector field is expressible uniquely in the form

$$V = v^j \frac{\partial}{\partial x^j}$$

where the components v^j are real-valued functions on some domain of $\mathbb{R}$. Then from axioms (1), (2), and (3) we have

$$\frac{\mathrm{D}V}{\mathrm{d}t} = \frac{\mathrm{d}v^j}{\mathrm{d}t}\frac{\partial}{\partial x^j} + v^j \frac{\mathrm{D}}{\mathrm{d}t}\left(\frac{\partial}{\partial x^j}\right)$$

$$= \frac{\mathrm{d}v^j}{\mathrm{d}t}\frac{\partial}{\partial x^j} + v^j \nabla_{\frac{\mathrm{d}c}{\mathrm{d}t}}\left(\frac{\partial}{\partial x^j}\right)$$

$$= \left(\frac{\mathrm{d}v^k}{\mathrm{d}t} + \Gamma^k_{ij}\frac{\mathrm{d}u^i}{\mathrm{d}t}v^j\right)\frac{\partial}{\partial x^k},$$

thus establishing uniqueness. Conversely if we use this equation as a definition of $\mathrm{D}V/\mathrm{d}t$, then clearly (1), (2), and (3) are satisfied, thus proving existence.

A vector field V along c is said to be *parallel* if the covariant derivative $\mathrm{D}V/\mathrm{d}t$ is identically zero. The importance of this concept is due to the following lemma.

Lemma (2.8.2) Let c be a given curve on M and let V_0 be a tangent vector to M at the point $c(0)$. Then there is one and only one parallel vector field V_t along c such that $V(0) = V_0$.

To prove the lemma we note that the components v^i of such a parallel vector field must satisfy the linear differential equations

$$\frac{\mathrm{d}v^k}{\mathrm{d}t} + \Gamma^k_{ij}\frac{\mathrm{d}u^i}{\mathrm{d}t}v^j = 0.$$

However a standard result from the theory of such equations (see Hurewicz (1958)) shows that the solutions v^k are uniquely determined by the initial values $v^k(0)$, and that the solutions are defined for all relevant values of t. The vector field V_t is said to be obtained from $V(0)$ by *parallel translation* along c.

The idea of parallel translation or parallel transport of vector fields is easily generalized to apply to parallel transport of tensor fields along c. For example if T is a tensor of type $(2, 3)$ we define

$$\left(\frac{\mathrm{D}}{\mathrm{d}t}T\right)^{hk}_{pqr} = \frac{\mathrm{d}}{\mathrm{d}t}T^{hk}_{pqr} + \Gamma^h_{ij}\frac{\mathrm{d}u^i}{\mathrm{d}t}T^{jk}_{pqr} + \Gamma^k_{ij}\frac{\mathrm{d}u^i}{\mathrm{d}t}T^{hj}_{pqr} - \Gamma^s_{ip}\frac{\mathrm{d}u^i}{\mathrm{d}t}T^{hk}_{sqr}$$

$$- \Gamma^s_{iq}\frac{\mathrm{d}u^i}{\mathrm{d}t}T^{hk}_{psr} - \Gamma^s_{ir}\frac{\mathrm{d}u^i}{\mathrm{d}t}T^{hk}_{pqs}$$

The idea of parallel translation is usually attributed to Levi-Civita, and gives some geometrical meaning to what was previously regarded as a purely analytical device.

Curvature and torsion tensors of Koszul connexions

We define the torsion tensor of a Koszul connexion by the relation

2.8 The Koszul connexion

$$T(X, Y) = \nabla_X Y - \nabla_Y X - [X, Y] \tag{2.6}$$

for any vector fields X, Y. To establish the tensorial nature of T we shall prove that T is a multilinear map of $\mathfrak{X}(M) \times \mathfrak{X}(M)$ into $\mathfrak{X}(M)$. Since T is skew-symmetric in X and Y, it is sufficient to establish linearity with respect to X. We have

$$\begin{aligned}T(f_1 X_1 + f_2 X_2, Y) &= \nabla_{f_1 X_1 + f_2 X_2} Y - \nabla_Y(f_1 X_1 + f_2 X_2) \\ &\quad - [f_1 X_1 + f_2 X_2, Y] \\ &= f_1 \nabla_{X_1} Y + f_2 \nabla_{X_2} Y - f_1 \nabla_Y X_1 - f_2 \nabla_Y X_2 - f_1 [X_1, Y] \\ &\quad - f_2 [X_2, Y] - (Y f_1) X_1 - (Y f_2) X_2 + (Y f_1) X_1 \\ &\quad + (Y f_2) X_2 \\ &= f_1 T(X_1, Y) + f_2 T(X_2, Y),\end{aligned}$$

and so we are through. In particular, if

$$T\left(\frac{\partial}{\partial x^i}, \frac{\partial}{\partial x^j}\right) = T_{ij}^k \frac{\partial}{\partial x^k}$$

then

$$T_{ij}^k \frac{\partial}{\partial x^k} = \Gamma_{ij}^k \frac{\partial}{\partial x^k} - \Gamma_{ji}^k \frac{\partial}{\partial x^k}$$

giving

$$T_{ij}^k = \Gamma_{ij}^k - \Gamma_{ji}^k$$

which agrees with the classical definition.

The curvature tensor R is defined by

$$R(X, Y)Z = \nabla_X \nabla_Y Z - \nabla_Y \nabla_X Z - \nabla_{[X, Y]} Z \tag{2.7}$$

where X, Y, and Z are smooth vector fields. We show that this expression is $\mathfrak{F}$-linear with respect to X, Y, and Z, and again, since it is skew in X and Y, it is sufficient to prove linearity with respect to X and Z. We have

$$\begin{aligned}R(f_1 X_1 + f_2 X_2, Y)Z &= f_1 \nabla_{X_1} \nabla_Y Z + f_2 \nabla_{X_2} \nabla_Y Z - \nabla_Y \{f_1 \nabla_{X_1} Z\} \\ &\quad - \nabla_Y \{f_2 \nabla_{X_2} Z\} \\ &\quad - \nabla_{f_1 [X_1, Y] + f_2 [X_2, Y] - (Y f_1) X_1 - (Y f_2) X_2} Z\end{aligned}$$

which simplifies to

$$f_1 R(X_1, Y)Z + f_2 R(X_2, Y)Z.$$

Similarly we may prove that
$$R(X, Y)(g_1 Z_1 + g_2 Z_2) = g_1 R(X, Y) Z_1 + g_2 R(X, Y) Z_2.$$
Perhaps, at first sight, it is surprising that no derivatives of Z are involved in the preceding calculation. As a consequence the value of $R(X, Y)Z$ at $p \in M$ depends on the value of Z at p, but not on the values of Z in a neighbourhood of p. If we write $X = \partial/\partial x^i$, $Y = \partial/\partial x^j$, $Z = \partial/\partial x^k$ we have

$$\begin{aligned} R(X, Y)Z &= \nabla_{\frac{\partial}{\partial x^i}} \left(\nabla_{\frac{\partial}{\partial x^j}} \frac{\partial}{\partial x^k} \right) - \nabla_{\frac{\partial}{\partial x^j}} \left(\nabla_{\frac{\partial}{\partial x^i}} \frac{\partial}{\partial x^k} \right) \\ &= \nabla_{\frac{\partial}{\partial x^i}} \left(\Gamma_{jk}^h \frac{\partial}{\partial x^h} \right) - \nabla_{\frac{\partial}{\partial x^j}} \left(\Gamma_{ik}^h \frac{\partial}{\partial x^h} \right) \\ &= \left(\frac{\partial \Gamma_{jk}^h}{\partial x^i} - \frac{\partial \Gamma_{ik}^h}{\partial x^j} + \Gamma_{jk}^p \Gamma_{ip}^h - \Gamma_{ik}^p \Gamma_{jp}^h \right) \frac{\partial}{\partial x^h} \\ &= R_{kij}^h \frac{\partial}{\partial x^h} \end{aligned}$$

which agrees with the classical formula. This shows that the Koszul treatment is consistent with the classical approach. The reader is again warned that roughly one half of the writers on differential geometry define the curvature tensor to be the **negative** of the definition of R given above. A good tip when reading any book or research paper on differential geometry is to observe the particular definition of curvature tensor used by that author.

2.9 CONNEXIONS AND MOVING FRAMES

Definition (2.9.1) *A frame at a point $p \in M$ is just a basis for the vector space M_p.*

Suppose now that we have over a neighbourhood U of M a set of n linearly independent vector fields X_i, such that at each point $p \in U$ the vectors $X_i(p)$ form a frame at p. Then the set of such frames constitutes a 'moving frame' in the sense of É. Cartan. The whole point of Cartan's approach is to choose a moving frame appropriate to each geometrical configuration, and this may well be impossible if one is restricted to coordinate frames of the form $(\partial/\partial x^i)$.

We have seen that a classical connexion is determined by n^3 functions Γ_{jk}^i which transform in a certain manner when the coordinates are changed. If we write

$$\omega_k^i = \Gamma_{jk}^i \, dx^j$$

2.9 Connexions and moving frames

we see that relative to a coordinate system a connexion determines a matrix whose elements are 1-forms.

Conversely, a matrix ω_k^i whose elements are 1-forms determines in each coordinate system a set of n^3 functions and hence a connexion. So we now have an alternative definition of connexion, namely, a connexion is a matrix of 1-forms relative to a moving frame. If we change from one moving frame $X = (X_1, \ldots, X_n)$ to another $X' = (X_1', \ldots, X_n')$ where

$$X' = Xa$$

and a is a matrix whose entries are functions, the law of transformation of the classical connexion components leads to

$$\omega' = a^{-1}\,da + a^{-1}\omega a.$$

So we are led to the following formal definition.

Definition (2.9.2) A connexion on a manifold M is an assignment of a matrix $\omega = (\omega_j^i)$ of 1-forms to every moving frame X such that the 1-forms assigned to X and the 1-forms assigned to the moving frame $X' = Xa$ are related by

$$\omega' = a^{-1}\,da + a^{-1}\omega a.$$

First we check that the above definition is consistent. Suppose $X'' = X'b = X(ab)$ is another moving frame. Then we have

$$\omega' = a^{-1}\,da + a^{-1}\omega a,$$
$$\omega'' = b^{-1}\,db - b^{-1}\omega' b;$$

then
$$\omega'' = b^{-1}\,db + b^{-1}(a^{-1}\,da + a^{-1}\omega a)b$$
$$= (b^{-1}a^{-1}(da)b + b^{-1}a^{-1}a(db)) + b^{-1}a^{-1}\omega ab$$
$$= (ab)^{-1}\,d(ab) + (ab)^{-1}\omega(ab),$$

thus confirming the consistency of the definition. Moreover, when

$$X = \left(\frac{\partial}{\partial x^1}, \ldots, \frac{\partial}{\partial x^n}\right),$$

$$X' = \left(\frac{\partial}{\partial x'^1}, \ldots, \frac{\partial}{\partial x'^n}\right),$$

then the relation

$$\omega' = a^{-1}\,da + a^{-1}\omega a$$

is none other than the classical law of transformation for the components of a connexion associated with a change of coordinates.

We now show how the torsion and the curvature of the connexion appear in the moving frames approach. Let X be a moving frame with respect to which the connexion forms are ω^i_j. Let θ be the moving frame of covectors dual to X, that is, let $\theta = (\theta^1, \ldots, \theta^n)$ where $\theta^i(X_j) = \delta^i_j$. We define a set of 2-forms Θ^i, $i = 1, \ldots, n$, and a matrix of 2-forms Ω^i_j by

$$d\theta + \omega \wedge \theta = \Theta, \qquad \text{i.e. } d\theta^i + \omega^i_j \wedge \theta^j = \Theta^i,$$
$$d\omega + \omega \wedge \omega = \Omega, \qquad \text{i.e. } d\omega^i_j + \omega^i_k \wedge \omega^k_j = \Omega^i_j.$$

The 2-forms Θ^i are called the *torsion* forms and the 2-forms Ω^i_j are called the *curvature* forms. We now justify our terminology. First we find the transformation formulas for Θ^i and Ω^i_j when the moving frames are changed. Let Θ^i and Ω^i_j correspond to the moving frame X and let Θ'^i, Ω'^i_j be the forms corresponding to the moving frame $X' = Xa$. We claim that

$$\Theta' = a^{-1}\Theta, \qquad \Omega' = a^{-1}\Omega a. \tag{2.8}$$

We have
$$\omega' = a^{-1}\,da + a^{-1}\omega a,$$
$$d\theta + \omega \wedge \theta = \Theta,$$
$$d\theta' + \omega' \wedge \theta' = \Theta',$$
$$d\omega + \omega \wedge \omega = \Omega,$$
$$d\omega' + \omega' \wedge \omega' = \Omega'.$$

Since $X' = Xa$, the dual bases are related by $\theta' = a^{-1}\theta$. Take the exterior derivative to get

$$d\theta' = da^{-1} \wedge \theta + a^{-1}\,d\theta.$$

Since $a^{-1}a = I$ we have $da^{-1}a + a^{-1}\,da = 0$. Hence

$$d\theta' = -a^{-1}\,da \wedge a^{-1}\theta + a^{-1}\,d\theta$$
$$= -a^{-1}\,da \wedge a^{-1}\theta + a^{-1}(\Theta - \omega \wedge \theta).$$

Also

$$d\theta' = \Theta' - \omega' \wedge \theta'$$
$$= \Theta' - ((a^{-1}\,da + a^{-1}\omega a) \wedge a^{-1}\theta)$$
$$= \Theta' - a^{-1}\,da \wedge a^{-1}\theta - a^{-1}\omega \wedge \theta.$$

Hence

$$\Theta' - a^{-1}\omega \wedge \theta = a^{-1}\Theta - a^{-1}\omega \wedge \theta$$

giving $\Theta' = a^{-1}\Theta$ as required. We now take the exterior derivative of the first equation to get

2.9 Connexions and moving frames

$$d\omega' = da^{-1} \wedge da + da^{-1} \wedge \omega a + a^{-1} d\omega a - a^{-1}\omega \wedge da$$
$$= -a^{-1} daa^{-1} \wedge da - a^{-1} daa^{-1} \wedge \omega a + a^{-1}(\Omega - \omega \wedge \omega) - a^{-1}\omega \wedge da.$$

Also
$$d\omega' = \Omega' - \omega' \wedge \omega' = \Omega' - (a^{-1} da + a^{-1}\omega a) \wedge (a^{-1} da + a^{-1}\omega a)$$
$$= \Omega' - a^{-1} da \wedge a^{-1} da - a^{-1} da \wedge a^{-1}\omega a - a^{-1}\omega a \wedge a^{-1} da$$
$$- a^{-1}\omega a \wedge a^{-1}\omega a$$
$$= \Omega' - a^{-1} da \wedge a^{-1} da - a^{-1} da \wedge a^{-1}\omega a - a^{-1}\omega \wedge da - a^{-1}\omega \wedge \omega a.$$

Comparing the two expressions for $d\omega'$ gives
$$a^{-1}\Omega a = \Omega'.$$

We now claim that the relations
$$\omega' = a^{-1} da + a^{-1}\omega a,$$
$$\Theta' = a^{-1}\Theta$$
$$\Omega' = a^{-1}\Omega a$$

enable us to recover ∇Y for vector fields, $T(X, Y)$ for tangent vectors X, Y, and $R(X, Y)Z$ for vector fields X, Y, Z. We can define
$$\nabla X = X\omega$$

that is,
$$\nabla X_j = \omega_j^i X_i,$$

or equivalently
$$\nabla_X X_j = \omega_j^i(X) X_i.$$

For an arbitary vector field $Y = b^j X_j$ we define
$$\nabla Y = db^j(X_j) + b^j \nabla X_j.$$

To show that this definition is meaningful, consider a change of frame $X' = Xa$. Then
$$\nabla X' = X'\omega' = Xa\omega'$$
$$= Xa(a^{-1} da + a^{-1}\omega a)$$
$$= X da + X(\omega a)$$
$$= X da + (X\omega)a$$
$$= X da + \nabla Xa$$

thus proving consistency. We now define
$$T(X_j, X_k) = \Theta^i(X_j, X_k) X_i.$$

Writing $X = b^j X_j$, $Y = c^k X_k$ we extend by linearity to get
$$T(X, Y) = b^j c^k \Theta^i(X_j, X_k) X_i$$
$$= b^j c^k T^i_{jk}.$$
In terms of the moving frame $X' = Xa$ we have
$$T(X'_i, X'_j) = \Theta'^k(X'_i, X'_j) X'_k$$
$$= a^{-1} \Theta^k(X'_i, X'_j) X'_k$$
$$= a^{-1} \Theta^k(a^r_i X_r, a^s_j X_s) a^t_k X_t$$
$$= a^r_i a^s_j \Theta^k(X_r, X_s) X_k$$
$$= a^r_i a^s_j T(X_r, X_s)$$
giving consistency with reference to moving frames. Finally define
$$R(X_k, X_l) X_j = \Omega^i_j(X_k, X_l) X_i,$$
and extend to arbitrary X, Y, Z by linearity. A change of frame $X' = Xa$ gives
$$R(X'_i, X'_j) = \Omega'^p_k(X'_i, X'_j) X'_p$$
$$= a_m^{-1p} \Omega^m_l a^l_k(a^r_i X_r, a^s_j X_s) a^t_p X_t$$
$$= \Omega^l_i(X_r, X_s) X_t a^l_k a^r_i a^s_j$$
$$= a^l_k a^r_i a^s_j R(X_r, X_s) X_l$$
again proving consistency.

If we now *define* Γ^i_{kj}, T^i_{jk}, R^i_{jkl} by the equations
$$\nabla_{X_k} X_j = \Gamma^i_{kj} X_i,$$
$$T(X_j, X_k) = T^i_{jk} X_i,$$
$$R(X_k, X_l) X_j = R^i_{jkl} X_i,$$
then we have

$$\Gamma^i_{kj} = \omega^i_j(X_k) \quad \text{or} \quad \omega^i_j = \Gamma^i_{kj} \theta^k,$$
$$T^i_{jk} = \Theta^i(X_j, X_k) \quad \text{or} \quad \Theta^i = \tfrac{1}{2} T^i_{jk} \theta^j \wedge \theta^k,$$
$$R^i_{jkl} = \Omega^i_j(X_k, X_l) \quad \text{or} \quad \Omega^i_j = \tfrac{1}{2} R^i_{jkl} \theta^k \wedge \theta^l.$$

Thus we have the structural equations
$$d\theta^i + \omega^i_j \wedge \theta^j = \Theta^i = \tfrac{1}{2} T^i_{jk} \theta^j \wedge \theta^k,$$
$$d\omega^i_j + \omega^i_k \wedge \omega^k_j = \Omega^i_j = \tfrac{1}{2} R^i_{jkl} \theta^k \wedge \theta^l,$$
$$\omega^i_j = \Gamma^i_{kj} \theta^k.$$

2.9 Connexions and moving frames

By reversing the argument we can obtain

$$T(X, Y) = \nabla_X Y - \nabla_Y X - [X, Y], \tag{2.9}$$

$$R(X, Y)Z = \nabla_X \nabla_Y Z - \nabla_Y \nabla_X Z - \nabla_{[X, Y]} Z, \tag{2.10}$$

thus showing that T and R are none other than the torsion and curvature tensors of the connexion as previously defined.

The entities θ, ω, Θ, and Ω are related by

$$d\Theta + \omega \wedge \Theta = \Omega \wedge \theta, \tag{2.11}$$

$$d\Omega + \omega \wedge \Omega - \Omega \wedge \omega = 0. \tag{2.12}$$

In fact these are none other than the first and second Bianchi identities! To prove the first identity we have

$$\begin{aligned} 0 = d(d\theta) &= d(-\omega \wedge \theta + \Theta) \\ &= -d\omega \wedge \theta + \omega \wedge d\theta + d\Theta \\ &= -(-\omega \wedge \omega + \Omega) \wedge \theta + \omega \wedge (-\omega \wedge \theta + \Theta) + d\Theta \\ &= -\Omega \wedge \theta + \omega \wedge \Theta + d\Theta, \end{aligned}$$

and we are home. Similarly

$$\begin{aligned} 0 = d(d\omega) &= d(-\omega \wedge \omega + \Omega) \\ &= -d\omega \wedge \omega + \omega \wedge d\omega + d\Omega \\ &= (-\omega \wedge \omega + \Omega) \wedge \omega + \omega \wedge (-\omega \wedge \omega + \Omega) + d\Omega \\ &= -\Omega \wedge \omega + \omega \wedge \Omega + d\Omega, \end{aligned}$$

and again we are through. Although there is no difficulty in establishing these identities, it is by no means obvious that they are the Bianchi identities in disguise. Again the reader is advised to take a large sheet of paper and write small! However, when the connexion has zero torsion, our first identity reduces to $\Omega \wedge \theta = 0$ so that

$$\Omega^i_j \wedge \theta^j = \tfrac{1}{2} R^i_{jkl} \theta^k \wedge \theta^l \wedge \theta^j = 0.$$

Apply this 3-form to the triad of vectors X_p, X_q, X_r to get

$$0 = R^i_{pqr} - R^i_{prq} + R^i_{qrp} - R^i_{qpr} + R^i_{rpq} - R^i_{rqp} = 0,$$

that is,

$$R^i_{pqr} + R^i_{qrp} + R^i_{rpq} = 0.$$

In a similar way when $\Theta = 0$ it is not difficult to show that the second identity reduces to

$$\nabla_l R^h_{ijk} + \nabla_j R^h_{ikl} + \nabla_k R^h_{ilj} = 0.$$

2.10 TENSORIAL FORMS

So far we have considered differential forms which map vector fields into functions. However, it is often useful to consider forms which take values in a fixed vector space or even in tensor fields.

Definition (2.10.1) *A tensorial p-form with values in tensor fields of type (r, s) is given by an alternating p-multilinear map of p vector fields into the space of tensor fields of type (r, s).*

Let T be such a tensorial form. With respect to a basis (e_i) of M, we have

$$T(X_1, \ldots, X_p) = T^{i_1 \cdots i_r}_{j_1 \cdots j_s k_1 \cdots k_p} X^{k_1} \cdots X^{k_p}$$

where the symbol $T^{i_1 \cdots i_r}_{j_1 \cdots j_s k_1 \cdots k_p}$ is skew-symmetric in the suffixes $k_1, \ldots, k_p$ but need not possess any other symmetry. Alternatively we can represent T as

$$T = t^{i_1 \cdots i_r}_{j_1 \cdots j_s k_1 \cdots k_p} e_{i_1} \otimes \cdots \otimes e_{i_r} \otimes \theta^{j_1} \otimes \cdots \otimes \theta^{j_s} \otimes \theta^{k_1} \wedge \cdots \wedge \theta^{k_p}.$$

We can regard T as a p-form with tensor coefficients, or equivalently as a tensor whose components are p-forms. We have already met an example of a vectorial 2-form, namely the torsion form Θ; the curvature form Ω is a tensorial 2-form of type $(1, 1)$.

If the space carries a connexion which we regard as a matrix of 1-forms ω^i_j relative to the basis (e_i), we can define an operator called covariant exterior differentiation, denoted by D, which maps tensorial p-forms of type (r, s) to tensorial $(p + 1)$-forms of type (r, s). Clearly for scalar-valued forms we require $D = d$, the exterior derivative operator already considered. For a tensorial p-form of type (r, s) we define

$$DT = dT^{i_1 \cdots i_r}_{j_1 \cdots j_s k_1 \cdots k_p} + \omega^{i_1}_h T^{h i_2 \cdots i_r}_{j_1 \cdots j_s k_1 \cdots k_p} + \omega^{i_2}_h T^{i_1 h i_3 \cdots i_r}_{j_1 \cdots j_s k_1 \cdots k_p}$$
$$+ \cdots - \omega^h_{j_1} T^{i_1 \cdots i_r}_{h j_2 \cdots j_s k_1 \cdots k_p} - \omega^h_{j_2} T^{i_1 \cdots i_r}_{j_1 h j_3 \cdots j_s k_1 \cdots k_p} - \cdots \quad (2.13)$$

A straightforward but tedious calculation shows that this definition is independent of the particular basis chosen, and does in fact map a tensorial p-form of type (r, s) to a tensorial $(p + 1)$-form of the same type.

In terms of this notation the torsion tensor satisfies

$$D\Theta = \Omega \wedge \theta, \quad (2.14)$$

while the second Bianchi identity becomes

$$D\Omega = 0. \quad (2.15)$$

2.11 EXERCISES

1. If a_{rs}, b_{rs} are the components of two symmetric covariant tensors such that
$$a_{rs}b_{tu} - a_{ru}b_{st} + a_{st}b_{ru} - a_{tu}b_{rs} = 0,$$
show that $a_{rs} = kb_{rs}$ for some real number k.

2. If $a_{\alpha\beta}$, b_α are respectively components of a symmetric covariant tensor and a covariant vector which satisfy the relation
$$a_{\beta\gamma}b_\alpha + a_{\gamma\alpha}b_\beta + a_{\alpha\beta}b_\gamma = 0$$
for $\alpha, \beta, \gamma = 1, 2, \ldots, n$, prove that either $a_{\alpha\beta} = 0$ or $b_\alpha = 0$.

3. Given any mixed tensor field h^i_j show that
$$H^i_{jk} = -\frac{1}{8}(h^p_j \partial_p h^i_k - h^p_k \partial_p h^i_j) + \frac{1}{8}(h^i_p \partial_j h^p_k - h^i_p \partial_k h^p_j)$$
is a tensor field, where $\partial_p h^i_k = \dfrac{\partial h^i_k}{\partial x^p}$.

[Hint: express the partial derivatives in terms of covariant derivatives with respect to an arbitrary symmetric affine connexion.]

4. Coefficients h^i_{jrs} are defined by
$$h^i_{jrs} = -\frac{1}{2}\partial_j H^i_{rs} + \frac{1}{2}h^i_p(h^q_j \partial_q H^p_{rs} - H^q_{rs}\partial_q h^p_j + H^p_{qs}\partial_r h^q_j - H^p_{qr}\partial_s h^q_j),$$
where $h^i_j h^j_k = -\delta^i_k$, and H^i_{jk} is defined in terms of the tensor h^i_j by the formula given in Exercise 3.

A. G. Walker has defined the *torsional derivative* $T^{ih\cdots}_{jk\cdots\|rs}$ of any tensor field with components $T^{ih\cdots}_{jk\cdots}$ by the formula
$$T^{ih\cdots}_{j\cdots\|rs} = H^p_{rs}\partial_p T^{ih\cdots}_{jk\cdots} + T^{ph\cdots}_{jk\cdots}h^i_{prs} + T^{ip\cdots}_{jk\cdots}h^h_{prs} + \cdots$$
$$- T^{ih\cdots}_{pk\cdots}h^p_{jrs} - T^{ih\cdots}_{jp\cdots}h^p_{krs} - \cdots,$$
where the dots indicate that there is a term for each suffix in $T^{ih\cdots}_{jk\cdots}$, as in the formula for covariant differentiation. Prove that $T^{ih\cdots}_{jk\cdots\|rs}$ are the components of a tensor of the type indicated by the suffixes.

5. Prove that $h^i_{j\|rs} = 0$ and also that $\delta^i_{j\|rs} = 0$.

6. Show that the tensors with components $H^i_{jk\|rs}$, $H^i_{jk\|ri}$ are not zero.

3
Riemannian manifolds

3.1 RIEMANNIAN METRICS

Let M be a differentiable manifold. We say that M carries a pseudo-Riemannian metric if there is a differentiable field $g = \{g_m\}$, $m \in M$, of non-degenerate symmetric bilinear forms g_m on the tangent spaces M_m of M. This makes the tangent space into an inner product space. Let Y, Z be differentiable vector fields defined over an open set U of M. By asserting that g is differentiable we mean that the function $g(Y, Z)$ is differentiable. Often we will write $\langle Y, Z \rangle$ instead of $g(Y, Z)$. If in addition the forms g_m are positive definite we call the metric Riemannian. A differentiable manifold with a (pseudo-)Riemannian metric is called a (pseudo-)Riemannian manifold.

Let M be a pseudo-Riemannian manifold with metric g. Let $X = (X_1, \ldots, X_n)$ be a moving frame over an open set U of M, and let $\theta = (\theta^1, \ldots, \theta^n)$ be the dual coframe so that $\theta^i(X_j) = \delta^i_j$. We define differentiable functions g_{ij} on U by

$$g_{ij}(m) = g_m(X_i, X_j).$$

Then g can be written as $g = g_{ij} \theta^i \otimes \theta^j$ on U, with $g_{ij} = g_{ji}$. In particular when X is a coordinate frame, $\theta^i = dx^i$ and we have $g = g_{ij} dx^i dx^j$, classically written as $ds^2 = g_{ij} dx^i dx^j$.

Definition (3.1.1) Let $\gamma: [a, b] \to M$ be a smooth curve. Then the length of γ is defined to be

$$L(\gamma) = \int_a^b |g_{\gamma(t)}(\dot{\gamma}(t), \dot{\gamma}(t))|^{\frac{1}{2}} dt.$$

If M is connected and Riemannian, we define the distance between the points x and y of M as the infimum of the lengths of sectionally smooth curves from x to y. We shall show that M is a metric space with respect to this distance function, and that the induced topology from this metric agrees with the original topology of the manifold.

Product manifolds

Let (M, g), (N, h) be two pseudo-Riemannian manifolds. Then the product manifold can be given a pseudo-Riemannian structure by defining the metric at $(p, q) \in M \times N$ as $g + h$ acting on the tangent space $(M \times N)_{p,q} = M_p \oplus N_q$. This product manifold with this metric will be denoted by $M \times N$.

Fundamental theorem

The fundamental theorem of Riemannian geometry is as follows.

Theorem (3.1.2) Let M be a pseudo-Riemannian manifold. Then there exists a uniquely determined connexion on M such that:

(i) parallel translation of tangent vectors along any curve γ in M preserves inner products;

(ii) the torsion tensor is zero.

To prove the theorem we note that (i) is equivalent to

$$Zg(X, Y) = g(\nabla_Z X, Y) + g(X, \nabla_Z Y)$$

for differentiable vector fields X, Y, Z; while (ii) gives

$$\nabla_X Y - \nabla_Y X = [X, Y].$$

Substitution gives

$$Zg(X, Y) = g(\nabla_Z X, Y) + g(X, \nabla_Y Z + [Z, Y]).$$

Cyclically permuting X, Y, Z twice gives a total of three equations. We add two and subtract the third to get

$$2g(\nabla_X Y, Z) = X \cdot g(Y, Z) + Y \cdot g(X, Z) - Z \cdot g(X, Y)$$
$$+ g([X, Y], Z) + g([Z, X], Y) + g([Z, Y], X),$$

which uniquely determines $\nabla_X Y$. In particular, if $X = \partial/\partial x^i$, $Y = \partial/\partial x^j$, $Z = \partial/\partial x^k$ we get

$$2g\left(\Gamma_{ij}^h \frac{\partial}{\partial x^h}, \frac{\partial}{\partial x^k}\right) = \frac{\partial g_{jk}}{\partial x^i} + \frac{\partial g_{ik}}{\partial x^j} - \frac{\partial g_{ij}}{\partial x^k},$$

that is,

$$2g_{hk}\Gamma_{ij}^h = g_{jk \cdot i} + g_{ik \cdot j} - g_{ij \cdot k},$$

where the dot denotes partial differentiation.

If g^{ij} is the reciprocal matrix of g_{ij} this gives

$$\Gamma_{ij}^h = \tfrac{1}{2} g^{hk} [g_{jk \cdot i} + g_{ik \cdot j} - g_{ij \cdot k}].$$

The Γ_{ij}^h are called Christoffel symbols, and are clearly uniquely determined. The connexion Γ has a curvature tensor R but the torsion tensor is zero. We note that g_{ij} are the components of a tensor field of type

(0, 2), and also that the g^{ij} are components of a tensor field of type (2, 0). Sometimes Christoffel symbols are denoted by $\{^i_{jk}\}$.

3.2 IDENTITIES SATISFIED BY THE CURVATURE TENSOR OF A RIEMANNIAN MANIFOLD

These are

$$R(X, Y)Z + R(Y, X)Z = 0, \qquad (3.1)$$

$$R(X, Y)Z + R(Y, Z)X + R(Z, X)Y = 0, \qquad (3.2)$$

$$\langle R(X, Y)Z, W\rangle + \langle R(X, Y)W, Z\rangle = 0, \qquad (3.3)$$

$$\langle R(X, Y)Z, W\rangle = \langle R(Z, W)X, Y\rangle. \qquad (3.4)$$

The first identity follows trivially from the definition of R. The second is another form of the first Bianchi identity for a torsion-free connexion.

To prove it using the suffix-free notation, we observe that because all terms of (3.2) are tensors it is sufficient to establish the identity in a coordinate system with $X = \partial/\partial x^i$, $Y = \partial/\partial x^j$, $Z = \partial/\partial x^k$ so that $[X, Y] = [Y, Z] = [Z, X] = 0$. So we have to prove that

$$\nabla_X(\nabla_Y Z) - \nabla_Y(\nabla_X Z) + \nabla_Y(\nabla_Z X) - \nabla_Z(\nabla_Y X) + \nabla_Z(\nabla_X Y) - \nabla_X(\nabla_Z Y) = 0.$$

However, because the torsion is zero we have $\nabla_Y Z - \nabla_Z Y = [Y, Z] = 0$ and the terms cancel in pairs (1, 6), (2, 3), and (4, 5).

Identities (3.3) and (3.4) involve the metric and so have no counterpart for the connexions considered in Chapter 2. To prove (3.3) we have to show that $\langle R(X, Y)Z, W\rangle$ is skew-symmetric in Z and W. It will be sufficient to show that $\langle R(X, Y)Z, Z\rangle = 0$ since by replacing Z by $(Z + W)$ this leads to the previous condition. Again we assume that $[X, Y] = 0$ giving

$$\langle R(X, Y)Z, Z\rangle = \langle \nabla_X(\nabla_Y Z) - \nabla_Y(\nabla_X Z), Z\rangle.$$

So we have to prove that $\langle \nabla_Y(\nabla_X Z), Z\rangle$ is symmetric in X and Y. We know that $YX\langle Z, Z\rangle$ is symmetric in X and Y since $[X, Y] = 0$. Moreover, since the connexion is compatible with the metric we have

$$X\langle Z, Z\rangle = \langle \nabla_X Z, Z\rangle + \langle Z, \nabla_X Z\rangle = 2\langle \nabla_X Z, Z\rangle.$$

Thus

$$YX\langle Z, Z\rangle = 2\langle \nabla_Y \nabla_X Z, Z\rangle + 2\langle \nabla_X Z, \nabla_Y Z\rangle.$$

3.2 The curvature tensor of a Riemannian manifold

Now the left-hand side is symmetric in X and Y, as is the second term on the right. It follows that $\langle \nabla_Y \nabla_X Z, Z \rangle$ must be symmetric in X and Y, and we are home.

The proof of (3.4) is rather subtle. From identity (3.2) we have

$$\langle R(X, Y)Z, W \rangle + \langle R(Y, Z)X, W \rangle + \langle R(Z, X)Y, W \rangle = 0. \quad (3.5)$$

We claim that

$$\langle R(X, Y)Z, W \rangle + \langle R(X, W)Y, Z \rangle + \langle R(Y, W)Z, X \rangle = 0. \quad (3.6)$$

The left-hand side is equal to

$$-\langle R(X, Y)W, Z \rangle \quad \text{(using (3.3))} \quad +\langle R(X, W)Y, Z \rangle$$
$$- \langle R(Y, W)X, Z \rangle = \langle R(Y, X)W + R(X, W)Y$$
$$+ R(W, Y)X, Z \rangle \quad \text{(using (3.1))} = 0 \quad \text{using (3.2)}.$$

In a similar way we prove

$$\langle R(Y, W)Z, X \rangle + \langle R(Y, Z)X, W \rangle + \langle R(Z, W)X, Y \rangle = 0, \quad (3.7)$$

$$\langle R(X, W)Y, Z \rangle + \langle R(Z, X)Y, W \rangle + \langle R(Z, W)X, Y \rangle = 0. \quad (3.8)$$

Adding the first two identities (3.5) and (3.6) and subtracting the last two (3.7) and (3.8) yields

$$2\langle R(X, Y)Z, W \rangle - 2\langle R(Z, W)X, Y \rangle = 0$$

and identity (3.4) is proved.

We now obtain a coordinate representation of the curvature tensor of a Riemannian manifold. We write

$$R(X, Y, Z, W) = \langle R(Z, W)Y, X \rangle$$

thus defining a covariant degree 4 tensor field also denoted by R. We recall that the components R^i_{jkl} of the curvature tensor are given by

$$R(X_k, X_l)X_j = R^i_{jkl} X_i.$$

Then

$$R\left(\frac{\partial}{\partial x^h}, \frac{\partial}{\partial x^i}, \frac{\partial}{\partial x^j}, \frac{\partial}{\partial x^k}\right) = \left\langle R\left(\frac{\partial}{\partial x^j}, \frac{\partial}{\partial x^k}\right) \frac{\partial}{\partial x^i}, \frac{\partial}{\partial x^h} \right\rangle$$

$$= \left\langle R^p_{ijk} \frac{\partial}{\partial x^p}, \frac{\partial}{\partial x^h} \right\rangle$$

$$= g_{hp} R^p_{ijk} = R_{hijk}.$$

Identity (3.2) now gives

$$R_{hijk} + R_{hjki} + R_{hkij} = 0,$$

which is the first Bianchi identity. In addition we have the second Bianchi identity which can be written

$$\nabla_l R_{hijk} + \nabla_j R_{hikl} + \nabla_k R_{hilj} = 0.$$

Identity (3.3) now gives

$$R_{kjhi} = -R_{jkhi},$$

and (3.4) gives

$$R_{kjhi} = R_{ihjk} = R_{hikj}.$$

3.3 SECTIONAL CURVATURE

Let P be a 2-plane, that is, a 2-dimensional subspace of the tangent space at m of a Riemannian manifold M, which is determined by an orthonormal basis X, Y. We define the sectional curvature $\sigma(P)$ of P at m to be given by $\sigma(P) = R(X, Y, X, Y)$.

We must first check that this definition is meaningful, that is, it is independent of the particular choice of basis X, Y. Any other orthonormal basis X', Y' will satisfy

$$X' = aX + bY, \qquad Y' = -bX + aY$$

where $a^2 + b^2 = 1$. Then

$$\begin{aligned} R(X', Y', X', Y') &= R(aX + bY, -bX + aY, aX + bY, -bX + aY) \\ &= R(aX, aY, aX, aY) + R(aX, aY, bY, -bX) \\ &\quad + R(bY, -bX, aX, aY) \\ &\quad + R(bY, -bX, bY, -bX) \\ &= a^4 R(X, Y, X, Y) - a^2 b^2 R(X, Y, Y, X) \\ &\quad - a^2 b^2 R(Y, X, X, Y) + b^4 R(Y, X, Y, X) \\ &= (a^2 + b^2)^2 R(X, Y, X, Y) \\ &\quad \text{using identities (3.1) and (3.2)} \\ &= R(X, Y, X, Y) \end{aligned}$$

so the definition makes sense.

We now prove a useful and perhaps rather surprising result, which says in effect that if you know the sectional curvature at m for all 2-planes P then you know the complete curvature tensor at m. Suppose that R and R' are two curvature tensors at m whose sectional curvatures are the same. We shall prove that $R = R'$. Due to the multilinearity

3.3 Sectional curvature

of the curvature tensor, it will be sufficient to prove that $R(X, Y, X, Y) = 0$ implies $R(X, Y, Z, W) = 0$, for all vectors X, Y, Z, W.

Let $R(X, Y, X, Y) = 0$. Then

$$R(X, Y + W, X, Y + W) = R(X, Y, X, W) + R(X, W, X, Y)$$
$$= 2R(X, Y, X, W)$$

using identity (3.4). Thus we have proved that $R(X, Y, X, W) = 0$ for all X, Y, W. Similarly we have

$$0 = R(X + Z, Y, X + Z, W)$$
$$= R(X, Y, Z, W) + R(Z, Y, X, W)$$
$$= R(X, Y, Z, W) + R(X, W, Z, Y) \quad \text{using (3.4),}$$
$$= R(X, Y, Z, W) - R(X, W, Y, Z) \quad \text{using (3.2).}$$

Thus

$$R(X, Y, Z, W) = R(X, W, Y, Z).$$

Replace Y, Z, W by Z, W, Y, respectively to give $R(X, Y, Z, W) = R(X, Z, W, Y)$. Thus

$$3R(X, Y, Z, W) = R(X, Y, Z, W) + R(X, W, Y, Z)$$
$$+ R(X, Z, W, Y) = 0$$

by the Bianchi identity. Thus $R(X, Y, Z, W) = 0$ so we have proved that the sectional curvatures determine the curvature tensor.

Suppose that we are given at the point m two vectors v_1, v_2 which span P but which are not orthonormal. Then using the Gram-Schmitt process we write

$$X = \frac{v_1}{\langle v_1, v_1 \rangle^{1/2}},$$

$$Y = \frac{\langle v_1, v_1 \rangle v_2 - \langle v_1, v_2 \rangle v_1}{[\langle v_1, v_1 \rangle (\langle v_1, v_1 \rangle \langle v_2, v_2 \rangle - \langle v_1, v_2 \rangle^2)]^{1/2}},$$

so that (X, Y) form an orthonormal basis. Then we have

$$\sigma(P) = \frac{R(v_1, \langle v_1, v_1 \rangle v_2 - \langle v_1, v_2 \rangle v_1, v_1, \langle v_1, v_1 \rangle v_2 - \langle v_1, v_2 \rangle v_1)}{\langle v_1, v_1 \rangle^2 [\langle v_1, v_1 \rangle \langle v_2, v_2 \rangle - \langle v_1, v_2 \rangle^2]}$$

$$= \frac{R(v_1, v_2, v_1, v_2)}{\langle v_1, v_1 \rangle \langle v_2, v_2 \rangle - \langle v_1, v_2 \rangle^2}.$$

Writing $v_1 = \lambda^i \partial/\partial x^i$, $v_2 = \mu^j \partial/\partial x^j$ we see that

$$\sigma(P) = \frac{R_{ijkl}\lambda^i\lambda^k\mu^j\mu^l}{(g_{pq}\lambda^p\lambda^q)(g_{rs}\mu^r\mu^s) - (g_{pr}\lambda^p\mu^r)^2}.$$

If, at the point m, $\sigma(P)$ is independent of the choice of P we say that M has *isotropic curvature* at m. If this is the case at all points of M and dim $M > 2$, then we have the result due to Schur that $\sigma(P)$ has the same value for all P and all m. To prove this we proceed as follows.

Suppose that M has the above property with respect to the Riemannian metric g. Consider the tensor field R' defined by

$$R'(W, Z, X, Y) = \langle W, X\rangle\langle Z, Y\rangle - \langle Z, X\rangle\langle Y, W\rangle.$$

It is easy to check that R' satisfies the identities (3.1), (3.2), (3.3), (3.4). In our proof that sectional curvature determines the curvature tensor, we effectively proved that if R and R' satisfy the identities (3.1), (3.2), (3.3), (3.4), and $R(X, Y, X, Y) = R'(X, Y, X, Y)$, then the two tensors are the same. Now we easily check that

$$R'(X, Y, X, Y) = \langle X, X\rangle\langle Y, Y\rangle - \langle X, Y\rangle^2,$$

so that $\sigma'(P) = 1$. Our assumption implies that $R = kR'$ for some real-valued function k on M.

It is instructive now to complete the proof of Schur's theorem by using three different methods, namely the suffix-free calculus, the classical tensor calculus, and the exterior calculus.

We give the first proof along the lines of that given by Kobayashi and Nomizu (1963) Vol. 1, page 202. We have, using our notation,

$$(\nabla_U R)(W, Z, X, Y) = (\nabla_U k)R'(W, Z, X, Y).$$

for any vector field U. This implies that for any X, Y, Z, U at m we have

$$[(\nabla_U R)(X, Y)]Z = (Uk)(g(Z, Y)X - g(Z, X)Y).$$

We now consider the cyclic sum of the above identity with respect to the permutation (U, X, Y) and use Bianchi's second identity on the left-hand side. We get

$$0 = (Uk)(g(Z, Y)X - g(Z, X)Y) + (Xk)(g(Z, U)Y - g(Z, Y)U) + (Yk)(g(Z, X)U - g(Z, U)X).$$

Now for given X, choose Y, Z and U so that X, Y, Z are mutually orthogonal and $U = Z$ with $\langle Z, Z\rangle = 1$. We can always do this when dim $M \geqslant 3$. We get

$$(Xk)Y - (Yk)X = 0.$$

Since X and Y are linearly independent, we must have $Xk = 0$ and $Yk = 0$ and hence k is a constant function and the result is established.

3.3 Sectional curvature

In terms of tensor calculus the equation $R = kR'$ gives immediately

$$R_{ijkl} = k(g_{ik}g_{jl} - g_{jk}g_{il}).$$

Taking the covariant derivative with respect to x^s gives

$$\nabla_s R_{ijkl} = (\nabla_s k) \cdot (g_{ik}g_{jl} - g_{jk}g_{il}).$$

Use Bianchi's second identity to get

$$0 = (\nabla_s k) \cdot (g_{ik}g_{jl} - g_{jk}g_{il}) + (\nabla_k k) \cdot (g_{il}g_{js} - g_{jl}g_{is})$$
$$+ (\nabla_l k) \cdot (g_{is}g_{jk} - g_{js}g_{ik}).$$

Multiply by g^{ik} to get

$$(\nabla_s k) \cdot n g_{jl} - (\nabla_s k) \cdot g_{jl} + (\nabla_l k) \cdot g_{js} - (\nabla_s k) \cdot g_{jl} + (\nabla_l k) g_{js}$$
$$- n(\nabla_l k) \cdot g_{js} = 0,$$

that is,

$$(n-2) \cdot ((\nabla_s k) \cdot g_{jl} - (\nabla_l k) g_{js}) = 0.$$

Multiply by g^{jl} to get

$$(n-1)(n-2)\nabla_s k = 0,$$

from which we deduce that if $\dim M > 2$ we must have $k =$ constant.

In terms of the exterior calculus we choose an orthonormal frame of tangent vectors (e_i) with dual coframe (θ^k). We have

$$R^i_{jkl} = R_{ijkl} = k(\delta_{ik}\delta_{jl} - \delta_{jk}\delta_{li}),$$

and hence

$$\Omega^i_j = \frac{1}{2}k[\theta^i \wedge \theta^j - \theta^j \wedge \theta^i] = k\theta^i \wedge \theta^j.$$

We take the covariant differential of each side to get

$$D\Omega^i_j = dk \wedge \theta^i \wedge \theta^j + kD\theta^i \wedge \theta^j - k\theta^i \wedge D\theta^j.$$

However $D\Omega^i_j = 0$ is just Bianchi's identity and $D\theta^i = 0$ because the metric has zero torsion. Thus we get

$$dk \wedge \theta^i \wedge \theta^j = 0.$$

If $\dim M = 2$ the 3-form vanishes identically and tells us nothing. But if $\dim M > 2$ we see immediately that $k =$ constant, and we are home.

The reader will see from this example that each method has its merits, and it illustrates our wish to switch from one method to another if this seems appropriate.

The Ricci Tensor

Let (e_i) be an orthonormal basis of $T(M)_m$ with respect to the Riemannian metric. We associate with R the symmetric tensor ρ of type $(0,2)$ given by

$$\rho(v_1, v_2) = \sum_{i=1}^{n} R(e_i, v_1, e_i, v_2).$$

We have to check that this definition makes sense, that is, it is independent of the particular basis chosen. If we write $e_i = p_i^{i'} e_{i'}$ where $(p_i^{i'})$ is an orthogonal matrix we have

$$\sum_i R(e_i, v_1, e_i, v_2) = \sum_{j'} \sum_{i'} \sum_i R(p_i^{i'} e_{i'}, v_1, p_i^{j'} e_{j'}, v_2)$$

$$= \sum_{j'} \sum_{i'} \sum_i p_i^{i'} p_i^{j'} R(e_{i'}, v_1, e_{j'}, v_2)$$

$$= \sum_{j'} \sum_{i'} \delta_{j'}^{i'} R(e_{i'}, v_1, e_{j'}, v_2)$$

$$= \sum_{i'} R(e_{i'}, v_1, e_{i'}, v_2)$$

as required. The new tensor is called the *Ricci tensor*.

From our definition it is clear that the Ricci tensor is symmetric. Moreover in terms of components we have

$$\rho_{jl} = R^i_{jil}.$$

An alternative notation R_{jl} is often used instead of ρ_{jl} for components of the Ricci tensor.

In terms of an orthonormal frame we define the scalar curvature by

$$\tau = \sum_{i=1}^{n} \rho(e_i, e_i).$$

The corresponding equation in components is given by

$$\tau = g^{rl} \rho_{rl}.$$

A substantial amount of modern research in differential geometry is devoted to investigating the consequences of imposing restrictions on one or more of these curvatures. In particular one is concerned with possible consequences on the topology on such manifolds. For example, one can ask whether a given differentiable manifold can be given a Riemannian metric whose curvatures are restricted in certain ways. We shall investigate these questions later on.

3.4 Geodesics

We now obtain an identity which has important applications in the general theory of relativity, namely,

$$\nabla_l \rho_j^l = \tfrac{1}{2} \nabla_j \tau, \quad \text{where } \rho_j^l = g^{il}\rho_{ij} = g^{il}R^h_{ihj} \text{ and } \tau = \rho_l^l.$$

We start with the second Bianchi identity which we write as

$$\nabla_l R_{hijk} + \nabla_j R_{hikl} + \nabla_k R_{hilj} = 0.$$

Multiply by g^{hp} to get

$$\nabla_l R^p_{ijk} - \nabla_j R^p_{ilk} + g^{hp}\nabla_k R_{hilj} = 0.$$

Contract for p and k to get

$$-\nabla_l \rho_{ij} + \nabla_j \rho_{il} + g^{hp}\nabla_p R_{hilj} = 0.$$

Multiply by g^{il} to get

$$-\nabla_l \rho_j^l + \nabla_j \tau - g^{hp}\nabla_p \rho_{hj} = 0,$$

that is,

$$2\nabla_l \rho_j^l = \nabla_j \tau$$

as required.

3.4 GEODESICS

Let M be a connected Riemannian manifold. We have the following definition.

Definition (3.4.1) *Let $\gamma: I \to M$, $t \to \gamma(t)$ be a parametrized path where I is an interval of the real numbers. We say that γ is a geodesic if*

$$\nabla_{\frac{d}{dt}}\left(\frac{d\gamma}{dt}\right) = 0.$$

Geometrically this means that the velocity vector field d/dt must be parallel along γ. Since

$$\nabla_{\frac{d}{dt}}\left\langle \frac{d\gamma}{dt}, \frac{d\gamma}{dt} \right\rangle = 2\left\langle \nabla_{\frac{d}{dt}} \frac{d\gamma}{dt}, \frac{d\gamma}{dt} \right\rangle = 0,$$

it follows that the length of the velocity vector $\|d\gamma/dt\| = \langle d\gamma/dt, d\gamma/dt \rangle^{\frac{1}{2}}$ must be constant along a geodesic.

The *arc length* function $s(t)$ is defined by

$$s(t) = \int^t \left\|\frac{d\gamma}{dt}\right\| dt + \text{constant}.$$

It follows that the parameter t is a linear function of the arc length along a geodesic.

In terms of local coordinates $x^1, \ldots, x^n$ the curve γ determines n functions $x^1(t), \ldots, x^n(t)$, and the equation for a geodesic takes the form

$$\frac{d^2 x^k}{dt^2} + \Gamma^k_{ij} \frac{dx^i}{dt} \frac{dx^j}{dt} = 0,$$

where we have used the summation convention. The existence of geodesics therefore depends upon solutions of a system of second-order differential equations, and we now state a theorem concerning such equations. Consider the system of differential equations of the form

$$\frac{d^2 u}{dt^2} = F\left(u, \frac{du}{dt}\right)$$

where $u = (u^1, \ldots, u^n)$, $F = (F^1, \ldots, F^n)$ and F^i is a C^∞-function defined throughout some neighbourhood U of a point (u, v) of $\mathbb{R}^{2n}$.

Then the following is a standard theorem from the theory of systems of differential equations.

Theorem (3.4.2) There exists a neighbourhood W of the point (u, v) of $\mathbb{R}^{2n}$ and a number $\varepsilon > 0$ such that, for each $(u_1, v_1) \in W$, the differential equation

$$\frac{d^2 u}{dt^2} = F\left(u, \frac{du}{dt}\right)$$

has a unique solution $t \to u(t)$ defined for $|t| < \varepsilon$ satisfying the initial conditions $u(0) = u_0$, $du/dt(0) = v_0$. Moreover the solution depends smoothly on the initial conditions. Thus the correspondence $(u_0, v_0, t) \to u(t)$ from $W \times (-\varepsilon, \varepsilon)$ to $\mathbb{R}^n$ is a C^∞-function of all the $2n + 1$ variables.

Applying this theorem to the differential equations for geodesics gives the following result.

Theorem (3.4.3) For every point p_0 on a Riemannian manifold M there exists a neighbourhood U of p_0 and a number $\varepsilon > 0$ such that for every $p \in U$ and each tangent vector $v \in TM(p)$ with length $< \varepsilon$, there is a unique geodesic $\gamma_v \colon (-2, 2) \to M$ such that $\gamma_v(0) = p$, $d\gamma_v/dt(0) = v$.

Proof. If instead of specifying the interval $(-2, 2)$ we replaced this by an arbitrarily small interval, then this result would follow from the previous theorem. We can certainly assert that there exists a neighbour-

3.4 Geodesics

hood U of p and numbers $\varepsilon_1, \varepsilon_2 > 0$ such that for each $p \in U$ and each $v \in TM(p)$ with $\|v\| < \varepsilon_1$, there is a unique geodesic

$$\gamma: (-2\varepsilon_2, 2\varepsilon_2) \to M$$

satisfying the required initial conditions. However, the following reasoning will allow us to extend the argument to the general case. Let $\gamma: t \to \gamma(t)$ be a geodesic and let c be any constant real number. Then, because of the form of the differential equation for geodesics, it follows that the parametrized curve $t \to \gamma_v(ct)$ is also a geodesic. Suppose now that ε is smaller than the product $\varepsilon_1 \varepsilon_2$.

Then $\|v\| < \varepsilon$ and $|t| < 2$ imply that $\|v/\varepsilon_2\| < \varepsilon_1$ and $|\varepsilon_2 t| < 2\varepsilon_2$.

So if we define $\gamma_v(t)$ to be $\gamma_{v/\varepsilon_2}(\varepsilon_2 t)$, the resulting geodesic satisfies our requirements.

The exponential map

Let $v \in TM(q)$ be a tangent vector, and suppose that there exists a geodesic $\gamma: [0, 1] \to M$ such that $\gamma(0) = q$, $d\gamma/dt(0) = v$. Then the point $\gamma(1) \in M$ will be denoted by $\exp_q(v)$, and the corresponding map $TM(q) \to M$ will be called the *exponential map*. The geodesic γ can be described by the formula

$$\gamma(t) = \exp_q(tv).$$

The reason for this terminology is as follows. If M is the Lie group of all $n \times n$ unitary matrices, then the tangent space to M at the identity I of the group can be identified with the space of $n \times n$ skew-Hermitian matrices (which is the Lie algebra of the unitary group). The function

$$\exp_I: TM(I) \to M$$

as defined above can be shown to be given by the exponential power series

$$\exp_I(A) = I + A + A^2/2! + A^3/3! + \cdots.$$

The reader will have observed that our analysis of geodesics so far could have been carried out for manifolds with an affine connexion. We have only used the concept of ratio of lengths along the same curve, and this makes sense for curves satisfying the equation for a geodesic when the functions Γ_{ij}^k are considered as connexion coefficients of a (not necessarily Riemannian) affine connexion. In fact many of the results of this section could be extended to apply to this more general case.

We have seen that $\exp_q(v)$ is also defined provided that $\|v\|$ is small enough, but it may not be defined for large values of $\|v\|$. However whenever $\exp_q(v)$ is defined the map is uniquely determined. We now prove that the exponential map is a local diffeomorphism. More specifically we prove the following theorem.

Theorem (3.4.4) For each point $p \in M$ there exists a neighbourhood W and a number $\varepsilon > 0$ such that:

(i) any two points of W are joined by a unique geodesic of M of length $< \varepsilon$;

(ii) this geodesic depends smoothly on the two points;

(iii) for each $q \in W$ the map $\exp_q$ maps the open ε-ball in $TM(q)$ diffeomorphically onto an open set $U_1 \supset W$.

Proof. Let TM be the tangent bundle of M. We recall that TM can be given a C^∞-structure as follows. Let U be a coordinate neighbourhood of M with coordinates $u^1, \ldots, u^n$. Then every tangent vector at $q \in U$ can be uniquely expressed as $t^i(\partial/\partial u^i)_q$ so that the functions $u^1, \ldots, u^n, t^1, \ldots, t^n$, constitute a coordinate system on the open set $TU \subset TM$. We have already seen that for each $p \in M$ the map

$$(q, v) \to \exp_q(v)$$

is defined throughout a neighbourhood V of the point $(p, 0) \in TM$ and is differentiable throughout V. Consider now the map $F: V \to M \times M$ defined by $F(q, v) = (q, \exp_q(v))$. Denote the induced coordinates on $U \times U \subset M \times M$ by (x^i, y^i). Then we have

$$F_*\left(\frac{\partial}{\partial u^i}\right) = \frac{\partial}{\partial x^i} + \frac{\partial}{\partial y^i},$$

$$F_*\left(\frac{\partial}{\partial t^i}\right) = \frac{\partial}{\partial y^i}.$$

Thus the Jacobian matrix of F at $(p, 0)$ has unit determinant and, from the implicit function theorem, it follows that F maps some neighbourhood V' of $(p, 0) \in TM$ diffeomorphically onto some neighbourhood of $(p, p) \in M \times M$. We can assume that the neighbourhood V' consists of all pairs (q, v) such that q belongs to a given neighbourhood U' of p and such that $\|v\| < \varepsilon$. Moreover we can always choose a smaller neighbourhood W to ensure that $F(V') \supset W \times W$. This completes the proof of the theorem.

Indeed a stronger result is valid. Whitehead (1932) showed that W can be chosen so that the geodesic joining any two points of W lies completely in W. Such a neighbourhood is called *simple convex* as distinct from our previous normal neighbourhoods which are simple. In fact Whitehead showed that M can be covered by a system of simple convex neighbourhoods.

We now prove a result known as the *Gauss lemma*.

3.4 Geodesics

Lemma (3.4.5) *Let U be a simple normal neighbourhood. Then the geodesics through q are orthogonal trajectories of the geodesic spheres given by*

$$\{\exp_q v : v \to TM(q), \|v\| = \text{constant}\}.$$

The lemma is named after Gauss since he first proved the result for two-dimensional surfaces. We see that this is a natural generalization of the obvious property of radii of a system of concentric circles in the plane. To prove the lemma, let $t \to v(t)$ denote any curve in $TM(q)$ with $\|v\| = 1$. We have to show that the corresponding curves

$$t \to \exp_q(r_0 v(t))$$

in U_q, where $0 < r_0 < c$, are orthogonal to the radial geodesics

$$r \to \exp_q(rv(t_0)).$$

In terms of the surface $f:(r, t) \to \exp_q(rv(t))$, $0 \leqslant r < \varepsilon$, we must prove that $\langle \partial f/\partial r, \partial f/\partial t \rangle = 0$ for all (r, t).

From the metric property of the connexion we have

$$\frac{\partial}{\partial r}\left\langle \frac{\partial f}{\partial r}, \frac{\partial f}{\partial t} \right\rangle = \left\langle \nabla_{\frac{\partial}{\partial r}} \frac{\partial f}{\partial r}, \frac{\partial f}{\partial t} \right\rangle + \left\langle \frac{\partial f}{\partial r}, \nabla_{\frac{\partial}{\partial r}} \frac{\partial f}{\partial t} \right\rangle.$$

Since the curves $t = $ constant are geodesics we have

$$\nabla_{\frac{\partial}{\partial r}} \frac{\partial f}{\partial r} = 0.$$

Moreover

$$\frac{\partial}{\partial t}\left\langle \frac{\partial f}{\partial r}, \frac{\partial f}{\partial r} \right\rangle = 2\left\langle \nabla_{\frac{\partial}{\partial t}} \frac{\partial f}{\partial r}, \frac{\partial f}{\partial r} \right\rangle$$

$$= 2\left\langle \nabla_{\frac{\partial}{\partial r}} \frac{\partial f}{\partial t}, \frac{\partial f}{\partial r} \right\rangle.$$

Since $\|\partial f/\partial r\| = \|v(t)\| = 1$, it follows that

$$\left\langle \nabla_{\frac{\partial}{\partial r}} \frac{\partial f}{\partial t}, \frac{\partial f}{\partial r} \right\rangle = 0,$$

and hence the previous equation reduces to

$$\frac{\partial}{\partial r}\left\langle \frac{\partial f}{\partial r}, \frac{\partial f}{\partial t} \right\rangle = 0.$$

Thus

$$\left\langle \frac{\partial f}{\partial r}, \frac{\partial f}{\partial t} \right\rangle$$

is independent of r. However, when $r = 0$ we have $f(0, t) = \exp_q(0) = q$ and hence $\partial f/\partial t(0, t) = 0$. Therefore $\langle \partial f/\partial r, \partial f/\partial t \rangle$ is identically zero and the lemma is proved.

We now prove that the shortest path joining two concentric geodesic spheres around q is a radial geodesic. More specifically we have the following result.

Lemma (3.4.6) *Let $w: [a, b] \to U_q \setminus \{q\}$ be a smooth curve lying in the punctured neighbourhood U_q, so that each point $w(t)$ can be expressed uniquely in the form $\exp_q(r(t)v(t))$ where $0 < r(t) < \varepsilon$, $\|v(t)\| = 1$, $v(t) \in TM(q)$. Then the length of w, given by*

$$\int_a^b \left\| \frac{dw}{dt} \right\| dt,$$

is greater than or equal to $|r(b) - r(a)|$. Moreover, equality holds only if the function $r(t)$ is monotone and $v(t)$ is a constant function.

To prove the lemma let

$$f(r, t) = \exp_q(rv(t)).$$

Then $w(t) = f(r(t), t)$, and

$$\frac{dw}{dt} = \frac{\partial f}{\partial r} \cdot \frac{dr}{dt} + \frac{\partial f}{\partial t}.$$

Taking the norm gives

$$\left\| \frac{dw}{dt} \right\|^2 = \left(\frac{dr}{dt} \right)^2 + \left\| \frac{\partial f}{\partial t} \right\|^2$$

where we have used the fact that the vectors $\partial f/\partial t$, $\partial f/\partial r$ are orthogonal (previous lemma!) and also that $\|\partial f/\partial r\| = 1$. Hence

$$\left\| \frac{dw}{dt} \right\|^2 \geq \left(\frac{dr}{dt} \right)^2,$$

equality holding only if $\partial f/\partial t = 0$, that is, if $dv/dt = 0$. Thus

$$\int_a^b \left\| \frac{dw}{dt} \right\| dt \geq \int_a^b \left| \frac{dr}{dt} \right| dt \geq |r(b) - r(a)|,$$

and equality holds only if $r(t)$ is a monotone function and $v(t)$ is constant.

We now use the preceding lemma to prove the next theorem.

Theorem (3.4.7) *Let (W, ε) be a convex normal neighbourhood and let $\gamma: [0, 1] \to M$ be a geodesic of length $< \varepsilon$ joining two points of W. Let $w: [0, 1] \to M$ be any other smooth curve joining the same two points. Then*

3.4 Geodesics

$$\int_0^1 \left\|\frac{d\gamma}{dt}\right\| dt \leq \int_0^1 \left\|\frac{dw}{dt}\right\| dt,$$

equality holding only if the set of points $w(t)$ and $\gamma(t)$ coincide. Thus $\gamma(t)$ is the shortest path joining its end points.

Proof. Let w be a smooth path from q to $p = \exp_q(rv) \in U_q$, where $0 < r < \varepsilon$, $\|v\| = 1$. Then for any positive number δ, the path w must contain a segment joining a point on the geodesic sphere, centre q of radius δ, to the geodesic sphere of radius r. The length of this segment must be greater than or equal to $r - \delta$ by the previous lemma. Since this must be true for all δ, we let $\delta \to 0$ to conclude that the length of w must be greater than or equal to r. Moreover, if $w([0, 1])$ does not coincide with $\gamma([0, 1])$, the inequality must be strict.

Corollary (3.4.8) *Suppose that a path $w: [0, l]$, parametrized by arc length, has a length less than or equal to the length of any other path from $w(0)$ to $w(l)$. Then w must be a geodesic.*

The proof follows immediately from the theorem.

Definition (3.4.9) *A geodesic $\gamma: [a, b] \to M$ will be called minimal if its length is less than or equal to the length of any other smooth path joining its end points.*

Clearly a sufficiently small geodesic arc is minimal. However it is intuitively evident that a great circle of the unit sphere in $\mathbb{R}^3$ of length greater than π is a geodesic but it is not minimal. Moreover, since there are an infinite number of semicircles joining the north pole to the south pole of this sphere, each of which is a minimal geodesic, it follows that a minimal geodesic joining two points is by no means unique.

We define the *distance* between two points p and q of M to be the greatest lower bound of the arc lengths of piecewise smooth paths joining p to q. This definition makes M a metric space. Moreover, from the previous theorem it follows that the topology determined by this metric is compatible with the topology used to define the manifold structure of M.

As a consequence of the theorem we have the following corollary.

Corollary (3.4.10) *Let K be a compact subset of M. Then there exists a number $\delta > 0$ such that any two points of K of distance apart less than δ are joined by a unique geodesic of length less than δ.*

Moreover this geodesic is minimal and depends differentiably on its end points.

To prove the corollary we cover K by a number of open sets $\{W_\alpha\}$. We now choose δ sufficiently small so that any two points with distance

Completeness

Definition (3.4.11) *The manifold M is said to be geodesically complete if $\exp_q(v)$ is defined for all $q \in M$ and for all $v \in TM(q)$.*

This is equivalent to M having the property that every geodesic segment $\gamma:[a,b] \to M$ can be extended to an infinitely long geodesic $\gamma: \mathbb{R} \to M$. In fact it is sufficient to assume that the manifold has this property for geodesics through a single point $q \in M$ but we shall not prove this. Clearly there must be some relation between geodesic completeness of M and completeness of M when considered as a metric space. The following result is known.

Theorem (3.4.12) *If M is geodesically complete, any two points can be joined by a minimal geodesic.*

From this, one can deduce that if M is geodesically complete, then every bounded subset of M as a metric space has compact closure. Hence M is complete as a metric space, that is, every Cauchy sequence converges. Conversely, it can be proved that if M is complete as a metric space, then it is geodesically complete. For a detailed treatment of these results the reader is referred to de Rham (1952) and Hopf and Rinow (1931).

The restriction of M to be complete seems natural geometrically. Compactness is a much stronger topological restriction which implies completeness, but the converse is clearly false as is shown by the plane $\mathbb{R}^2$ with the Euclidean metric.

3.5 NORMAL COORDINATES

We have seen that $\exp_q v$ is defined provided that $\|v\|$ is small enough. Equivalently, if $\|v\| = 1$, then $\exp rv$ is defined for sufficiently small r. Let (u_i) be an orthonormal basis of the tangent space $TM(q)$. We choose for v one of the basis vectors. We write

$$x^j\left(\exp_q\left(\sum_{i=1}^n t^i u_i\right)\right) = t^j.$$

Then the functions x^i form a coordinate system for some neighbourhood of q. These are called *normal coordinates* centred at q. Relative to this system the equations of geodesics through q are given by

$$x^i = a^i s$$

3.5 Normal coordinates

where s is the arc length and $a = \Sigma a^i \partial/\partial x^i$ is a unit vector. Since the geodesics must satisfy the equation

$$\frac{d^2 x^i}{ds^i} + \Gamma^i_{jk} \frac{dx^j}{ds} \frac{dx^k}{ds} = 0,$$

it follows that $\Gamma^i_{jk}(q) a^j a^k = 0$ at q for all choices of a^j. Hence $\Gamma^i_{jk}(q) = 0$. Thus normal coordinates have a property already met in the previous chapter.

Normal coordinates are important because they are the nearest system of coordinates that one can find in a Riemannian manifold to rectangular Cartesian coordinates in Euclidean geometry of $\mathbb{R}^n$. However they have their limitations: if one is given a non-normal system of coordinates and wishes to transform this to a normal system, the problem is usually very difficult and sometimes impossible. It usually involves the explicit solution of the differential equations for a geodesic in the given system of coordinates; hence the difficulty.

However, it is often useful to express geometrical objects, such as tensors, by expressing their components as power series in normal coordinates. This idea goes back to Riemann but a modern and powerful method of achieving this has been developed by Gray (1973).

Gray defines his curvature operator, which we denote by S, so that it has opposite sign to our R, that is, $S(X, Y) = R(Y, X)$. This implies in components that $S^h_{ijk} = -R^h_{ijk}$. He defines the covariant curvature tensor by $S(X, Y, Z, W) = \langle S(X, Y,) Z, W \rangle$, and the Ricci tensor as the trace of the endomorphism $V \to S(X, V)Y$. The net result is that $S(X, Y, Z, W) = R(X, Y, Z, W)$ so that the covariant curvature tensors and the Ricci tensors corresponding to R and S are identical. Note, however, that $\rho_{ij} = S^h_{ijh}$, that is, contraction is made with the third lower suffix.

Let $q \in M$, and let $(x^1, \ldots, x^n)$ be a normal coordinate system defined in a neighbourhood U_q of q with $x^1(q) = \cdots = x^n(q) = 0$. We shall say that X is a *coordinate vector field at* q if there exist constants $a^1, \ldots, a^n$ such that in a neighbourhood of q we have

$$X = \sum a^i \frac{\partial}{\partial x^i}.$$

Coordinate vector fields will be denoted by $X, Y, \ldots$ and their corresponding integral curves by $\alpha, \beta, \ldots$. We normalize so that $\alpha(0) = \beta(0) = \cdots = q$. Thus $\alpha, \beta, \ldots$ are geodesics starting at q, such that $\alpha'(t) = X_{\alpha(t)}$ whenever $\alpha(t)$ is defined. We use the notation

$$\nabla^p_{X \cdots X} = \nabla_X \cdots \nabla_X$$

where ∇ is the Riemannian connexion.

Lemma (3.5.1) Let X, Y be coordinate vector fields and let α be an integral curve of X passing through q. Then
$$(\nabla^p_{X\ldots X} X)_{\alpha(t)} = 0, \quad \text{for } p = 1, 2, \ldots;$$
$$(\nabla_X Y)_q \quad = 0.$$

We note that since α is a geodesic
$$(\nabla_X X)_{\alpha(t)} = 0.$$

Then the first assertion follows by induction because $(\nabla^p_{X\ldots X} X)_{\alpha(t)}$ depends only on the values of $X_{\alpha(t)}$ and $(\nabla^{p-1}_{X\ldots X} X)_{\alpha(t)}$. To prove the second assertion, it follows, because X and Y are coordinate vector fields, that $[X, Y] = 0$. Since ∇ has zero torsion this implies $\nabla_X Y - \nabla_Y X = 0$. However, if we replace X in $(\nabla_X X)_q = 0$ by $X + Y$ we obtain
$$(\nabla_X Y)_q + (\nabla_Y X)_q = 0.$$

This, with the previous equation, gives $(\nabla_X Y)_q = 0$ as required.

Lemma (3.5.2) For all p we have
$$(\nabla^p_{YX\ldots X} X)_q = (\nabla^p_{XYX\ldots X} X)_q = \cdots = (\nabla^p_{X\ldots XYX} X)_q$$
where X, Y are coordinate vector fields.

Proof. Since $[X, Y] = 0$ we get
$$R_{YX} = -\nabla_X \nabla_Y + \nabla_Y \nabla_X.$$

Let $A_k = (\nabla^p_{X\ldots Y\ldots X} X)_q$ where Y occurs in the kth place. Then
$$A_k - A_{k-1} = (\nabla^{k-2}_{X\ldots X} R_{YX} \nabla^{p-k}_{X\ldots X} X)_q.$$

We can expand the right-hand member in terms of covariant derivatives of R at q. However, if $2 \leq k < p$, each term contains a factor of the form $(\nabla^i_{X\ldots X} X)_q$, and each of these is zero. Hence $A_k = A_{k-1}$ for $2 \leq k < p$ and the lemma is proved.

Lemma (3.5.3) If X, Y are coordinate vector fields, then
$$2(\nabla^p_{X\ldots XY} X)_q + (p-1)(\nabla^p_{X\ldots XYX} X)_q = 0, \quad p = 1, 2, \ldots.$$

This follows from linearization of the relation
$$(\nabla^p_{X\ldots X} X)_{\alpha(t)} = 0,$$
using the relation $\nabla_X Y = \nabla_Y X$ and one of the lemmas.

Lemma (3.5.4) If X and Y are coordinate vector fields, then
$$(\nabla^p_{X\ldots X} Y)_q = -\left(\frac{p-1}{p+1}\right)(\nabla^{p-2}_{X\ldots X} R_{YX} X)_q \quad p = 2, 3, \ldots.$$

3.5 Normal coordinates

From the relation
$$R_{YX} = \nabla_Y \nabla_X - \nabla_X \nabla_Y$$
and the previous lemma we have
$$(\nabla_X^{p-2}\!\ldots_X R_{YX} X)_q = -(\nabla_X^p\!\ldots_{XY} X)_q + (\nabla_X^p\!\ldots_{XYX} X)_q$$
$$= -\left(\frac{p+1}{p-1}\right)(\nabla_X^p\!\ldots_X Y)_q,$$
and the lemma is proved.

We follow Gray's treatment, and fix the point q and a coordinate vector field X at q. We define operators D and Q by
$$DY = \nabla_X Y, \qquad QY = R_{YX} X$$
where Y is any coordinate vector field at q. Then
$$D^p Y = \nabla_X^p\!\ldots_X Y \quad \text{and} \quad D^p(Q) Y = D^p(R)_{YX} X.$$
For consistency we put $D^0 Y = Y$ and $D^0(Q) = Q$. We shall assume that all these vector fields are evaluated at q. In terms of this notation the previous lemma becomes
$$D^p Y = -\left(\frac{p-1}{p+1}\right) D^{p-2}(QY).$$

Lemma (3.5.5) *We have the identity*
$$D^p Y = -\left(\frac{p-1}{p+1}\right) D^{p-2}(Q) \cdot Y$$
$$+ \left(\frac{p-1}{p+1}\right) \sum_{i=1}^{[p/2]-1} \sum_{\substack{0 \leq k_1 + \cdots k_i \leq p-2i-2 \\ k_j \geq 0}} (-1)^{i+1}$$
$$\left[\prod_{j=1}^{i} t_p(k_1, \ldots, k_j) D^{k_j}(Q)\right]$$
$$D^{p-k_1-\cdots-k_i-2i-2}(Q) Y,$$

where
$$t_p(k_1, \ldots, k_j) = \left(\frac{p - k_1 - \cdots - k_j - 2j - 1}{p - k_1 - \cdots - k_j - 2j + 1}\right) \cdot$$
$$\left(\frac{p - k_0 - \cdots - k_{j-1} - 2j}{k_j}\right),$$

and $k_0 = 0$.

The proof follows from the previous lemma and the Leibnitz rule

$$D^k(QY) = \sum_{s=0}^{k} \binom{k}{s} D^s(Q) D^{k-s}Y$$

The above lemma is indeed correct as a general result, but in practice it is easier to compute $D^p Y$ for $1 \leq p \leq 8$ directly from the preceding lemma. We find

$$D^1 Y = 0, \quad D^2 Y = -\frac{1}{3} QY, \quad D^3 Y = -\frac{1}{2} D(Q) Y,$$

$$D^4 Y = \{-\frac{3}{5} D^2(Q) + \frac{1}{5} Q^2\} Y,$$

$$D^5 Y = \{-\frac{2}{3} D^3(Q) + \frac{2}{3} D(Q)Q + \frac{1}{3} QD(Q)\} Y,$$

$$D^6 Y = \{-\frac{5}{7} D^4(Q) + \frac{10}{7} D^2(Q)Q + \frac{3}{7} QD^2(Q) + \frac{10}{7} D(Q)^2 - \frac{1}{7} Q^3\} Y,$$

$$D^7 Y = \{-\frac{3}{4} D^5(Q) + \frac{5}{2} D^3(Q)Q + \frac{1}{2} QD^3(Q) + \frac{15}{4} D^3(Q)D(Q)$$
$$+ \frac{9}{4} D(Q)D^2(Q) - \frac{3}{4} D(Q)Q^2 - \frac{1}{2} QD(Q)Q - \frac{1}{4} Q^2 D(Q)\} Y,$$

$$D^8 Y = \{-\frac{7}{9} D^6(Q) + \frac{35}{9} D^4(Q)Q + \frac{5}{9} QD^4(Q) + \frac{70}{9} D^3(Q)D(Q)$$
$$+ \frac{28}{9} D(Q)D^3(Q) + 7 D^2(Q)^2 - \frac{7}{3} D^2(Q)Q^2$$
$$- \frac{10}{9} QD^2(Q)Q - \frac{1}{3} Q^2 D^2(Q) - \frac{28}{9} D(Q)^2 Q - \frac{14}{9} D(Q)QD(Q)$$
$$- \frac{10}{9} QD(Q)^2 + \frac{1}{9} Q^4\} Y.$$

Remark (3.5.6) We note that if in a Riemannian manifold we have $\nabla R = 0$, then

$$D^{2p+1} Y = 0 \quad \text{for } p = 0, 1, \ldots;$$

$$D^{2p} Y = (-1)^p Q^p / (2p+1) \quad \text{for } p = 1, 2, \ldots.$$

Note that, in the general case, $D^p(Q)$ and $D^p Y$ are polynomials of degree $(p+2)$ in X. For example,

$$D^2 Y = (\nabla^2_{XX} Y)_q = -\frac{1}{3} (R_{YX} X)_q.$$

We may linearize this formula and obtain the relation

$$(\nabla_{ZX} Y)_q + (\nabla_{XZ} Y)_q = -\frac{1}{3} (R_{YZ} X)_q - \frac{1}{3} (R_{YX} Z)_q.$$

Similarly numerous other formulae may be obtained by linearization of $D^p Y$.

3.5 Normal coordinates

Applications

Let us assume now that M is an analytic manifold. Let W be an analytic tensor field, covariant of order r, defined in a neighbourhood of q. Assume that $X_1, \ldots, X_n$ are coordinate vector fields that are orthonormal at q. Let $(x^1, \ldots, x^n)$ denote the corresponding normal coordinate system, and write

$$W(X_{\alpha_1}, \ldots, X_{\alpha_r}) = W_{\alpha_1 \cdots \alpha_r}.$$

From Taylor's theorem we have

$$W_{\alpha_1 \cdots \alpha_r} = \sum_{k=0}^{\infty} \sum_{i_1, \ldots, i_k = 1}^{n} \frac{1}{k!} (X_{i_1} \cdots X_{i_k} W_{\alpha_1 \cdots \alpha_r})(q) x^{i_1} \cdots x^{i_k}.$$

Now, for any coordinate vector field X and normal coordinate system we have

$$(X^p W_{\alpha_1 \cdots \alpha_r})_q = \sum_{\substack{\nu_1 + \cdots + \nu_{r+1} = p \\ \nu_i \geq 0}} \frac{p!}{\nu_1! \cdots \nu_n!} \nabla^{\nu_{r+1}}_{X \cdots X}(W)$$

$$(\nabla^{\nu_1}_{X \cdots X} X_{\alpha_1}, \ldots, \nabla^{\nu_r}_{X \cdots X} X_{\alpha_r})(q). \quad (3.9)$$

Since the coefficients in the Taylor series are symmetric in $X_{i_1}, \ldots, X_{i_k}$ it follows that these coefficients may be determined by linearizing the left-hand side of (3.9). On the other hand the right-hand side of (3.9) can be computed by means of the previous lemma. In principle it should be possible to write down a general expression for the coefficients of the Taylor series in terms of covariant derivatives of W at q and the curvature operator of M at q. However, such a formula would be extremely complicated and we content ourselves by computing the coefficients up to the fourth order. To simplify the notation we write

$$\nabla^p_{X_i \cdots X_j} = \nabla^p_{i \cdots j}, \quad \langle R_{X_j X_i} X_k, X_l \rangle = R_{ijkl}$$

$$R_{jk} = \sum_{i=1}^{n} R_{ijik}.$$

We find after some computation

$$W_{\alpha_1 \cdots \alpha_r} = W_{\alpha_1 \cdots \alpha_r}(q) + \sum_{i=1}^{n} (\nabla_i W_{\alpha_1 \cdots \alpha_r})(q) x^i$$

$$+ \frac{1}{2} \sum_{i,j=1}^{n} \left\{ \nabla^2_{ij} W_{\alpha_1 \cdots \alpha_r} - \frac{1}{3} \sum_{s=1}^{n} \sum_{a=1}^{r} R_{i\alpha_a js} W_{\alpha_1 \cdots \alpha_{a-1} s \alpha_{a+1} \cdots \alpha_r} \right\} (q) x^i x^j$$

78   3   *Riemannian manifolds*

$$+ \frac{1}{6} \sum_{i,j,k=1}^{n} \left\{ \nabla^3_{ijk} W_{\alpha_1 \cdots \alpha_r} - \sum_{s=1}^{n} \right.$$

$$\sum_{a=1}^{r} R_{i\alpha_a js} \nabla_k W_{\alpha_1 \cdots \alpha_{a-1} s \alpha_{a+1} \cdots \alpha_r}$$

$$\left. - \frac{1}{2} \sum_{s=1}^{n} \sum_{a=1}^{r} \nabla_i R_{j\alpha_a ks} W_{\alpha_1 \cdots \alpha_{a-1} s \alpha_{a+1} \cdots \alpha_r} \right\} (q) x^i x^j x^k$$

$$+ \frac{1}{24} \sum_{i,j,k,l=1}^{n} \left\{ \nabla^4_{ijkl} W_{\alpha_1 \cdots \alpha_r} - 2 \sum_{s=1}^{n} \right.$$

$$\sum_{a=1}^{r} R_{i\alpha_a js} \nabla^2_{kl} W_{\alpha_1 \cdots \alpha_{a-1} s \alpha_{a+1} \cdots \alpha_r}$$

$$- 2 \sum_{s=1}^{n} \sum_{a=1}^{r} \nabla_i R_{j\alpha_a ks} \nabla_l W_{\alpha_1 \cdots \alpha_{a-1} s \alpha_{a+1} \cdots \alpha_r}$$

$$- \frac{3}{5} \sum_{s=1}^{n} \sum_{a=1}^{r} \nabla^2_{ij} R_{k\alpha_a ls} W_{\alpha_1 \cdots \alpha_{a-1} s \alpha_{a+1} \cdots \alpha_r}$$

$$+ \frac{1}{5} \sum_{s,t=1}^{n} \sum_{a=1}^{r} R_{i\alpha_a js} R_{kslt} W_{\alpha_1 \cdots \alpha_{a-1} t \alpha_{a+1} \cdots \alpha_r}$$

$$\left. + \frac{2}{3} \sum_{s,t=1}^{n} \sum_{\substack{a,b=1 \\ a<b}}^{r} R_{i\alpha_a js} R_{k\alpha_b lt} W_{\alpha_1 \cdots \alpha_{a-1} s \alpha_{a+1} \cdots \alpha_{b-1} t \alpha_{b+1} \cdots \alpha_r} \right\}$$

$$(q) x^i x^j x^k x^l$$
$$+ \cdots$$

Corollary (3.5.7) Assume that W is parallel, that is, all its covariant derivatives are zero. Then we have

$$W_{\alpha_1 \cdots \alpha_r} = W_{\alpha_1 \cdots \alpha_r}(q) - \frac{1}{6} \sum_{i,j,s=1}^{n} \sum_{a=1}^{r} (R_{i\alpha_a js} W_{\alpha_1 \cdots \alpha_{a-1} s \alpha_{a+1} \cdots \alpha_r})$$

$$(q) x^i x^j$$

$$- \frac{1}{12} \sum_{i,j,k,s=1}^{n} \sum_{a=1}^{r} (\nabla_i R_{j\alpha_a ks} W_{\alpha_1 \cdots \alpha_{a-1} s \alpha_{a+1} \cdots \alpha_r}) (q) x^i x^j x^k$$

$$+ \frac{1}{24} \sum_{i,j,k,l=1}^{n} \left\{ -\frac{3}{5} \sum_{s=1}^{n} \sum_{a=1}^{r} \nabla^2_{ij} R_{k\alpha_a ls} W_{\alpha_1 \cdots \alpha_{a-1} s \alpha_{a+1} \cdots \alpha_r} \right.$$

$$+ \frac{1}{5} \sum_{s,t=1}^{n} \sum_{a=1}^{r} R_{i\alpha_a js} R_{kslt} W_{\alpha_1 \cdots \alpha_{a-1} t \alpha_{a+1} \cdots \alpha_r}$$

3.6 *Volume form* 79

$$+\frac{2}{3}\sum_{s,t=1}^{n}\sum_{\substack{a,b=1\\a<b}}^{r} R_{i\alpha_a js}R_{k\alpha_b lt}W_{\alpha_1\cdots\alpha_{a-1}s\alpha_{a+1}\cdots\alpha_{b-1}t\alpha_{b+1}\cdots\alpha_r}\Bigg\}$$

$(q)x^i x^j x^k x^l$

$$+\frac{1}{120}\sum_{i,j,k,l,h=1}^{n}\Bigg\{-\frac{2}{3}\sum_{s=1}^{n}\sum_{a=1}^{r}\nabla^3_{ijk}R_{l\alpha_a hs}W_{\alpha_1\cdots\alpha_{a-1}s\alpha_{a+1}\cdots\alpha_r}$$

$$+\sum_{s,t=1}^{n}\sum_{a=1}^{r}\left(\frac{1}{3}\nabla_i R_{j\alpha_a ks}R_{lsht}+\frac{2}{3}R_{i\alpha_a js}\nabla_k R_{lsht}\right)$$

$W_{\alpha_1\cdots\alpha_{a-1}t\alpha_{a+1}\cdots\alpha_r}$

$$+\frac{5}{3}\sum_{s,t=1}^{n}\sum_{\substack{a,b=1\\a<b}}^{r}(\nabla_i R_{j\alpha_a ks}R_{l\alpha_b ht}+R_{i\alpha_a js}\nabla_k R_{l\alpha_b ht})$$

$$\cdot W_{\alpha_1\cdots\alpha_{a-1}s\alpha_{a+1}\cdots\alpha_{b-1}t\alpha_{b+1}\cdots\alpha_r}\Bigg\}(q)x^i x^j x^k x^l x^h+\cdots.$$

In particular, if we take for W the metric tensor, with $g_{pq}=\langle X_p, X_q\rangle$, we get

$$g_{pq}=\delta_{pq}-\frac{1}{3}\sum_{i,j=1}^{n} R_{ipjq}(q)x^i x^j-\frac{1}{6}\sum_{i,j,k=1}^{n}\nabla_i R_{jpkq}(q)x^i x^j x^k$$

$$+\frac{1}{120}\sum_{i,j,k,l=1}^{n}\Bigg\{-6\nabla^2_{ij}R_{kplq}+\frac{16}{3}\sum_{s=1}^{n} R_{ipjs}R_{kqls}\Bigg\}(q)x^i x^j x^k x^l$$

$$+\frac{1}{90}\sum_{i,j,k,l,h=1}^{n}\Bigg\{-\nabla^3_{ijk}R_{lphq}+2\sum_{s=1}^{n}(\nabla_i R_{jpks}R_{lqhs}+\nabla_i R_{jqks}R_{lphs})\Bigg\}$$

$(q)x^i x^j x^k x^l x^h+\cdots.$

Before proceeding with further applications, we deal with a section involving the volume element of a Riemannian manifold.

3.6 VOLUME FORM

Let M be a Riemannian manifold and let U be a coordinate neighbourhood of a point $q\in M$. Let $e_1,\ldots,e_n$ be a collection of C^∞-vector fields on U which form an orthonormal frame at each point of U. Let $\omega^1,\ldots,\omega^n$ be the corresponding covectors with respect to the natural isomorphism of $T(M)$ and $T^*(M)$ induced by the metric. Then at each point q, $(\omega^1,\ldots,\omega^n)$ is itself an orthonormal basis for the space $T^*M(q)$.

Let $\omega^{1'}, \ldots, \omega^{n'}$ be a similar orthonormal coframe over some neighbourhood U'. Then on $U \cap U'$ we have

$$\omega^1 \wedge \cdots \wedge \omega^n = \det(\sigma) \omega^{1'} \wedge \cdots \wedge \omega^{n'}$$

where σ is an orthogonal matrix whose entries are C^∞-functions over $U \cap U'$. Thus

$$\omega^1 \wedge \cdots \wedge \omega^n = \pm \omega^{1'} \wedge \cdots \wedge \omega^{n'}.$$

We now assume that M is oriented. This enables us to choose a local oriented n-coframe field at each point of M, so that the corresponding n-forms $\omega = \omega^1 \wedge \cdots \wedge \omega^n$ agree on the overlaps and therefore determine a nowhere vanishing n-form ω over the whole of M. This is called the *volume form* of (M, g), and its integral over M is called the *volume* of M.

In terms of a general coordinate system (x^i) over U, write

$$\frac{\partial}{\partial x^i} = \sum_j \alpha_i^j e_j.$$

Then

$$\omega\left(\frac{\partial}{\partial x^1}, \ldots, \frac{\partial}{\partial x^n}\right) = \det(\alpha_i^j) \omega(e_1, \ldots, e_n) = \det(\alpha_i^j).$$

Moreover,

$$g_{ij}(q) = \left\langle \frac{\partial}{\partial x^i}, \frac{\partial}{\partial x^j} \right\rangle$$

$$= \left\langle \sum_k \alpha_i^k e_k, \sum_h \alpha_j^h e_h \right\rangle$$

$$= \sum \alpha_i^k \alpha_j^h \delta_{kh}$$

$$= \sum_k \alpha_i^k \alpha_j^k \big|_q.$$

Thus $\det(g_{ij}(q)) = (\det \alpha_i^n)^2$, and we have

$$\omega_q\left(\frac{\partial}{\partial x^1}, \ldots, \frac{\partial}{\partial x^n}\right) = (\det g_{ij}(q))^{1/2}.$$

We can now obtain an expansion of the volume element in terms of normal coordinates. Let $X_1, \ldots, X_n$ be coordinate vector fields which are orthonormal at q. We write $\omega_{1 \cdots n} = \omega(X_1, \ldots, X_n)$ and apply the previous expansion to give

3.6 Volume form

$$\omega_{1\cdots n} = 1 - \frac{1}{6}\sum_{i,j=1}^{n} R_{ij}(q)x^i x^j - \frac{1}{12}\sum_{i,j,k=1}^{n} \nabla_i R_{jk}(q) x^i x^j x^k$$

$$+ \frac{1}{24}\sum_{i,j,k,l=1}^{n} \left\{ -\frac{3}{5}\nabla_{ij}^2 R_{kl} + \frac{1}{3} R_{ij} R_{kl} - \frac{2}{15}\sum_{a,b=1}^{n} R_{iajb} R_{kalb} \right\}$$

$$(q) x^i x^j x^k x^l$$

$$+ \frac{1}{120}\sum_{i,j,k,l,h=1}^{n} \left\{ -\frac{2}{3}\nabla_{ijk}^3 R_{lh} + \frac{5}{3}\nabla_i R_{jk} R_{lh} \right.$$

$$\left. - \frac{2}{3}\sum_{a,b=1}^{n} \nabla_i R_{jakb} R_{lahb} \right\} (q) x^i x^j x^k x^l x^h$$

$$+ \frac{1}{720}\sum_{i,j,k,l,h,g=1}^{n} \left\{ -\frac{5}{7}\nabla_{ijkl}^4 R_{hg} + 3\nabla_{ij}^2 R_{kl} R_{hg} + \frac{5}{2}\nabla_i R_{jk} \nabla_l R_{hg} \right.$$

$$- \frac{8}{7}\sum_{a,b=1}^{n} \nabla_{ij}^2 R_{kalb} R_{hagb} - \frac{5}{9} R_{ij} R_{kl} R_{hg} - \frac{15}{14}\sum_{a,b=1}^{n} \nabla_i R_{jakb} \nabla_l R_{hagb}$$

$$- \frac{16}{63}\sum_{a,b,c=1}^{n} R_{iajb} R_{kblc} R_{bcga}$$

$$\left. + \frac{2}{3} R_{ij} \sum_{a,b=1}^{n} R_{kalb} R_{hagb} \right\} (q) x^i x^j x^k x^l x^h x^g + \cdots .$$

In a similar way we obtain the following expansion of the components of the metric tensor:

$$g_{pq} = \delta_{pq} - \frac{1}{3}\sum_{i,j=1}^{n} R_{ipjq}(q) x^i x^j - \frac{1}{6}\sum_{i,j,k=1}^{n} \nabla_i R_{jpkq}(q) x^i x^j x^k$$

$$+ \frac{1}{120}\sum_{i,j,k,l=1}^{n} \left\{ -6\nabla_{ij}^2 R_{kplq} + \frac{16}{3}\sum_{s=1}^{n} R_{ipjs} R_{kqls} \right\} (q) x^i x^j x^k x^l$$

$$+ \frac{1}{90}\sum_{i,j,k,l,h=1}^{n} \left\{ -\nabla_{ijk}^3 R_{lphq} + 2\sum_{s=1}^{n} (\nabla_i R_{jpks} R_{lqhs} + \nabla_i R_{jqks} R_{lphs}) \right\}$$

$$(q) x^i x^j x^k x^l x^h + \cdots .$$

Let M be an analytic Riemannian manifold, and suppose r small enough to ensure that $\exp_q$ is defined in a ball of radius r in the tangent space M_q. We write

$$S_q(r) = \text{volume of } \{\exp_q(x) \mid \|x\| = r\},$$
$$V_q(r) = \text{volume of } \{\exp_q(x) \mid \|x\| \leq r\}.$$

Here $S_q(r)$ denotes the $(n-1)$-dimensional volume of the sphere and $V_q(r)$ denotes the n-dimensional volume of the ball. We write

$$\tau(R) = \sum_{i=1}^{n} R_{ii}, \qquad \|R\|^2 = \sum_{i,j,k,l=1}^{n} R_{ijkl}^2,$$

Write

$$\alpha_n = 2\Gamma\left(\frac{1}{2}\right)^n \Gamma\left(\frac{n}{2}\right)^{-1},$$

$$\|\rho(R)\|^2 = \sum_{i,j=1}^{n} R_{ij}^2,$$

$$\Delta R = \text{Laplacian of } R = \sum_{i=1}^{n} \nabla_{ii}^2 \tau(R).$$

Then by applying the previous techniques we obtain the formulas:

$$S_q(r) = \alpha_n r^{n-1} \left\{ 1 - \frac{\tau(R)}{6n} r^2 + \frac{1}{360n(n+2)} \left(-3\|R\|^2 + 8\|\rho(R)\|^2 \right.\right.$$
$$\left.\left. + 5\tau(R)^2 - 18\Delta R \right) r^4 + O(r^6) \right\}_q,$$

$$V_q(r) = \frac{\alpha_n r^n}{n} \left\{ 1 - \frac{\tau(R)}{6(n+2)} r^2 + \frac{1}{360(n+2)(n+4)}\right.$$
$$\left. \left(-3\|R\|^2 + 8\|\rho(R)\|^2 + 5\tau(R)^2 - 18\Delta R\right) r^4 + O(r^6) \right\}_q.$$

We shall make applications of these formulas when studying special metrics on manifolds.

3.7 THE LIE DERIVATIVE

We have already seen that when a differential operator (covariant differentiation, for example) is defined for scalars and for vector fields, its domain of definition can be uniquely extended to tensor fields of type (r, s) provided that the operator satisfies the Leibnitz law and commutes with contraction. Another example of such a differential operator, which depends upon a given vector field X, is called Lie derivation. We denote this operator by L_X and define its operation on scalars and vector fields, respectively, by

$$L_X f = Xf, \qquad L_X Y = [X, Y].$$

3.7 The Lie derivative

It is easily verified that
$$L_X(gY) = (L_X g)Y + g(L_X Y)$$
and hence its domain can be extended. For example, if ω is a covariant vector field, the relation
$$L_X(\omega(Y)) = (L_X \omega)(Y) + \omega(L_X(Y))$$
gives the required action of L_X on covariant vector fields, namely,
$$(L_X \omega)(Y) = X(\omega(Y)) - \omega([X, Y]).$$

In terms of coordinates we have

$$L_X f = X^i \frac{\partial f}{\partial x^i},$$

$$(L_X Y)^j = X^i \frac{\partial Y^j}{\partial x^i} - Y^i \frac{\partial X^j}{\partial x^i},$$

$$(L_X T)^{ij} = X^k \frac{\partial T^{ij}}{\partial x^k} - T^{kj} \frac{\partial X^i}{\partial x^k} - T^{ik} \frac{\partial X^j}{\partial x^k},$$

$$(L_X \omega)_j = X^k \frac{\partial \omega_j}{\partial x^k} + \omega_k \frac{\partial X^k}{\partial x^j},$$

$$(L_X T)_{ij} = X^k \frac{\partial T_{ij}}{\partial x^k} + T_{kj} \frac{\partial X^k}{\partial x^i} + T_{ik} \frac{\partial X^k}{\partial x^j},$$

$$(L_X T)^i_j = X^k \frac{\partial T^i_j}{\partial x^k} - T^k_j \frac{\partial X^i}{\partial x^k} + T^i_k \frac{\partial X^k}{\partial x^j}.$$

Clearly we have extended the domain of operation so that
$$L_X(S \otimes T) = (L_X S) \otimes T + S \otimes L_X T.$$
In particular the action of L_X on the algebra of exterior forms has many nice properties, including the following which we ask the reader to verify:

(1) $L_X(u \wedge v) = L_X u \wedge v + u \wedge L_X v$ where $u, v \in \Lambda T^*(M)$;
(2) on $\Lambda T^*(M)$ we have $L_X = i(X) \circ d + d \circ i(X)$;
(3) L_X commutes with d;
(4) let $\omega \in \Lambda^p T^*(M)$ and let $Y_0, \ldots, Y_p$ be vector fields on M. Then

$$L_{Y_0}(\omega(Y_1, \ldots, Y_p)) = (L_{Y_0} \omega)(Y_1, \ldots, Y_p) \\ + \sum_{i=1}^{p} \omega(Y_1, \ldots, Y_{i-1}, L_{Y_0} Y_i, \ldots, Y_p).$$

Finally we note that the Jacobi identity

$$[X, [Y, Z]] + [Y, [Z, X]] + [X, [Y, Z]] = 0$$

can be written in the form

$$L_{[X,Y]} Z = [L_X, L_Y] Z$$

where the bracket on the right-hand side denotes the commutator $(L_X L_Y - L_Y L_X)$. The algebraically minded reader will see that this gives a homomorphism of the Lie algebra of vector fields on M to the algebra of Lie derivations over M.

3.8 THE HODGE STAR OPERATOR

We recall that an inner product space V is a linear space with a real-valued function on $V \times V$, denoted by $\langle \Box, \Box \rangle$ which is:

(1) linear in each variable;
(2) symmetric: $\langle \alpha, \beta \rangle = \langle \beta, \alpha \rangle$;
(3) non-degenerate, that is, if for fixed α, $\langle \alpha, \beta \rangle = 0$ for all β, then $\alpha = 0$.

An orthonormal basis of V is a basis $\sigma^1, \ldots, \sigma^n$ such that

$$\langle \sigma^i, \sigma^j \rangle = \pm \delta^{ij}.$$

If there are r plus signs and s minus signs, then $r + s = n$ and $r - s$ is called the *signature* of the inner product. A standard result is that the signature does not depend on the choice of basis. Another standard result is that each inner product space possesses an orthonormal basis. A further property of inner product spaces is the following lemma.

Lemma (3.8.1) Let f be a linear functional on V. Then there is a uniquely determined β in V such that $f(\alpha) = \langle \alpha, \beta \rangle$.

To prove this we take an orthonormal basis $\sigma^1, \ldots, \sigma^n$. Set $b_i = f(\sigma^i)$, and take for β

$$\beta = \sum \pm b_j \sigma^j = \sum \langle \sigma^j, \sigma^k \rangle b_k \sigma^j.$$

Then

$$\langle \sigma^i, \beta \rangle = \sum \langle \sigma^i, \Sigma \langle \sigma^j, \sigma^k \rangle b_k \sigma^j \rangle$$

$$= \sum \delta^{ij} \delta^{jk} b_k$$

$$= b_i = f(\sigma^i)$$

3.8 The Hodge star operator

and the lemma is proved.

The inner product on V extends in a natural way to each of the spaces of exterior powers $\Lambda^p V$ of V. For if $\lambda = \alpha_1 \wedge \cdots \wedge \alpha_p$ and $\mu = \beta_1 \wedge \cdots \wedge \beta_p$ we set

$$\langle \lambda, \mu \rangle = \det(\langle \alpha_i, \beta_j \rangle).$$

Condition (1) is satisfied because the determinant is an alternating multilinear function of the α's and the β's. Condition (2) is satisfied because interchanging the rows and columns of a determinant does not alter its value. Condition (3) is checked by considering the products of elements formed from an orthonormal basis of V.

We choose for V a definite orientation, that is, we select one basis of V and consider only other bases related to this one by a matrix of positive determinant. The space V has two orientation — we select one and keep to it.

We now define an operation $*$ called the Hodge star operator. This is a linear map of $\Lambda^p V$ onto $\Lambda^{n-p} V$. The operator depends on the inner product and also on the orientation of V. Reversing orientation will change the sign of $*$.

Let $(\sigma^1, \ldots, \sigma^n)$ be an orthonormal basis of V and hence it determines a basis of $\Lambda^n V$. Let λ be a fixed element of $\Lambda^p V$. Then the mapping

$$\mu \to \lambda \wedge \mu$$

is a linear transformation on $\Lambda^{n-p} V$ onto the one-dimensional space $\Lambda^n V$. We may write

$$\lambda \wedge \mu = f_\lambda(\mu) \sigma^1 \wedge \cdots \wedge \sigma^n$$

so that f_λ is a linear functional on $\Lambda^{n-p} V$. From the previous lemma it follows that there is a unique $(n-p)$-vector, which we denote by $*\lambda$, such that

$$\lambda \wedge \mu = \langle *\lambda, \mu \rangle \sigma$$

where $\sigma = \sigma^1 \wedge \cdots \wedge \sigma^n$. This defines the operator $*$.

In order to compute $*\lambda$ for generators of $\Lambda^p V$ it is sufficient to compute $*\lambda$ where $\lambda = \sigma^1 \wedge \cdots \wedge \sigma^p$, where $\sigma^1, \ldots, \sigma^n$ form an orthonormal basis of V. We let the index K run over sets of $(n-p)$ indices. Then

$$\lambda \wedge \sigma^K = \langle *\lambda, \sigma^K \rangle \sigma.$$

The only non-zero entries on the left-hand side occur when $K = \{p+1, \ldots, n\}$, and hence

$$*\lambda = c\sigma^{p+1} \wedge \cdots \wedge \sigma^n$$

for some constant c. To find c we have

$$\sigma = \lambda \wedge \sigma^K = c \langle \sigma^K, \sigma^K \rangle \sigma$$

from which we deduce that

$$c = \langle \sigma^K, \sigma^K \rangle = \pm 1.$$

Thus

$$*\lambda = \langle \sigma^K, \sigma^K \rangle \sigma^K.$$

If we set $H = \{1, \ldots, p\}$, $K = \{p+1, \ldots, n\}$, we have

$$*\sigma^H = \langle \sigma^K, \sigma^K \rangle \sigma^K$$

Since $\sigma^K \wedge \sigma^H = (-1)^{p(n-p)} \sigma^H \wedge \sigma^K$ we deduce that

$$*\sigma^K = (-1)^{p(n-p)} \langle \sigma^H, \sigma^H \rangle \sigma^H.$$

Hence

$$\begin{aligned} *(*\sigma^H) &= (-1)^{p(n-p)} \langle \sigma^H, \sigma^H \rangle \langle \sigma^K, \sigma^K \rangle \sigma^H, \\ &= (-1)^{p(n-p)} \langle \sigma^H, \sigma^H \rangle \sigma^H \\ &= (-1)^{p(n-p)+s} \sigma^H, \end{aligned}$$

where s is the number of negative eigenvalues of the metric tensor. Thus, for any p-vector α we have

$$**\alpha = (-1)^{p(n-p)+s} \alpha.$$

We now prove the following lemma.

Lemma (3.8.2) *If $\alpha, \beta \in \Lambda^p V$ then $\alpha \wedge *\beta = \beta \wedge *\alpha$.*

To prove this we note that when $\beta = \sigma^H$, the only generator α for which both sides do not vanish is $\alpha = \sigma^H$. Then

$$\begin{aligned} \alpha \wedge *\beta = \sigma^H \wedge \langle \sigma^K, \sigma^K \rangle \sigma^K &= \langle \sigma^K, \sigma^K \rangle \sigma \\ &= \langle \sigma^H, \sigma^H \rangle (-1)^s \sigma \\ &= (-1)^s \langle \alpha, \beta \rangle \sigma, \end{aligned}$$

which is clearly symmetric in α and β, and the lemma is proved.

We note that the volume element can be written as $*1$ provided that s is even.

We now obtain two further identities satisfied by the $*$ operator. Let $\alpha, \beta \in \Lambda^p V$. Then

$$\langle \alpha, \beta \rangle = (-1)^s \langle *\alpha, *\beta \rangle.$$

To prove this we note that

3.8 The Hodge star operator

$$\langle *\alpha, *\beta \rangle \sigma = (-1)^s * \alpha \wedge **\beta$$
$$= (-1)^{2s}(-1)^{p(n-p)} * \alpha \wedge \beta$$
$$= \beta \wedge *\alpha$$
$$= (-1)^s \langle \alpha, \beta \rangle \sigma$$

giving the result.

Also we have the identity

$$\langle *\alpha, \beta \rangle = \langle \alpha \wedge \beta, \sigma \rangle,$$

for all $\alpha \in \Lambda^p V$ and $\beta \in \Lambda^{n-p} V$. To prove the identity write $\beta = *\gamma$ for $\gamma \in \Lambda^p V$. Then the left-hand side becomes

$$\langle *\alpha, \beta \rangle = \langle *\alpha, *\gamma \rangle$$
$$= (-1)^s \langle \alpha, \gamma \rangle.$$

The right-hand side becomes

$$\langle \alpha \wedge *\gamma, \sigma \rangle = \langle \langle \alpha, \gamma \rangle \sigma, \sigma \rangle$$
$$= \langle \alpha, \gamma \rangle \langle \sigma, \sigma \rangle$$
$$= (-1)^s \langle \alpha, \gamma \rangle$$

and so the identity is proved.

Summarising, we see that $*$ satisfies:

(1) $\alpha \wedge *\beta = \beta \wedge *\alpha = \langle \alpha, \beta \rangle \sigma$, $\alpha, \beta \in \Lambda^p Y$;
(2) $*1 = (-1)^s \sigma$;
(3) $**\alpha = (-1)^{p(n-p)+s} \alpha$;
(4) $\langle \alpha, \beta \rangle = (-1)^s \langle *\alpha, *\beta \rangle$;
(5) $\langle *\alpha, \beta \rangle = \langle \alpha \wedge \beta, \sigma \rangle$ for $\alpha \in \Lambda^p V$, $\beta \in \Lambda^{n-p} V$.

The coderivative

Let M be a compact oriented pseudo-Riemannian manifold with metric (g_{ij}) having s minus signs when expressed as a sum of squares. Let $e_1, \ldots, e_n$ be an oriented orthonormal basis defined over some coordinate neighbourhood U of M, and let $\sigma^1, \ldots, \sigma^n$ be the dual basis of 1-forms. The volume element is intrinsic, that is, independent of the choice of basis. This follows because if $\bar{\sigma}^1, \ldots, \bar{\sigma}^n$ is a corresponding dual basis over a neighbourhood $\bar{U}$, then on $U \cap \bar{U}$ we have

$$\bar{\sigma}^i = \sum a^i_j \sigma^j,$$

where (a_j^i) is an orthogonal matrix of determinant $+1$. Then

$$\bar{\sigma}^1 \wedge \cdots \wedge \bar{\sigma}^n = \det(a_j^i)\sigma^1 \wedge \cdots \wedge \sigma^n = \sigma^1 \wedge \cdots \wedge \sigma^n.$$

The algebraic operator $*$ can now be applied to a p-form defined over U. Locally we have

$$*(\sigma^1 \wedge \cdots \wedge \sigma^p) = \sigma^{p+1} \wedge \cdots \wedge \sigma^n$$

We recall that $**\omega = (-1)^{p(n-p)+s}\omega$.

We now define a new operator, called the *coderivative*, denoted by δ, where

$$\delta\omega = (-1)^{np+n+1+s}*d*\omega.$$

We turn the space of p-forms over M into an infinite-dimensional inner product space as follows. If ω and η are p-forms then $\omega \wedge *\eta$ is an n-form and we define

$$\langle \omega, \eta \rangle = \int_m \omega \wedge *\eta.$$

To show that $\langle \ , \ \rangle$ so defined is indeed an inner product, we already know that $\omega \wedge *\eta = \eta \wedge *\omega$ which implies that $\langle \omega, \eta \rangle = \langle \eta, \omega \rangle$. The product is linear in each factor. There remains just to prove nondegeneracy. One way of doing this is to use the coordinate representation of $*$. Our definition implies that if

$$\eta = \frac{1}{p!} \sum_{i_1 \cdots i_p} n_{i_1 \cdots i_p} dx^{i_1} \wedge \cdots \wedge dx^{i_p}$$

then

$$*\eta = \frac{1}{(n-p)!} \sum_{i_{p+1} \cdots i_n} \left[\frac{1}{p!} \sum_{i_1 \cdots i_p} \varepsilon_{i_1 \cdots i_n} |\det(g_{kl})|^{1/2} \right.$$

$$\left. \sum_{j_1 \cdots j_p} \eta_{j_1 \cdots j_p} g^{i_1 j_1} \cdots g^{i_p j_p} \right] dx^{i_{p+1}} \wedge \cdots \wedge dx^{i_n}.$$

In terms of coordinates we get

$$\int \omega *\eta = \frac{1}{p!} \sum \int \omega_{i_1 \cdots i_p} \eta_{j_1 \cdots j_p} g^{i_1 j_1} \cdots g^{i_p j_p} |\det(g_{kl})|^{1/2} dx^1 \wedge \cdots \wedge dx^n$$

from which it follows that $\langle \omega, \eta \rangle = 0$ for all η implies that $\omega = 0$.

Example (3.8.3) Let V be 4-space with coordinates x^1, x^2, x^3, t normalized so that dx^1, dx^2, dx^3, dt is an orthonormal basis with $\langle dx^i, dx^i \rangle = 1$, $i = 1, 2, 3$; $\langle dt, dt \rangle = -1$. Then for a 2-form:

3.8 The Hodge star operator

$$*(dx^i \wedge dt) = dx^j \wedge dx^k \quad \text{where } i, j, k \text{ are in cyclic order;}$$
$$*(dx^i \wedge dx^j) = -dx^k \wedge dt.$$

Clearly $\delta(dx^i \wedge dx^j) = 0$.

Example (3.8.4) Let V be E with the usual Euclidean metric, and let f, g be real-valued functions; then

$$*df = \frac{\partial f}{\partial x} dy \wedge dz + \frac{\partial f}{\partial y} dz \wedge dx + \frac{\partial f}{\partial z} dx \wedge dy;$$

$$df \wedge *dg = \left(\frac{\partial f}{\partial x} \frac{\partial g}{\partial x} + \frac{\partial f}{\partial y} \frac{\partial g}{\partial y} + \frac{\partial f}{\partial z} \frac{\partial g}{\partial z} \right) dx \wedge dy \wedge dz;$$

$$\delta df = - \left(\frac{\partial f^2}{\partial x^2} + \frac{\partial^2 f}{\partial y^2} + \frac{\partial^2 f}{\partial z^2} \right).$$

Theorem (3.8.5) Let ω be a p-form and η be a $(p+1)$-form. Then

$$\int d\omega \wedge *\eta = \int \omega \wedge *\delta\eta,$$

that is,

$$\langle d\omega, \eta \rangle = \langle \omega, \delta\eta \rangle.$$

Thus δ is the formal adjoint operator of d with respect to the inner product $\langle \ , \ \rangle$.

To prove the theorem we integrate over M the relation

$$d(\omega \wedge *\eta) = d\omega \wedge *\eta + (-1)^p \omega \wedge d*\eta.$$

We get

$$\int_M d\omega * \eta + (-1)^p \int_M \omega \wedge d*\eta = \int_M d(\omega \wedge *\eta) = \int_{\partial M} \omega \wedge *\eta = 0$$

since M is assumed to have zero boundary. Thus

$$\langle d\omega, \eta \rangle = (-1)^{p-1} \int_M \omega \wedge d*\eta.$$

Since $d*\eta$ is an $(n-p)$-form we have

$$**(d*\eta) = *(*d*\eta) = (-1)^{(p+1)n+n+1+s} *\delta\eta.$$

But η is a $(p+1)$-form so

$$**(d*\eta) = (-1)^s (-1)^{(n-p)p} d*\eta$$

hence

$$d*\eta = (-1)^{np+2n+1+s-s-pn+p^2}*\delta\eta$$
$$= (-1)^{p^2+1}*\delta\eta.$$

Thus
$$\int_M d\omega \wedge *\eta = (-1)^{2+p(p+1)} \int \omega \wedge *\delta\eta = \int \omega \wedge *\delta\eta,$$

so $\langle d\omega, \eta \rangle = \langle \omega, \delta\eta \rangle$, and the theorem is proved.

We now make use of the operator δ to define the generalized Laplacian operator Δ, where
$$\Delta = -(d\delta + \delta d).$$

In the particular case of a function f (0-form) we find
$$\Delta f = \frac{1}{|g|^{1/2}} \sum_{j,k} \frac{\partial}{\partial x^k} \left(|g|^{1/2} g^{jk} \frac{\partial f}{\partial x^j} \right),$$

agreeing with a formula previously obtained.

We denote by b the isomorphism which maps a vector field to a 1-form by means of the metric tensor. We denote the inverse map sending a 1-form to its corresponding vector field by #. Then we have
$$\text{div } v = -\delta \flat v = \frac{1}{|g|^{1/2}} \sum_{k=1}^n \frac{\partial}{\partial x^k}(|g|^{1/2} v^k),$$

giving the divergence of the vector field v.

Similarly in E^3 we have
$$\text{curl } v = \#*d\flat v = \frac{1}{|g|^{1/2}} \sum_{i,j,k} \left[\frac{\partial}{\partial x^j} (v^l g_{li} \varepsilon_{jik}) \right] \frac{\partial}{\partial x^k}.$$

Of course our generalized Laplacian operates on p-forms, sending a p-form to another p-form. A form ω is said to be *harmonic* if
$$\Delta\omega = 0.$$

Naturally if $d\omega = 0$ and $\delta\omega = 0$, then $\Delta\omega = 0$. However the converse is true provided that M is compact, orientable, and the metric is positive definite. To prove this we note that for any p-form
$$\langle \Delta\omega, \omega \rangle = \langle d\delta\omega, \omega \rangle + \langle \delta d\omega, \omega \rangle$$
$$= \langle \delta\omega, \delta\omega \rangle + \langle d\omega, d\omega \rangle.$$

If ω is harmonic then $\Delta\omega = 0$ and hence
$$\langle \delta\omega, \delta\omega \rangle + \langle d\omega, d\omega \rangle = 0.$$

3.8 The Hodge star operator

Since the metric is positive definite it follows that $d\omega = 0$ and $\delta\omega = 0$, and the claim is justified.

We have already proved that d and δ are adjoint operators with respect to the inner product $\langle \ , \ \rangle$. We now observe that Δ is self-adjoint with respect to $\langle \ , \ \rangle$, that is,

$$\langle \Delta\omega, \eta \rangle = \langle \omega, \Delta\eta \rangle.$$

This follows because each side is equal to

$$\langle d\omega, d\eta \rangle + \langle \delta\omega, \delta\eta \rangle.$$

Since $\langle \Delta\omega, \omega \rangle \geq 0$ with equality only when $\Delta\omega = 0$, Δ is a positive definite (elliptic) self-adjoint differential operator.

Hodge's theorem concerning harmonic forms is one of the most powerful tools in global Riemannian geometry. It relates differential forms to topological properties of the manifold. The main result is a deep theorem which we state but do not prove. This is the Hodge decomposition theorem.

If ω is any p-form on M, then there is a $(p-1)$-form α, a $(p+1)$-form β, and a harmonic p-form γ such that

$$\omega = d\alpha + \delta\beta + \gamma.$$

The forms $d\alpha$, $\delta\beta$, and γ are uniquely determined.

A corollary is Hodge's theorem:

Theorem (3.8.6) The harmonic p-forms over a compact, orientable manifold M form a vector space whose finite dimension is equal to the pth Betti number over the reals.

We remind the reader that over any compact manifold of dimension n there are $(n+1)$ cohomology classes with $(n+1)$ corresponding integers β_i, $i = 0, 1, \ldots, n$ called Betti numbers. These are topological invariants and are independent of the particular Riemannian metric chosen on M. However, as often happens in mathematics, it sometimes pays off to introduce subsidiary material which appears to be irrelevant but which nevertheless enables one to obtain a result. We have already made use of this idea when we wished to establish whether a very complicated expression involving partial derivatives are components of a tensor field. We introduced a torsion-free connexion and expressed the partial derivatives in terms of covariant derivatives and connexion coefficients. If all the connexion coefficients cancel then the given expression represents components of a tensor field.

Hodge's theorem can be regarded as a refinement of the theorem of De Rham which we now describe. A *closed* form is a differential p-form ω such that $d\omega = 0$. Since $d^2 = 0$ it seems reasonable that there should exist a $(p-1)$-form η such that $\omega = \mathrm{d}\eta$. Indeed an *exact* form is a differential form ω such that $\omega = \mathrm{d}\eta$ for some $(p-1)$-form η. A theorem due to Poincaré, known as *Poincaré's lemma*, states that *if the domain U is star-shaped, then a p-form is closed if and only if it is exact*. The set of closed p-forms defined over a general manifold forms an abelian group (of large dimension!), denoted by $\mathfrak{Z}^p$. The exact forms form a subgroup, denoted by $\mathfrak{B}^p$. The theorem of De Rham states that *the quotient group $\mathfrak{Z}^p/\mathfrak{B}^p$ is isomorphic to the pth cohomology group, which itself is a vector space whose dimension is equal to the pth Betti number*.

The great advantage of Hodge's theorem is that it is more useful than De Rham's theorem in making computations. An apparently irrelevant Riemannian metric is introduced which enables one to talk about harmonic forms, and a consideration of these leads to topological invariants which are independent of the choice of metric.

To convince one's self that not every closed form is exact, consider the following standard example. Let M be the Euclidean plane minus the origin. Let ϕ be the 1-form

$$\phi = \frac{1}{(x^2+y^2)}(x\mathrm{d}y - y\mathrm{d}x).$$

Then

$$\begin{aligned}\mathrm{d}\phi &= \mathrm{d}\left(\frac{x}{x^2+y^2}\right)\wedge \mathrm{d}y - \mathrm{d}\left(\frac{y}{x^2+y^2}\right)\wedge \mathrm{d}x\\ &= \frac{\mathrm{d}x\wedge \mathrm{d}y}{x^2+y^2} - \frac{2x^2\mathrm{d}x\wedge \mathrm{d}y}{(x^2+y^2)^2} - \frac{\mathrm{d}y\wedge \mathrm{d}x}{(x^2+y^2)} + \frac{2y^2\mathrm{d}y\wedge \mathrm{d}x}{(x^2+y^2)^2}\\ &= \frac{2\mathrm{d}x\wedge \mathrm{d}y}{x^2+y^2} - \frac{2\mathrm{d}x\wedge \mathrm{d}y}{x^2+y^2} = 0,\end{aligned}$$

so that ϕ is closed. However if ϕ were exact, there must be some function f such that $\mathrm{d}f = \phi$. This would involve

$$\frac{\partial f}{\partial x} = -\frac{y}{(x^2+y^2)}, \quad \frac{\partial f}{\partial y} = \frac{x}{(x^2+y^2)}.$$

The solution of these differential equations, modulo an arbitrary constant, is

$$f(x,y) = \tan^{-1}(y/x)$$

which is the polar angle. However f is not continuous along a path which crosses the negative x-axis. Thus ϕ is closed but not exact.

3.9 MAXWELL'S EQUATIONS

We make a diversion at this point to show how Maxwell's field equations in classical electromagnetic theory can be simplified by using the operators d and δ. We recall that the theory involves the electric field **E**, the magnetic field **H**, magnetic induction **B**, electric current density **J**, dielectric displacement **D**, and charge density ρ. These are functions of the space variables x^1, x^2, x^3 and the time t. In vector notation Maxwell's equations are:

$$\operatorname{curl} \mathbf{E} = -\frac{1}{c}\frac{\partial \mathbf{B}}{\partial t} \qquad (3.10)$$

which is Faraday's law of induction;

$$\operatorname{curl} \mathbf{H} = \frac{4\pi}{c}\mathbf{J} + \frac{1}{c}\frac{\partial \mathbf{D}}{\partial t} \qquad (3.11)$$

which is Ampere's law;

$$\operatorname{div} \mathbf{D} = 4\pi\rho \qquad (3.12)$$

giving the law of continuity; and

$$\operatorname{div} \mathbf{B} = 0. \qquad (3.13)$$

showing the non-existence of true magnetism.

Here c represents the speed of light. We put these equations in the notation of differential forms.

We set

$$F = (E_1 dx^1 + E_2 dx^2 + E_3 dx^3) \wedge c\,dt$$
$$+ (B_1 dx^2 \wedge dx^3 + B_2 dx^3 \wedge dx^1 + B_3 dx^1 \wedge dx^2),$$
$$G = -(H_1 dx^1 + H_2 dx^2 + H_3 dx^3) \wedge c\,dt$$
$$+ (D_1 dx^2 \wedge dx^3 + D_2 dx^3 \wedge dx^1 + D_3 dx^1 \wedge dx^2),$$
$$L = (J_1 dx^2 dx^3 + J_2 dx^3 \wedge dx^1 + J_3 dx^1 \wedge dx^2) \wedge dt - \rho\,dx^1 \wedge dx^2 \wedge dx^3.$$

Straightforward computation shows that equations (3.10) and (3.13) become

$$dF = 0.$$

Equations (3.11) and (3.12) become

$$dG + 4\pi L = 0.$$

Taking the exterior derivative of each side of this equation gives

$$dL = 0$$

which in vector notation is

$$\operatorname{div} \mathbf{J} + \frac{\partial \rho}{\partial t} = 0.$$

Using Poincaré's lemma, it follows from the equation $dF = 0$ that at least in a star-shaped region there exists a 1-form λ such that

$$d\lambda = F.$$

We introduce a vector potential $\mathbf{A}$ and a scalar A_0 by writing

$$\lambda = A_1 dx^1 + A_2 dx^2 + A_3 dx^3 + A_0 c dt,$$

The equation $d\lambda = F$ is now equivalent to the vector equations

$$\operatorname{curl} \mathbf{A} = \mathbf{B},$$

$$\operatorname{grad} A_0 - \frac{1}{c} \frac{\partial \mathbf{A}}{\partial t} = \mathbf{E}.$$

We now simplify matters by assuming that we are dealing with free space, and then

$$\mathbf{E} = \mathbf{D}, \quad \mathbf{H} = \mathbf{B}, \quad \mathbf{J} = \mathbf{O}, \quad \rho = 0.$$

Then Maxwell's equations reduce to the form

$$\operatorname{curl} \mathbf{E} = -\frac{1}{c}\frac{\partial \mathbf{H}}{\partial t}, \qquad \operatorname{div} \mathbf{E} = 0,$$

$$\operatorname{curl} \mathbf{H} = \frac{1}{c}\frac{\partial \mathbf{E}}{\partial t}, \qquad \operatorname{div} \mathbf{H} = 0.$$

We now introduce the Lorentz metric into 4-space so that $dx^1, dx^2, dx^3, c dt$ is an orthonormal basis. Then $s = 1$ and we have

$$*(dx^1 \wedge dx^2) = -dx^3 \wedge c dt, \qquad *(dx^1 \wedge c dt) = dx^2 \wedge dx^3, \qquad \text{etc.}$$

A short calculation shows that

$$\begin{aligned} F &= (E_1 dx^1 + E_2 dx^2 + E_3 dx^3) \wedge c dt \\ &\quad + (H_1 dx^2 \wedge dx^3 + H_2 dx^3 \wedge dx^1 + H_3 dx^1 \wedge dx^2) \\ G &= -(H_1 dx^1 + H_2 dx^2 + H_3 dx^3) \wedge c dt \\ &\quad + (E_1 dx^2 \wedge dx^3 + E_2 dx^3 \wedge dx^1 + E_3 dx^1 \wedge dx^2) \end{aligned}$$

and $G = *F$.

Thus Maxwell's equations in free space are simply

3.10 Consequences of the theorems of Stokes and of Hodge

$$dF = 0, \quad d*F = 0.$$

Alternatively this may be written

$$dF = 0 \quad \text{and} \quad \delta F = 0,$$

so that F is a harmonic 2-form.

A detailed treatment of classical electromagnetic theory by differential forms is given by Parrott (1987).

3.10 CONSEQUENCES OF THE THEOREMS OF STOKES AND OF HODGE

In this section we find it convenient to use the suffix notation of classical tensor calculus, although no doubt each step could be formulated in invariant notation. Indeed, I have good reason to believe that several differential geometers first obtain results with suffixes but their published papers ultimately appear in suffix-free form. We follow the treatment of Lichnerowicz (1958), except that our Laplacian differs from his in sign.

We assume that we are dealing with a proper Riemannian manifold M (positive definite metric) of dimension m. Let α be a p-form on M so that we may write

$$\alpha = \frac{1}{p!} \alpha_{i_1 \ldots i_p} dx^{i_1} \wedge \cdots \wedge dx^{i_p}$$

where summation is taken over repeated indices. Then, as components of $d\alpha$ and $\delta\alpha$, respectively, we have

$$(d\alpha)_{j_1 \ldots j_{p+1}} = \frac{1}{p!} \varepsilon_{j_1 \ldots j_{p+1}}^{k i_1 \ldots i_p} \nabla_k \alpha_{i_1 \ldots i_p},$$

$$(\delta\alpha)_{i_2 \ldots i_p} = -\nabla_l \alpha^l_{i_2 \ldots i_p}.$$

It follows that

$$(\delta d\alpha)_{j_1 \ldots j_p} = -\frac{1}{p!} \varepsilon_{l j_1 \ldots j_p}^{k i_1 \ldots i_p} \nabla^l \nabla_k \alpha_{i_1 \ldots i_p}.$$

The term with $k = l$ is given by

$$-\nabla^k \nabla_k \alpha_{j_1 \ldots j_p}.$$

If k takes the value of one of the i's, we get

$$\frac{1}{(p-1)!} \varepsilon_{j_1 \ldots j_p}^{k i_2 \ldots i_p} \nabla^l \nabla_k \alpha_{l i_2 \ldots i_p}.$$

It follows that

$$(\delta d\alpha)_{j_1\ldots j_p} = -\nabla^k\nabla_k\alpha_{j_1\ldots j_p} + \frac{1}{(p-1)!}\varepsilon^{ki_2\ldots i_p}_{j_1\ldots j_p}\nabla_l\nabla_k\alpha^l_{i_2\ldots i_p}.$$

Moreover

$$(d\delta\alpha)_{j_1\ldots j_p} = -\frac{1}{(p-1)!}\varepsilon^{ki_2\ldots i_p}_{j_1\ldots j_p}\nabla_k\nabla_l\alpha^l_{i_2\ldots i_p}.$$

Thus, remembering our sign convention for Δ we have

$$-(\Delta\alpha)_{j_1\ldots j_p} = -\nabla^k\nabla_k\alpha_{j_1\ldots j_p}$$
$$+ \frac{1}{(p-1)!}\varepsilon^{ki_2\ldots i_p}_{j_1\ldots j_p}(\nabla_l\nabla_k - \nabla_k\nabla_l)\alpha^l_{i_2\ldots i_p}.$$

The term in the second bracket can now be evaluated with the aid of the Ricci identities, to give

$$\nabla_l\nabla_k\alpha^l_{i_2\ldots i_p} - \nabla_k\nabla_l\alpha^l_{i_2\ldots i_p} = R^l_{rlk}\alpha^r_{i_2\ldots i_p} - \sum_{s=2}^{p} R^r_{i_s lk}\alpha^l_{i_2\ldots i_{s-1}ri_{s+1}\ldots i_p}.$$

After further computation, making use of the identity

$$R_{klmn} + R_{kmnl} + R_{knlm} = 0,$$

Lichnerowicz obtains the result

$$-(\Delta\alpha)_{j_1\ldots j_p} = -\nabla^k\nabla_k\alpha_{j_1\ldots j_p} + \frac{1}{(p-1)!}\varepsilon^{ki_2\ldots i_p}_{j_1\ldots j_p}R_{kl}\alpha^l_{i_1\ldots i_p}$$
$$- \frac{1}{2(p-2)!}\varepsilon^{kli_3\ldots i_p}_{j_1\ldots j_p}R_{klmn}\alpha^{mn}_{i_3\ldots i_p}.$$

For our purposes it will be sufficient to restrict attention to the special case when α is a 1-form, and then the calculations become much simpler. We have

$$-(\Delta\alpha)_j = -\nabla^k\nabla_k\alpha_j + \nabla_l\nabla_j\alpha^l - \nabla_j\nabla_l\alpha^l$$
$$= -\nabla^k\nabla_k\alpha_j + R_{kj}\alpha^k.$$

We now calculate the value of $\Delta|\alpha|^2$ for a 1-form α. We have

$$|\alpha|^2 = \langle\alpha,\alpha\rangle.$$

Then

$$-\Delta|\alpha|^2 = \delta d(|\alpha|^2)$$

where

3.10 Consequences of the theorems of Stokes and of Hodge

$$|\alpha|^2 = \alpha^j \alpha_j.$$

By covariant differentiation we get

$$\frac{1}{2}d(|\alpha|^2)_k = \alpha^j \nabla_k \alpha_j.$$

Consequently

$$-\frac{1}{2}\Delta|\alpha|^2 = -\nabla^k \alpha^j \nabla_k \alpha_j - \alpha^j \nabla^k \nabla_k \alpha_j.$$

Substituting for the term $\nabla^k \nabla_k \alpha_j$ we get

$$\frac{1}{2}\Delta|\alpha|^2 = \langle \alpha, \Delta\alpha \rangle + \nabla^k \alpha^j \nabla_k \alpha_j + R_{kj} \alpha^k \alpha^j.$$

We integrate over M to get

$$\frac{1}{2}\int \Delta|\alpha|^2 * 1 = \int \{\langle \alpha, \Delta\alpha \rangle + \nabla^k \alpha_j \nabla_k \alpha^j + R_{kj} \alpha^k \alpha^j\} * 1.$$

We now make use of the important fact that *for any smooth function f, the integral of Δf over an oriented compact manifold without boundary is zero;* that is,

$$\int_m \Delta f * 1 = 0.$$

This follows from Green's theorem: *in a compact orientable Riemannian manifold without boundary we have, for any 1-form α,*

$$\int_M \delta\alpha * 1 = 0,$$

that is,

$$\int_M (g^{ij}\nabla_j \alpha_i) * 1 = 0.$$

In particular if α is the differential df of a function f we have

$$\int_M \Delta f * 1 = 0.$$

Applying this result to the function $f = |\alpha|^2$ we get

$$\int_M \{\langle \alpha, \Delta\alpha \rangle + \nabla^k \alpha^j \nabla_k \alpha_j + R_{jk} \alpha^j \alpha^k\} * 1 = 0.$$

Suppose now that α is a harmonic 1-form, that is, $\Delta\alpha = 0$. The above equation then gives

$$\int_M (R_{jk} \alpha^j \alpha^k) * 1 = -\int_M (\nabla^k \alpha^j \nabla_k \alpha_j) * 1 \leqslant 0.$$

Suppose now that the Ricci tensor determines a positive definite quadratic form. Then the above inequality forces the conclusion $\alpha = 0$. Thus we have proved the following result.

Theorem (3.10.1) (Myers–Bochner) *If a compact orientable manifold admits a Ricci curvature everywhere strictly positive, then its first Betti number $b_1(M)$ is zero.*

Green's theorem which we have used above follows as a corollary to Stokes' theorem.

The vanishing of the integral of Δf also leads to the next result.

Lemma (3.10.2) (Bochner's Lemma) *If $\Delta f \geq 0$ (or $\Delta f \leq 0$) in M, then f is constant in M.*

To prove the lemma we note that $\Delta f = g^{ij}\nabla_j\nabla_i f$. Thus if $\Delta f \geq 0$ (or $\Delta f \leq 0$) in M we must have $\Delta f = 0$. Also

$$\frac{1}{2}\Delta f^2 = g^{ij}\nabla_j(f\nabla_i f) = f\Delta f + |\nabla f|^2.$$

Hence

$$\int_M \{f\Delta f + |\nabla f|^2\} * 1 = 0.$$

Since our hypothesis implies that $\Delta f = 0$, this gives

$$\int |\nabla f|^2 * 1 = 0.$$

Thus $\nabla f = 0$ and f is a constant function.

We can now prove that, with our definition of Δ, all the eigenvalues of Δ are negative. That is, if $\Delta f = cf$ then c must be negative. This follows on integrating the equation

$$\frac{1}{2}\Delta f^2 = f\Delta f + |\nabla f|^2,$$

to get

$$\int_M \{f\Delta f + |\nabla f|^2\} * 1 = 0,$$

that is,

$$\int_M \{cf^2 + |\nabla f|^2\} * 1 = 0$$

from which we deduce that c must be negative. It is for this reason that some mathematicians define the Laplacian to be the negative of ours, so that all the eigenvalues would then be positive. However the price to pay for this is that we would then have

$$\Delta f = -\left(\frac{\partial^2 f}{\partial x^2} + \frac{\partial^2 f}{\partial y^2} + \frac{\partial^2 f}{\partial z^2}\right),$$

which would offend some engineers and physicists! It is a 'no-win' situation!

Another consequence of the integral of Δf being zero is the following theorem.

Theorem (3.10.3) On a compact orientable manifold, the vanishing of the rth-order covariant derivative of a tensor field implies the vanishing of the first-order covariant derivative.

To illustrate the proof we consider what happens when we have a tensor field with components (t_{ij}) which satisfy the relation

$$\nabla^k \nabla_k t_{ij} = 0.$$

In particular this will be satisfied if the second covariant derivative vanishes. From our hypothesis we have

$$-\nabla^k(t^{ij}\nabla_k t_{ij}) = -\nabla^k t^{ij} \nabla_k t_{ij}.$$

We write $|t|^2 = \tfrac{1}{2} t^{ij} t_{ij}$ so that

$$\Delta |t|^2 = \nabla^k t^{ij} \nabla_k t_{ij}$$

and hence

$$\int_M (\nabla^k t^{ij} \nabla_k t_{ij}) * 1 = 0.$$

It follows that the first covariant derivative of t_{ij} must be zero. The theorem now follows by recurrence. The same proof, slightly adapted, applies to a tensor field of type (p, q).

The book by Yano (1970) contains a large number of theorems, nearly all of which depend on the vanishing of the integral of Δf on a compact orientable Riemannian manifold and, more specifically, on Bochner's lemma.

3.11 THE SPACE OF CURVATURE TENSORS

We first obtain some algebraic results concerning the Riemannian curvature tensor. Let V be an n-dimensional vector space over $\mathbb{R}$ with inner product $\langle \ , \ \rangle$. Let Λ^2 denote the space of 2-vectors of V with inner product

$$\langle u_1 \wedge u_2, v_1 \wedge v_2 \rangle = \det(\langle u_i, v_j \rangle), \qquad u_i, v_i \in V, \ i = 1, 2.$$

We extend the inner product to the whole of Λ^2 by linearity. By a *curvature tensor on V* we mean a symmetric linear transformation $R: \Lambda^2 \to \Lambda^2$. The set of curvature tensors on V forms a vector space $\mathfrak{R}$ with inner product given by $\langle R, S \rangle = \text{trace}(RS)$. Using the usual isomorphism $\#$ and $\flat$ defined by the inner product (corresponding to the raising and lowering of suffixes in standard tensor calculus) a curvature tensor on V may be regarded as a 2-form on V with values in the space of skew-symmetric endomorphisms of V. The *sectional curvature* σ_R of $R \in \mathfrak{R}$ is the real-valued function defined on the Grassmann manifold G of oriented two-dimensional subspaces of V by

$$\sigma_R(P) = \langle RP, P \rangle, \qquad (P \in G \subset \Lambda^2).$$

where G is identified with the set of decomposable 2-vectors of length 1. The orthogonal group $O(V)$ acts isometrically on Λ^2 by

$$g(u \wedge v) = (gu) \wedge (gv), \qquad g \in O(V),\ u, v \in V.$$

The group $O(V)$ also acts isometrically on $\mathfrak{R}$ by

$$g(R) = g^t R g = g^{-1} R g : \Lambda^2 \to \Lambda^2$$

since we regard R as a skew-symmetric endomorphism of V.

Consider now the following three maps:

Bianchi map. This is the map $b: \mathfrak{R} \to \mathfrak{R}$ defined by

$$[b(R)](u_1, u_2)(u_3) = -\sum_\alpha R(u_{\alpha(1)}, u_{\alpha(2)}) u_{\alpha(3)}$$

where α runs through all cyclic permutations of $(1, 2, 3)$.

Ricci contraction. This is the map $r: \mathfrak{R} \to \mathfrak{J}$ where $\mathfrak{J}$ is the space of symmetric linear transformations of V defined by

$$\langle r(R)(v), w \rangle = \text{trace}\{u \to R(u, v)w\}.$$

The trace functional. $\text{tr}: \mathfrak{R} \to \mathbb{R}$.

Each of these maps is equivariant with respect to the $O(V)$ action on the spaces involved, and hence the kernel of each of these maps is an $O(V)$-invariant subspace of $\mathfrak{R}$. Since the group $O(V)$ acts isometrically on $\mathfrak{R}$, orthogonal complements of invariant subspaces are also invariant. We set

$$\mathfrak{R}_1 = (\text{Ker } b)^\perp, \qquad \mathfrak{R}_2 = (\text{Ker tr})^\perp,$$
$$\mathfrak{R}_3 = (\text{Ker } r) \cap (\text{Ker } b), \quad \mathfrak{R}_4 = (\text{Ker } r)^\perp \cap (\text{Ker tr}).$$

We have the following fundamental theorem.

Theorem (3.11.1)

$$\mathfrak{R} = \sum_{i=1}^{4} \oplus \mathfrak{R}_i.$$

3.11 The space of curvature tensors

Moreover:

(i) $R \in \mathcal{R}_1 \Leftrightarrow$ the sectional curvature of R is identically zero;
(ii) $R \in \mathcal{R}_1 \oplus \mathcal{R}_2 \Leftrightarrow$ the sectional curvature of R is constant;
(iii) $R \in \mathcal{R}_1 \oplus \mathcal{R}_3 \Leftrightarrow$ the Ricci tensor of R is zero;
(iv) $R \in \mathcal{R}_1 \oplus \mathcal{R}_2 \oplus \mathcal{R}_3 \Leftrightarrow$ the Ricci curvature of R is a scalar multiple of the identity;
(v) $R \in \mathcal{R}_1 \oplus \mathcal{R}_3 \oplus \mathcal{R}_4 \Leftrightarrow$ the scalar curvature of R is zero;
(vi) $R \in \mathcal{R}_2 \oplus \mathcal{R}_3 \oplus \mathcal{R}_4 \Leftrightarrow R$ satisfies the first Bianchi identity;

Proof. (i) $R \in \mathcal{R}_1 \Rightarrow \operatorname{tr} RS = \langle R, S \rangle = 0$ for all $S \in \mathcal{R}$ with $b(S) = 0$. Let P be a 2-plane belonging to the Grassmannian G. Define the tensor $S_P \in \mathcal{R}$ by $S_P(P) = P$ and $S_P(Q) = 0$ whenever $\langle P, Q \rangle = 0$. Then $b(S_P) = 0$ and hence $0 = \operatorname{tr}(RS) = \sigma_R(P)$. Conversely, if $R \in \mathcal{R}$ is such that $\sigma_R = 0$ then $R = R_1 + R'$ where $R_1 \in \mathcal{R}_1$ and $b(R') = 0$. Then $\sigma_{R'} = \sigma_R - \sigma_{R_1} = 0$. Thus the sectional curvature is everywhere zero. This fact together with the Bianchi identity implies that the curvature tensor is everywhere zero.

(ii) We have

$$\langle R(u,v)w, u \rangle = \frac{1}{2}[\langle R(u, v+w)(v+w), u \rangle$$
$$- \langle R(u,v)v, u \rangle - \langle R(u,w)w, u \rangle],$$

and since $R \in \mathcal{R}_1$ each term on the right is zero. Thus we can conclude

$$\mathcal{R}_1 \subset \operatorname{Ker} r.$$

Moreover $\operatorname{tr} r(R) = 2 \operatorname{tr} R$ so we have

$$\mathcal{R}_1 \subset \operatorname{Ker} r \subset \operatorname{Ker} \operatorname{tr}.$$

These inequalities ensure that $\mathcal{R}$ is the orthogonal direct sum $\mathcal{R} = \Sigma_{i=1}^{4} \oplus \mathcal{R}_i$. Moreover since Kernel (trace) has dimension 1 then the dimension of $\mathcal{R}_2 = 1$ and hence $\mathcal{R}_2$ consists of all scalar multiples of the identity map $I: \Lambda^2 \to \Lambda^2$. But since $b(R) = 0$ we have $R = aI$ if and only if the space has constant sectional curvature, that is, $\sigma_R \equiv a$.

(iii) and (vi) These follow directly since $\mathcal{R}_1 \subset \operatorname{Ker} r \subset \operatorname{Ker} \operatorname{tr}$.

(iv) This follows since r maps $\mathcal{R}_2$ isomorphically onto multiples of the identity in $\mathcal{I}$ and $\operatorname{Ker} r = \mathcal{R}_1 \oplus \mathcal{R}_3$.

(v) This is straightforward because of the trace-free requirement.

It can be proved that the subspaces $\mathcal{R}_i$ are the minimal invariant subspaces of $\mathcal{R}$ under the action of $O(V)$.

We are primarily concerned with curvature tensors coming from a

Riemannian metric. These satisfy the Bianchi identity so that $R \in \mathcal{R}_1^\perp$. Since from (i) the $\mathcal{R}_1$ component contributes nothing to the sectional curvature function, in Riemannian geometry there will be no loss of generality in assuming that all our curvature tensors lie in $\mathcal{R}_2 \oplus \mathcal{R}_3 \oplus \mathcal{R}_4$. Let $\mathcal{E} = \mathcal{R}_2 \oplus \mathcal{R}_3$. It follows from (iv) that a Riemannian manifold has curvature tensor in $\mathcal{E}$ at each point if and only if it is an Einstein space, that is, a space for which the Ricci tensor is proportional to the metric tensor. We have the following $\mathcal{R}_2 \oplus \mathcal{R}_4$ theorem.

Theorem (3.11.2) If $n > 2$ the Ricci contraction map r sends $\mathcal{R}_2 \oplus \mathcal{R}_4$ isomorphically onto $\mathfrak{J}$. The inverse map $s: \mathfrak{J} \to \mathcal{R}_2 \oplus \mathcal{R}_4$ is given by

$$s(T)(u_1, u_2)(u_3) = -\frac{1}{(n-2)}[\langle u_1, u_3 \rangle T u_2 - \langle u_2, u_3 \rangle T u_1 + \langle T u_1, u_3 \rangle u_2 - \langle T u_2, u_3 \rangle u_1]$$
$$+ \frac{1}{(n-1)(n-2)}(\operatorname{tr} T)(\langle u_1, u_3 \rangle u_2 - \langle u_2, u_3 \rangle u_1), \quad \text{for } T \in \mathfrak{J}, u_i \in V.$$

Proof. We observe that since $\mathcal{R}_2 \oplus \mathcal{R}_4$ is the orthogonal complement of Ker r, the map r sends $\mathcal{R}_2 \oplus \mathcal{R}_4$ isomorphically onto the image of r. Moreover, straightforward substitution shows that $r \circ s$ is the identity and hence the image of r is $\mathfrak{J}$ itself. It remains to check that, for $T \in \mathfrak{J}$, we have $s(T) \perp \mathcal{R}_1 \oplus \mathcal{R}_3$. Singer and Thorpe (1969) do this by choosing an orthonormal basis $\{e_1, \ldots, e_n\}$ consisting of eigenvectors of T. Let $\lambda_1, \ldots, \lambda_n$ be the corresponding eigenvalues. Then we have

$$s(T)(e_i \wedge e_j) = -\frac{1}{(n-2)}\left[\lambda_i + \lambda_j - \frac{1}{(n-1)}\operatorname{tr}(T)\right] e_i \wedge e_j \quad (3.14)$$

so that $\{e_i \wedge e_j | i < j\}$ is an orthonormal basis of Λ^2 consisting of eigenvectors of $s(T)$. Compute $\langle s(T), R \rangle$ using $r(R) = 0$ for $R \in \mathcal{R}_1 \oplus \mathcal{R}_3$ to get $\langle s(T), R \rangle = \operatorname{trace} s(T)R = 0$ for all $R \in \mathcal{R}_1 \oplus \mathcal{R}_3$. This completes the proof.

Note that under the isomorphism $r: \mathcal{R}_2 \oplus \mathcal{R}_4 \to \ker r^\perp$, $\mathcal{R}_2$ is mapped into scalar multiples of the identity I, while $\mathcal{R}_4$ is mapped onto $\{T \in \mathfrak{J} | \operatorname{tr} T = 0\}$.

Let us assume now that R satisfies the first Bianchi identity so we may write $R \in \mathcal{R}_2 \oplus \mathcal{R}_3 \oplus \mathcal{R}_4$. The theorem says that $r: \mathcal{R} \to \mathfrak{J}$ when restricted to $\mathcal{R}_2 \oplus \mathcal{R}_4$ is an isomorpohism and gives a formula for the inverse map $s: \mathfrak{J} \to \mathcal{R}_2 \oplus \mathcal{R}_4$. So if $\mathcal{R} = \mathcal{R}_2 + \mathcal{R}_3 + \mathcal{R}_4$ are the components of R, that is, the projections of R on $\mathcal{R}_2$, $\mathcal{R}_3$, and $\mathcal{R}_4$, then

$$s(r(\mathcal{R}_2 + \mathcal{R}_4)) = \mathcal{R}_2 + \mathcal{R}_4.$$

3.11 The space of curvature tensors

But $r(\mathcal{R}_3) = 0$ so that $\mathcal{R}_2 + \mathcal{R}_4 = s(r\mathcal{R})$. Hence $\mathcal{R}_3 = \mathcal{R} - s(r)(\mathcal{R})$. When we evaluate this expression we find that its components in the classical tensor calculus notation are given by

$$W^h_{ijk} = R^h_{ijk} - \frac{1}{(n-2)}(\delta^h_j R_{ik} - \delta^h_k R_{ij} + g_{ik}R^h_j - g_{ij}R^h_k)$$

$$- \frac{\tau}{(n-1)(n-2)}(\delta^h_k g_{ij} - \delta^h_j g_{ik}).$$

Thus the $\mathcal{R}_3$ component of the curvature tensor R is none other than the conformal curvature tensor introduced by Weyl (1918). This tensor is invariant under a change of metric of the form $g \to e^\phi g$ for some function ϕ.

The special case when $n = 4$

Let V be given a definite orientation. We recall that the Hodge star operator $*: \Lambda^2 \to \Lambda^2$ satisfies $\langle *\alpha, \beta \rangle \omega = \alpha \wedge \beta$ where ω is the generator for Λ^4 determined by the inner product and the orientation. If $P \in G$ then $*P$ is the oriented orthogonal complement $P^\perp$. The map $*$ is symmetric and satisfies $*^2 = I$. The six-dimensional space Λ^2 decomposes into an orthogonal direct sum $\Lambda^2 = \Lambda^+ \oplus \Lambda^-$ where $\Lambda^\pm$ are the three-dimensional eigenspaces corresponding to the eigenvalues ± 1 of $*$. We have the following theorem.

Theorem (3.11.3) *Let V be a four-dimensional oriented vector space. Then:*

(i) $\mathcal{R}_1$ *consists of all scalar multiples of* $*$;
(ii) $\mathcal{R}_2$ *consists of all scalar multiples of I;*
(iii) $R \in \mathcal{R}_3$ *if and only if* $*R = R*$, $\operatorname{tr} R$ *and* $\operatorname{tr} *R = 0$;
(iv) $R \in \mathcal{R}_4$ *if and only if* $*R = -R*$.

In particular $R \in \mathcal{E}$ if and only if both $R* = *R$ and $b(R) = 0$.

Proof. If we restrict $*$ to Λ^+ we get an isomorphism of

$$\mathcal{Q} = \{R \in \mathcal{R} \mid *R = -R*\}$$

with $\operatorname{Hom}(\Lambda^+, \Lambda^-)$. Since the dimension of this latter space of 3×3 matrices is 9, it follows that

$$\dim \mathcal{Q} = 9 = \dim \mathfrak{I} - 1 = \dim \mathcal{R}_4.$$

From formula (3.14) we have $s(T)* = -*s(T)$ for all $T \in \mathfrak{I}$ with trace $T = 0$. Hence $\mathcal{R}_4 \subset \mathcal{Q}$ and hence $\mathcal{Q} = \mathcal{R}_4$. Thus $R \in \mathcal{R}_4$ if and only if

$*R = -R*$. It follows that $\mathcal{R}_4^{\perp} = \{R \in \mathcal{R} \mid *R = R*\}$. The remainder of the statements of the theorem follow by using the relation $\operatorname{tr} *R = 2b(R)$.

We now have the following interesting result.

Theorem (3.11.4) Let M be a four-dimensional Riemannian manifold. Then M is an Einstein space if and only if at each tangent 2-plane P the sectional curvature is equal to that at $P^{\perp}$, that is, $\sigma(P^{\perp}) = \sigma(P)$.

Proof. We have

$$\sigma_R(P^{\perp}) = \sigma_R(*P) = \langle R*P, *P \rangle = \langle *R*P, P \rangle.$$

Hence $\sigma_R(P^{\perp}) = \sigma_R(P)$ if and only if

$$\langle (*R* - R)P, P \rangle = 0.$$

We have already seen that the Einstein property is characterized by the relation $R* = *R$ which is independent of the chosen orientation. Clearly if $R* = *R$ this condition is satisfied by all P since $*^2 = I$. Conversely if this condition is satisfied by all P, then the curvature tensor $*R* - R$ must have sectional curvature zero. Then

$$b(*R* - R) = \frac{1}{2}\operatorname{tr}[*(*R* - R)] = \frac{1}{2}\operatorname{tr}[(R* - *R)] = 0.$$

Thus $*R* - R = 0$, that is, $R* = *R$.

We may observe that a very similar proof shows that a Riemannian four-dimensional manifold has its curvature tensor in $\mathcal{R}_4$ at each point if and only if $\sigma(P) = -\sigma(P^{\perp})$ for each 2-plane P.

As an application of this we obtain the following result due to M. Berger (1961).

Theorem (3.11.5) The Euler–Poincaré characteristic of a compact four-dimensional Einstein space is non-negative and is zero only if M is flat.

To prove the theorem we use the fact that the Euler–Poincaré characteristic of a compact four-dimensional manifold is given by the Gauss–Bonnet formula

$$\chi(M) = \frac{3}{4\pi^2} \int_M \operatorname{tr}(*R)^2 \, dV.$$

For an Einstein space we have $R* = *R$ and hence

$$\operatorname{tr}(*R)^2 = \operatorname{tr} R^2 = \langle R, R \rangle \geq 0.$$

The result now follows immediately.

3.12 CONFORMAL METRICS AND CONFORMAL CONNEXIONS

For the first part of this section we follow the treatment given in Eisenhart's Riemannian Geometry (1926), remembering that his definition of the Ricci tensor has the opposite sign to ours so the resulting equations will differ from his. Let M be an n-dimensional C^∞-differentiable manifold with metric tensors g and $\bar{g}$ where, relative to a neighbourhood U with local coordinates x^i, we have

$$\bar{g}_{ij} = e^{2\sigma} g_{ij} \tag{3.15}$$

where σ is a C^∞-function of the coordinates x^i. Clearly the angle between any two directions at points of U is independent of the choice of metric g or $\bar{g}$. We say that the manifolds (M, g), $(M, \bar{g})$ are conformally related. From (3.15) it follows that

$$\bar{g}^{ij} = e^{-2\sigma} g^{ij}. \tag{3.16}$$

A straightforward calculation shows that the Christoffel symbols are related by

$$[\overline{ij, k}] = e^{2\sigma}([ij, k] + g_{ik}\sigma_{,j} + g_{jk}\sigma_{,i} - g_{ij}\sigma_{,k}), \tag{3.17}$$

$$\bar{\Gamma}^l_{ij} = \Gamma^l_{ij} + \delta^l_i \sigma_{,j} + \delta^l_j \sigma_{,i} - g_{ij} g^{lm} \sigma_{,m} \tag{3.18}$$

where we have written $\sigma_{,i} = \partial\sigma/\partial x^i$ and $[ij, k] = g_{hk} \Gamma^h_{ij}$.

In Eisenhart's notation, the covariant form of the curvature tensor has components

$$R_{hijk} = \frac{\partial}{\partial x^j}[ik, h] - \frac{\partial}{\partial x^k}[ij, h] + \Gamma^l_{ij}[hk, l] - \Gamma^l_{ik}[hj, l]. \tag{3.19}$$

If we substitute for the analogous expression derived from $\bar{g}$ we find

$$e^{-2\sigma}\bar{R}_{hijk} = R_{hijk} + g_{hk}\sigma_{ij} + g_{ij}\sigma_{hk} - g_{hj}\sigma_{ik} - g_{ik}\sigma_{hj}$$
$$+ (g_{hk}g_{ij} - g_{hj}g_{ik})\Delta_1\sigma, \tag{3.20}$$

where we have written

$$\Delta_1 \sigma = g^{ij} \sigma_{,i} \sigma_{,j} \tag{3.21}$$

and

$$\sigma_{ij} = \sigma_{,ij} - \sigma_{,i}\sigma_{,j}, \tag{3.22}$$

the first term on the right-hand side of (3.22) being the second covariant derivative with respect to the Levi-Civita connexion of g. To get the corresponding relations between the Ricci tensors we have, remembering that $R_{ij} = R^k_{ikj}$, we find

106   3 *Riemannian manifolds*

$$\bar{R}_{ij} = R_{ij} - (n-2)\sigma_{ij} - g_{ij}(\Delta\sigma + (n-2)\Delta_1\sigma), \quad (3.23)$$

where $\Delta\sigma = g^{ij}\sigma_{,ij}$.

The scalar curvatures τ and $\bar{\tau}$ are related by

$$\bar{\tau} = e^{-2\sigma}[\tau - 2(n-1)\Delta\sigma - (n-1)(n-2)\Delta_1\sigma]. \quad (3.24)$$

The reader is now invited to take a large sheet of paper and, writing small, to check that the tensor with components

$$W^h_{ijk} = R^h_{ijk} - \frac{1}{(n-2)}(\delta^h_j R_{ik} - \delta^h_k R_{ij} + g_{ik}R^h_j - g_{ij}R^h_k)$$
$$- \frac{\tau}{(n-1)(n-2)}(\delta^h_k g_{ij} - \delta^h_j g_{ik}) \quad (3.25)$$

satisfies the relation

$$\bar{W}^h_{ijk} = W^h_{ijk}. \quad (3.26)$$

Thus the tensor W is conformally invariant. This is the Weyl conformal curvature tensor discovered by him (Weyl (1918)). At first sight the formula (3.25) seems very surprising. However, as we have seen, it turns out to be nothing but the formula for expressing the projection of the curvature tensor into one of the three irreducible components of the space of curvature tensors considered from the point of view of representations of the special orthogonal group $SO(n)$.

From the point of view of *local* differential geometry, when $\dim M = 2$ the tensor W has little interest since any quadratic form is reducible to the form $\lambda[(dx^1)^2 + (dx^2)^2]$, so that locally (*but not globally!*) all two-dimensional Riemannian manifolds are conformally equivalent. The case $\dim M = 3$ is rather exceptional because then the conformal curvature tensor W vanishes identically. Another way of saying this is that when $\dim M = 3$, the Riemann curvature tensor is algebraically related to the Ricci tensor and the scalar curvature – in short, the Ricci tensor determines the Riemann curvature tensor.

To prove this we proceed as follows. When $n = 3$, (3.25) reduces to

$$W^h_{ijk} = R^h_{ijk} - (\delta^h_j R_{ik} - \delta^h_k R_{ij} + g_{ik}R^h_j - g_{ij}R^h_k) - \frac{\tau}{2}(\delta^h_k g_{ij} - \delta^h_j g_{ik}), \quad (3.27)$$

or in covariant form

$$W_{lijk} = R_{lijk} - (g_{lj}R_{ik} - g_{lk}R_{ij} + g_{ik}R_{lj} - g_{ij}R_{lk}) - \frac{\tau}{2}(g_{lk}g_{ij} - g_{lj}g_{ik}). \quad (3.28)$$

Our proof, the idea of which we owe to Dr. F. Pedit, depends only on linear algebra and does not involve the Bianchi identities. We show that

3.12 Conformal metrics and conformal connexions

W_{lijk} vanishes at an arbitrary point m – in fact, we evaluate the three parts of the right-hand side of (3.28) at m.

Since R_{ij} and g_{ij} are symmetric and g_{ij} is non-singular we may choose coordinates so that, *at m*, the tangent space V at m has an orthonormal basis $\mathbf{e}_1, \mathbf{e}_2, \mathbf{e}_3$ such that

$$R_{ij} = \lambda_i \delta_{ij} \quad \text{(not summed for } i\text{)}, \quad \lambda_i \in \mathbb{R},$$

$$g_{ij} = \delta_{ij},$$

$$\tau = \lambda_1 + \lambda_2 + \lambda_3.$$

All the following quantities are expressed in terms of this orthogonal basis. We compute now, bit by bit, the three expressions on the right-hand side of (3.28).

The first expression. We have

$$R_{1213} = R_{1213} + R_{2223} + R_{3233} = R_{23} = 0.$$

Similarly $R_{1232} = R_{13} = 0$ and $R_{1323} = 0$. Thus the only non-zero components of R_{lijk} are R_{1212}, R_{1313}, and R_{2323}.

The second expression. We denote this by $-B_{lijk}$. To compute this we have only to consider the three indices combinations 1212, 1313, 2323, otherwise these terms are trivially zero. We have

$$B_{1212} = R_{11} + R_{22} = \lambda_1 + \lambda_2,$$

$$B_{1313} = R_{11} + R_{33} = \lambda_1 + \lambda_3,$$

$$B_{2323} = R_{22} + R_{33} = \lambda_2 + \lambda_3.$$

Now

$$R_{11} = R_{1212} + R_{1313} = \lambda_1, \qquad R_{1212} = \tfrac{1}{2}(\lambda_1 + \lambda_2 - \lambda_3),$$

$$R_{22} = R_{1212} + R_{2323} = \lambda_2, \qquad R_{1313} = \tfrac{1}{2}(\lambda_1 - \lambda_2 + \lambda_3),$$

$$R_{33} = R_{1313} + R_{2323} = \lambda_3. \qquad R_{2323} = \tfrac{1}{2}(-\lambda_1 + \lambda_2 + \lambda_3).$$

The third expression. We denote this by C_{lijk}. We have

$$C_{1212} = C_{1313} = C_{2323} = \tfrac{1}{2}(\lambda_1 + \lambda_2 + \lambda_3).$$

We substitute in the equation for W_{lijk} to find that every term vanishes. For example we have

$$W_{1212} = R_{1212} - B_{1212} + C_{1212} = \tfrac{1}{2}(\lambda_1 + \lambda_2 - \lambda_3)$$
$$- (\lambda_1 + \lambda_2) + \tfrac{1}{2}(\lambda_1 + \lambda_2 + \lambda_3) = 0.$$

This completes the proof that $W = 0$ identically in a three-dimensional Riemannian manifold.

It is interesting to give an alternative treatment using differential forms. Let (M, g) be an n-dimensional differentiable manifold, and let (e_i) be an orthonormal frame with respect to g with dual frame (ω^i) defined over some coordinate neighbourhood of a point $m \in M$. Any conformally related metric is given by $\bar{g} = \phi^2 g$ where we have replaced e^σ by ϕ.

Let $\bar{\omega}^i = \phi \omega^i$, $\bar{e}_i = \phi \bar{e}_i$. Then $(\bar{e}_i)$, $(\bar{\omega}^i)$ form orthonormal frames and coframes with respect to $\bar{g}$. The equations of structure give

$$d\omega^i + \omega^i_j \wedge \omega^j = 0, \qquad d\bar{\omega}^i + \bar{\omega}^i_j \wedge \bar{\omega}^j = 0,$$

$$\omega^i_j + \omega^j_i = 0, \qquad \bar{\omega}^i_j + \bar{\omega}^j_i = 0,$$

where ω^i_j and $\bar{\omega}^i_j$ are connexion forms of the corresponding Levi-Civita connexions. Since $\bar{\omega}^i = \phi \omega^i$ we have

$$d\bar{\omega}^i = d\phi \wedge \omega^i + \phi \, d\omega^i.$$

Thus

$$-\bar{\omega}^i_j \wedge \bar{\omega}^j = d\phi \wedge \omega^i - \phi \omega^i_j \wedge \omega^j.$$

We write

$$d\phi = c_j \omega^j$$

to get

$$-\bar{\omega}^i_j \wedge \bar{\omega}^j = c_j \omega^j \wedge \omega^i - \phi \omega^i_j \wedge \omega^j,$$

which can be written as

$$(\phi \bar{\omega}^i_j - c_j \omega^i - \phi \omega^i_j) \wedge \omega^j = 0.$$

This in turn can be written

$$(\phi \bar{\omega}^i_j - c_j \omega^i + c_i \omega^j - \phi \omega^i_j) \wedge \omega^j = 0,$$

since

$$c_i \omega^j \wedge \omega^j = 0.$$

The 1-form in brackets is skew-symmetric in i and j. We write

$$\phi \bar{\omega}^i_j - c_j \omega^i + c_i \omega^j - \phi \omega^i_j = F^i_{jk} \omega^k,$$

and see that $F^i_{jk} \omega^k \wedge \omega^j = 0$. Thus $F^i_{jk} = F^i_{kj}$. However

$$F^i_{jk} = -F^j_{ik} = -F^j_{ki} = F^k_{ji} = F^k_{ij} = -F^i_{kj}$$

and hence $F^i_{jk} = 0$. Thus we conclude that

$$\phi \bar{\omega}^i_j = \phi \omega^i_j + c_j \omega^i - c_i \omega^j,$$

giving the relation between the connexion forms. Writing $\omega^i_j = \Gamma^i_{kj} \omega^k$, $\bar{\omega}^i_j = \bar{\Gamma}^i_{kj} \bar{\omega}^k$ we find

3.12 Conformal metrics and conformal connexions

$$\phi^2 \bar{\Gamma}^i_{kj} \omega^k = \phi \Gamma^i_{kj} \omega^k + c_j \omega^i - c_i \omega^j,$$

that is,

$$\omega^k (\phi^2 \bar{\Gamma}^i_{kj} - \phi \Gamma^i_{kj} - c_j \delta^i_k + c_i \delta^j_k) = 0,$$

so that

$$\phi^2 \bar{\Gamma}^i_{kj} = \phi \Gamma^i_{kj} + c_j \delta^i_k - c_i \delta^j_k,$$

giving the relation between the connexion coefficients.

Let $X = X^i \mathbf{e}_i = \bar{X}^i \bar{\mathbf{e}}_i$, $Y = Y^i \mathbf{e}_i = \bar{Y}^i \bar{\mathbf{e}}_i$. Then $\bar{X}^i = \phi X^i$, $\bar{Y}^i = \phi Y^i$. We now compute $\bar{\nabla}_X Y$ and $\nabla_X Y$. We have

$$\nabla_X Y = \nabla_{X^i \mathbf{e}_i}(Y^j \mathbf{e}_j) = X^i \{\mathbf{e}_i(Y^j) \mathbf{e}_j + Y^j \Gamma^k_{ij} \mathbf{e}_k \}.$$

$$\bar{\nabla}_X Y = \bar{\nabla}_{\bar{X}^i \bar{\mathbf{e}}_i}(\bar{Y}^j \bar{\mathbf{e}}_j) = \bar{X}^i \{\bar{\mathbf{e}}_i(\bar{Y}^j) \bar{\mathbf{e}}_j + \bar{Y}^j \bar{\Gamma}^k_{ij} \bar{\mathbf{e}}_k \}$$

$$= \phi X^i \{\phi^{-1} \mathbf{e}_i(\phi Y^j) \phi^{-1} \mathbf{e}_j + \phi Y^j \bar{\Gamma}^k_{ij} \phi^{-1} \mathbf{e}_k \}$$

$$= X(\phi Y^j) \phi^{-1} \mathbf{e}_j + \phi X^i Y^j \bar{\Gamma}^k_{ij} \mathbf{e}_k$$

$$= X(\phi) Y^j \phi^{-1} \mathbf{e}_j + X(Y^j) \mathbf{e}_j + \phi X^i Y^j \bar{\Gamma}^k_{ij} \mathbf{e}_k.$$

Then

$$\bar{\nabla}_X Y - \nabla_X Y = \phi X^i Y^j \bar{\Gamma}^k_{ij} \mathbf{e}_k + X(\phi) Y^j \phi^{-1} \mathbf{e}_j - X^i Y^j \Gamma^k_{ij} \mathbf{e}_k$$

$$= X^i Y^j \{\Gamma^k_{ij} + c_j \delta^k_i \phi^{-1} - c_k \delta^j_i \phi^{-1}\} \mathbf{e}_k + \phi^{-1} X(\phi) Y$$
$$\quad - X^i Y^j \Gamma^k_{ij} \mathbf{e}_k$$

$$= \phi^{-1} X^i Y^j \{c_j \mathbf{e}_i - c_k \delta^j_i \mathbf{e}_k\} + \phi^{-1} X(\phi) Y$$

$$= X(\phi^{-1} c_j Y^j) - g(X,Y) c_k \phi^{-1} \mathbf{e}_k + \phi^{-1} X(\phi) Y.$$

Let α be the 1-form given by

$$\alpha = \phi^{-1} c_j \omega^j = \phi^{-1} d\phi = d(\log \phi).$$

Then $\alpha(Y) = \phi^{-1} c_j \omega^j(Y^k \mathbf{e}_k) = \phi^{-1} c_k Y^k$. Hence

$$\bar{\nabla}_X Y - \nabla_X Y = \alpha(Y) X + \phi^{-1} X(\phi) Y - g(X,Y) \alpha^\#$$

$$= \alpha(Y) X + \phi^{-1} d\phi(X) Y - g(X,Y) \alpha^\#$$

$$= \alpha(Y) X + \alpha(X) Y - g(X,Y) \alpha^\#,$$

where $\alpha^\#$ denotes the contravariant form of α with respect to g. Thus we have obtained the relation

$$\bar{\nabla}_X Y - \nabla_X Y = \alpha(Y) X + \alpha(X) Y - g(X,Y) \alpha^\# \tag{3.29}$$

and the 1-form α is closed. In fact, in terms of the function σ we have $\alpha = d\sigma$.

110   3 *Riemannian manifolds*

So far this is quite straightforward. However, in the course of lecturing at Durham in 1986, Dr. F. Pedit made the following interesting generalization.

Suppose we are given a conformal manifold $(M, [g])$, that is, a C^∞-manifold with a conformal class of metrics. Consider the following triple product $T: TM \times TM^* \times TM \to TM$ given by

$$T(X, \alpha, Y) = T_\alpha(X, Y) = \alpha(X)Y + \alpha(Y)X - g(X, Y)\alpha^\#, \quad (3.30)$$

a mapping which is motivated by (3.29). The essential difference here is that the 1-form α is no longer required to be closed. Clearly T satisfies the identity $T(X, \alpha, Y) = T(Y, \alpha, X)$. But we also have a further identity given by the next theorem.

Theorem (3.12.1) Let $T'(\alpha, Z, \beta)$ be a 1-form defined for 1-forms α, β and vector field Z by

$$T'(\alpha, Z, \beta)(W) = \alpha(T(Z, \beta, W)), \quad (3.31)$$

where W is any vector field. Then the following identity is satisfied:

$$T(X, \alpha, T(Y, \beta, Z)) - T(X, T'(\alpha, Z, \beta), Y) = \\ T(T(X, \alpha, Y), \beta, Z) - T(T(X, \beta, Z), \alpha, Y). \quad (3.32)$$

Proof. The only way I have succeeded in establishing this identity is by the time-honoured method of tediously showing that the left-hand side and the right-hand side when expanded do in fact lead to identical expressions. We leave the ready to verify that this is the case.

Following Pedit, we define an equivalence class on $\mathcal{C}_0(M)$ the set of torsion-free connexions on M. We assert that $\nabla \approx \tilde{\nabla}$ if and only if there exists a 1-form α such that $\tilde{\nabla} = \nabla + T_\alpha$. Then $\mathcal{C}_0(M)/\sim$ is the set of *Conformal connexions* on M. Thus a conformal connexion is an equivalence class of connexions.

Theorem (3.12.2) Let $\nabla \sim \tilde{\nabla}$; then $\nabla T = \tilde{\nabla} T$.
Proof. We note that $\nabla \sim \tilde{\nabla}$ implies that

$$\tilde{\nabla} = \nabla + T_\beta$$

for some 1-form β. Then

$$(\tilde{\nabla}_Z T)(X, \alpha, Y) = \tilde{\nabla}_Z T(X, \alpha, Y) - T(\tilde{\nabla}_Z X, \alpha, Y) \\ - T(X, \tilde{\nabla}_Z \alpha, Y) - T(X, \alpha, \tilde{\nabla}_Z Y) \\ = (\nabla_Z T)(X, \alpha, Y) + T(Z, \beta, T(X, \alpha, Y)) \\ - T(T(Z, \beta, X), \alpha, Y) + T(X, T'(\alpha, Z, \beta), Y)$$

3.12 Conformal metrics and conformal connexions

$$- T(X, \alpha, T(Z, \beta, Y)$$
$$= (\nabla_Z T)(X, \alpha, Y)$$

by Theorem (3.12.1) and this completes the proof.

Theorem (3.12.3) *If $\nabla \sim \nabla'$ where $\nabla' = \nabla_g$, the Levi-Civita connexion of g, then $\nabla T = 0$.*

To prove this theorem we note that

$$(\nabla_Z T)(X, \alpha, Z) = -(\nabla_Z g)(X, Y)\alpha^\# + g(X, Y)[(\nabla_Z \alpha)^\# - \nabla_Z \alpha^\#].$$

From Theorem (3.12.2) it now follows that

$$\nabla_Z T = \nabla'_Z T.$$

Moreover, our assumption implies that the right-hand side of the above identity vanishes. Hence $\nabla T = 0$ and we are home.

We now prove a result which can be regarded as the converse of theorem (3.12.2).

Theorem (3.12.4) *If $\nabla T = \tilde{\nabla} T$ for $\nabla, \tilde{\nabla} \in \mathcal{C}_0(M)$, then $\nabla \sim \tilde{\nabla}$.*

Proof. Define $D: TM \times TM \to TM$ by $D = \nabla - \tilde{\nabla}$. We have to show that $D = T_\omega$ for some 1-form ω. We note that our hypothesis implies that D is symmetric. Moreover

$$0 = (\nabla - \tilde{\nabla})T$$

implies, by using the expression of ∇T in the proof of Theorem (3.12.3) and the formula

$$g((\nabla_Z \dot{\alpha})^\# - \nabla_Z \alpha^\#, -) = (\nabla_Z g)(\alpha^\#, -),$$

the following identity:

$$[g(D(Z, X), Y) + g(X, D(Z, Y))]g(U, W) =$$
$$[g(D(Z, U), W) + g(U, D(Z, W))]g(X, Y).$$

If we choose an orthonormal basis $e = (e_1, \ldots, e_n)$ with respect to g of the tangent space at a fixed point, the above equation becomes

$$(D^j_{lk} + D^l_{jk})\delta_{rs} = (D^s_{kr} + D^r_{ks})\delta_{lj},$$

where we have written

$$D^j_{kl} = g(D(e_k, e_l), e_j).$$

Putting $r = s$ and $l \neq j$ we obtain $D^j_{kl} = -D^l_{kj}$. Putting $r = s$ and $l = j$ we obtain

$$D^j_{kj} = D^r_{kr}$$

which we denote by ω_k. From this we conclude that $D_{kl}^s = 0$ if s, k, l are all distinct. Also $D_{kj}^j = -D_{jj}^k$ if $j \neq k$. If we put $X = X^i\mathbf{e}_i$, $Y = Y^j\mathbf{e}_j$, $Z = Z^k\mathbf{e}_j$, we get

$$g(D(X,Y),Z) = \sum X^iY^jZ^kD_{ij}^k.$$

Moreover, the right-hand side may be written

$$\sum_i X^iY^iZ^iD_{ii}^i + \sum_{\substack{i=j \\ k \neq i}} X^iY^iZ^kD_{ii}^k + \sum_{\substack{i=k \\ i \neq j}} X^iY^jZ^iD_{ij}^i + \sum_{\substack{j=k \\ i \neq j}} X^iY^jZ^jD_{ij}^j$$

$$= \sum_i X^iY^iZ^iD_{ii}^i + \sum_{j,k; k \neq i} X^iY^iZ^kD_{ki}^i + \sum_{i,j; i \neq j} X^iY^jZ^iX_{ji}^i + \sum_{i,j; i \neq j} X^iY^jZ^jD_{ij}^j$$

$$= \sum_{i,j} X^iY^jZ^i\omega_j + \sum_{i,j} X^iY^jZ^j\omega_i - \sum_{i,k} X^iY^iZ^k\omega_k$$

$$= g(X,Z)\omega(Y) + g(Y,Z)\omega(X) - g(X,Y)\omega(Z),$$

where $\omega = \omega_k\sigma^k$ ((σ^k dual to (e_k)) is a well-defined 1-form. But this is just another way of writing

$$g(D(X,Y),Z) = g(T_\omega(X,Y),Z)$$

and so the theorem is proved.

Theorem (3.12.5) *Let $\nabla T = 0$ for $\nabla \in \mathcal{C}_0(M)$. Then $\nabla \sim \nabla_g$ where ∇_g is the Levi-Civita connexion for g.*

Proof. Because $\nabla_g T = 0$ (see Theorem (3.12.3)) we conclude from the previous theorem that $\nabla_g \sim \nabla$.

We conclude that in the realm of conformal geometry there is a unique *class* of torsion-free connexions satisfying $\nabla T = 0$. This is the analogue of the theorem in Riemannian geometry which states that there is a uniquely determined torsion-free connexion which satisfies $\nabla g = 0$.

Earlier in this section we remarked that in a three-dimensional manifold the Ricci curvature determines the Riemann curvature tensor. In particular for a 3-manifold an Einstein metric has constant sectional curvature. Use was made of this by Hamilton (1982) who proved the following remarkable result.

Theorem (3.12.6) *Let M be a compact 3-manifold which admits a Riemannian metric g_0 with positive Ricci curvature. Then this metric can be deformed to a metric of constant positive sectional curvature.*

This result is particularly interesting because, since the only simply connected 3-manifold with constant sectional curvature is the sphere S^3, if one can prove that a simply connected 3-manifold has a metric with positive Ricci curvature then one would have a proof of the famous Poincaré conjecture.

To prove the theorem the author proves that the metric g_0 can be deformed to an Einstein metric which in the case of three-dimensions has constant sectional curvature. The deformation is obtained by solving the heat equation

$$\frac{\partial g}{\partial t} = 2\left(\frac{rg}{3} - \text{Ric}(g)\right)$$

with initial condition $g(0) = g_0$. Here r is the average scalar curvature of the metric g. He shows that, when the dimension of M is 3, the solution of this equation exists for all time t and as $t \to \infty$ the metric $g(t)$ converges to an Einstein metric with positive curvature. The proof of this depends on the fact that the manifold has dimension 3. We refer the reader to Hamilton's paper for the details.

3.13 EXERCISES

1. A Riemannian manifold is called *recurrent* if the curvature tensor R satisfies the equation

   $$\nabla R = R \otimes \omega$$

   for some 1-form ω. It is assumed that the metric is positive definite and that $R \neq 0$. Prove that ω is exact. In the notation of classical tensor calculus the recurrence condition becomes

   $$\nabla_h R_{ijkl} = R_{ijkl} \kappa_h$$

   for some covariant vector κ_h. Prove that κ_h is a gradient.

2. Investigate if the result still holds if the metric is no longer positive definite.

3. Let M be an n-dimensional differentiable manifold with a torsion-free connexion Γ. We wish to prove that this induces a pseudo-Riemannian metric g' on the covariant tangent bundle $T^*(M)$. Let U be a coordinate neighbourhood of M with coordinates x^α, $\alpha = 1, \ldots, n$, and let Γ have components $\Gamma^\gamma_{\alpha\beta}$ with respect to this system. A covariant vector field over U is represented by coordinates (x^α, ξ_α). Define the matrix (g_{ij}), $i, j = 1, \ldots, 2n$, by

   $$g_{ij} = \begin{pmatrix} g_{\alpha\beta}, & I \\ I, & 0 \end{pmatrix}$$

where $g_{\alpha\beta} = -2\Gamma^{\gamma}_{\alpha\beta}\xi_{\gamma}$ and I is the unit $n \times n$ matrix. Prove that (g_{ij}) defines a pseudo-Riemannian metric over $T^*(M)$.

4. Prove that the curvature tensor on $T^*(M)$ so defined is zero if and only if the connexion Γ on M has zero curvature.

5. Prove that the curvature tensor on $T^*(M)$ has zero covariant derivative if and only if the curvature tensor of Γ has zero covariant derivative.

6. Give a coordinate-free description of how a torsion-free connexion on M determines a pseudo-Riemannian metric on $T^*(M)$.

4
Submanifold theory

4.1 SUBMANIFOLDS

In this chapter we shall consider the immersion of a manifold M into a Riemannian manifold $(N, \tilde{g})$, and examine the structures on M induced from the given structure on N. This is a natural generalization of the study of surfaces in Euclidean 3-space with properties induced from the Euclidean metric — which was the origin of the classical theory of surfaces.

Induced connexion and second fundamental form

Let $f: (M, g) \to (N, \tilde{g})$ be an isometric immersion of Riemannian manifolds, so that the metric g of M can be regarded as the pull-back of the metric $\tilde{g}$ of N. Let $U \subset M$ be a neighbourhood with coordinates $x: U \to \mathbb{R}^m$, and let $V \subset N$ be a neighbourhood with coordinates $y: V \to \mathbb{R}^n$. We shall work locally so we may assume that $f: M \to N$ is an embedding. The mapping f is given locally by equations

$$y^A = f^A(x^i)$$

where $A = 1, \ldots, n$ and $i = 1, \ldots, m$. Then we have

$$g_{ij} = \frac{\partial y^A}{\partial x^i} \frac{\partial y^B}{\partial x^j} \tilde{g}_{AB}. \tag{4.1}$$

Let X be a vector field defined on U. Then $f_* X$ is a vector field on $f(U) \subset V$. For simplicity we shall identify $f_* X$ with X and treat M as a subset of N. We want to operate on a vector field defined on the whole of V, so we consider an extension of $f_*(X)$ which we denote by $\tilde{X}$. The relation between X and $\tilde{X}$ is shown by the following commutative diagram.

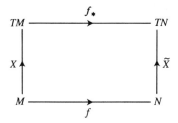

116   4 Submanifold theory

Let X, Y be two vector fields over U and let $\tilde{X}, \tilde{Y}$ be any extensions of X and Y.

Proposition (4.1.1) The restriction of $[\tilde{X}, \tilde{Y}]$ to $f(U)$ is independent of the choice of the extensions.

The proposition follows from a result previously proved in Chapter 1, namely, that if $\tilde{X}, \tilde{Y}$ and X, Y, respectively, are f-related then so is $[\tilde{X}, \tilde{Y}]$, that is,

$$[\tilde{X}, \tilde{Y}] \circ f = f_*[X, Y]. \tag{4.2}$$

However we give another proof here in terms of coordinates. Let $\tilde{X} = \tilde{\xi}^A \partial/\partial y^A$, $\tilde{Y} = \tilde{\eta}^B \partial/\partial y^B$. Then $\tilde{X} \circ f = f_* X$ implies

$$\tilde{\xi}^A \circ f = \frac{\partial f^A}{\partial x^i} \xi^i = \frac{\partial (y^A \circ f)}{\partial x^i} \xi^i = f_i^A \xi^i.$$

Similarly $\tilde{\eta}^B \circ f = f_i^B \eta^i$. We have

$$\frac{\partial (\tilde{\eta}^A \circ f)}{\partial x^i} = \left(\frac{\partial \tilde{\eta}^A \circ f}{\partial y^B}\right) \left(\frac{\partial (y^B \circ f)}{\partial x^i}\right) \tag{4.3}$$

by the chain rule, and hence

$$(\tilde{\xi}^B \circ f) \left(\frac{\partial \tilde{\eta}^A \circ f}{\partial y^B}\right) = \frac{\partial (y^B \circ f)}{\partial x^i} \left(\frac{\partial \tilde{\eta}^A \circ f}{\partial y^B}\right) \xi^i = \frac{\partial (\tilde{\eta}^A \circ f)}{\partial x^i} \xi^i$$

$$= \frac{\partial^2 (y^A \circ f)}{\partial x^i \partial x^j} \xi^i \eta^j + \frac{\partial (y^A \circ f)}{\partial x^j} \frac{\partial \eta^j}{\partial x^i} \xi^i.$$

Similarly

$$(\tilde{\eta}^B \circ f) \left(\frac{\partial \tilde{\xi}^A \circ f}{\partial y^B}\right) = \frac{\partial^2 (y^A \circ f)}{\partial x^j \partial x^i} \xi^i \eta^j + \frac{\partial (y^A \circ f)}{\partial x^j} \frac{\partial \xi^j}{\partial x^i} \eta^i.$$

Now

$$[\tilde{X}, \tilde{Y}] \circ f = \left[(\tilde{\xi}^B \circ f) \left(\frac{\partial \tilde{\eta}^A \circ f}{\partial y^B}\right) - (\tilde{\eta}^B \circ f) \left(\frac{\partial \tilde{\xi}^A \circ f}{\partial y^B}\right)\right] \left(\frac{\partial}{\partial y^A} \circ f\right).$$

Hence

$$[\tilde{X}, \tilde{Y}] \circ f = \left(\xi^i \frac{\partial \eta^j}{\partial x^i} - \eta^i \frac{\partial \xi^j}{\partial x^i}\right) \frac{\partial (y^A \circ f)}{\partial x^j} \left(\frac{\partial}{\partial y^A} \circ f\right)$$

$$= \left(\xi^i \frac{\partial \eta^j}{\partial x^i} - \eta^i \frac{\partial \xi^j}{\partial x^i}\right) f_* \left(\frac{\partial}{\partial x^j}\right)$$

$$= f_*[X, Y]$$

and the proposition is proved.

4.1 Submanifolds

Proposition (4.1.2) *The vector field $(\tilde{\nabla}_{\tilde{X}} \tilde{Y})$ when restricted to M is independent of the choice of the particular extensions of X and Y.*

To prove this we write

$$\tilde{X} = \tilde{\xi}^A \frac{\partial}{\partial y^A}, \qquad \tilde{Y} = \tilde{\eta}^B \frac{\partial}{\partial y^B};$$

then

$$(\tilde{\nabla}_{\tilde{X}} \tilde{Y}) \circ f = (\tilde{\xi}^A \circ f) \left\{ \left[\frac{\partial \tilde{\eta}^B}{\partial y^A} \circ f \right] \left[\frac{\partial}{\partial y^B} \circ f \right] \right.$$

$$\left. + (\tilde{\Gamma}^C_{AB} \circ f)(\tilde{\eta}^B \circ f) \left[\frac{\partial}{\partial y^C} \circ f \right] \right\}.$$

Using the relation (4.3) and that $\tilde{X}$ is an extension of X we get

$$(\tilde{\nabla}_{\tilde{X}} \tilde{Y}) \circ f = \left\{ \xi^i \frac{\partial (\tilde{\eta}^C \circ f)}{\partial x^i} + \tilde{\Gamma}^C_{AB} \frac{\partial (y^A \circ f)}{\partial x^i} \frac{\partial (y^B \circ f)}{\partial x^j} \xi^i \eta^j \right\} \left[\frac{\partial}{\partial y^C} \circ f \right]$$

$$= \xi^i \left\{ \frac{\partial}{\partial x^i} \left[\frac{\partial y^C \circ f}{\partial x^k} \eta^k \right] \right.$$

$$\left. + \tilde{\Gamma}^C_{AB} \frac{\partial (y^A \circ f)}{\partial x^i} \frac{\partial (y^B \circ f)}{\partial x^j} \eta^j \right\} \left[\frac{\partial}{\partial y^C} \circ f \right].$$

Since this is clearly independent of the choice of the extensions of X and Y, the claim is proved. We can therefore use the notation $\tilde{\nabla}_X Y$ to denote $\tilde{\nabla}_{\tilde{X}} \tilde{Y}$.

Since f is an immersion it follows that $f_*(T_p M) \subset T_{f(p)} N$ is an m-dimensional vector subspace. We denote by $(T_p M)^\perp$ the *normal space* at p of M in N where

$$(T_p M)^\perp = \{\xi \in T_{f(p)} N : \tilde{g}(\xi, v) = \tilde{g}(\xi, f_* v) = 0, \forall v \in T_p M\} \quad (4.4)$$

Any vector $\xi \in (T_p M)^\perp$ is a *normal vector* of M at p. The set

$$TM^\perp = \bigcup_{p \in M} (T_p M)^\perp$$

has the structure of a vector bundle over M and is called the *normal bundle* of M in N.

The tangent bundle TN splits up over M into the (Whitney) sum of the tangent bundle TM and the normal bundle $TM^\perp$, that is,

$$TN|_M = TM \oplus TM^\perp.$$

In particular the vector field $\tilde{\nabla}_X Y$ can be decomposed into its components in TM and $TM^\perp$, respectively, as follows:

$$\tilde{\nabla}_X Y = \nabla_X Y + h(X, Y). \quad (4.5)$$

This is called the *equation of Gauss* (not to be confused with the Gauss equations which we shall obtain shortly), who first obtained it for surfaces embedded in Euclidean space $\mathbb{R}^3$. Here the notation $\nabla_X Y$ is just a symbolic expression for the tangential component of $\tilde{\nabla}_X Y$, but as the notation suggests, we shall prove that it is the covariant derivative of the induced Riemannian connexion from the metric $g = f^*(\tilde{g})$. Moreover, from the connexion properties we have

$$\tilde{\nabla}_{\alpha X} Y = \alpha \tilde{\nabla}_X Y,$$
$$\tilde{\nabla}_X (\beta Y) = (X\beta) Y + \beta \tilde{\nabla}_X Y$$

for real-valued functions α, β restricted to M. Therefore

$$\tilde{\nabla}_{\alpha X} Y = \nabla_{\alpha X} Y + h(\alpha X, Y) = \alpha \nabla_X Y + \alpha h(X, Y)$$
$$\tilde{\nabla}_X (\beta Y) = (X\beta) Y + \beta \{\nabla_X Y + h(X, Y)\} = \nabla_X (\beta Y) + h(X, \beta Y).$$

By taking the tangential and normal parts we deduce that

$$\nabla_{\alpha X} Y = \alpha \nabla_X Y,$$
$$\nabla_X (\beta Y) = (X\beta) Y + \beta \nabla_X Y,$$
$$h(\alpha X, Y) = \alpha h(X, Y),$$
$$h(X, \beta Y) = \beta h(X, Y).$$

Since the additivity properties are easily established we deduce that ∇ defines a connexion over M, and also that h is a bilinear mapping of X, Y into $TM^\perp$. Since $\tilde{\nabla}$ has zero torsion we have

$$0 = \tilde{\nabla}_X Y + \tilde{\nabla}_Y X - [X, Y]$$
$$= \nabla_X Y - \nabla_Y X - [X, Y] + h(X, Y) - h(Y, X).$$

Equating the tangential part to zero shows that ∇ has zero torsion. Equating the normal part to zero shows that the bilinear form h is symmetric, that is, $h(X, Y) = h(Y, X)$.

Since $\tilde{\nabla}$ is the Levi-Civita connexion we have, for any vector fields X, Y, Z on M,

$$\nabla_X (g(Y, Z)) = \tilde{\nabla}_X (\tilde{g}(Y, Z)) = \tilde{g}(\tilde{\nabla}_X Y, Z) + \tilde{g}(Y, \tilde{\nabla}_X Z)$$
$$= \tilde{g}(\nabla_X Y + h(X, Y), Z) + \tilde{g}(Y, \nabla_X Z + h(X, Z))$$
$$= \tilde{g}(\nabla_X Y, Z) + \tilde{g}(Y, \nabla_X Z)$$
$$= g(\nabla_X Y, Z) + g(Y, \nabla_X Z).$$

Thus the connexion ∇ is a metric connexion with zero torsion, and because of uniqueness we can identify it with the Levi-Civita connexion of the pulled-back metric. We shall call ∇ the *induced connexion* and h the *second fundamental form* of the submanifold M.

4.1 Submanifolds

Let (e_i), $i = 1, \ldots, m$ be an orthonormal frame for the tangent space to M at p. Then we define

$$H = \frac{1}{m}\operatorname{trace} h = \frac{1}{m}\sum_{i=1}^{m} h(e_i, e_i) \qquad (4.6)$$

to be the *mean curvature vector* at p.

Let ξ be a unit normal vector field on M, and let X be a vector field tangent to M. Then we may write

$$\tilde{\nabla}_X \xi = -(A_\xi X) + \nabla^\perp_X \xi \qquad (4.7)$$

where $-(A_\xi X)$ and $\nabla^\perp_X \xi$ are respectively the tangential and the normal components. This is called the *equation of Weingarten*, after the mathematician who first obtained the equation for surfaces in Euclidean space.

Proposition (4.1.3) The function $A_\xi X$ is bilinear in ξ and X, and hence its value at the point $p \in M$ depends only on the values ξ_p and X_p.

To prove this we have, for functions α and β on M,

$$\tilde{\nabla}_{\alpha X}(\beta \xi) = \alpha\{(X\beta)\xi + \beta \tilde{\nabla}_X \xi\}$$
$$= \alpha(X\beta)\xi + \alpha\beta\nabla^\perp_X \xi - \alpha\beta A_\xi X$$
$$= -A_{\beta\xi}(\alpha X) + \nabla^\perp_{\alpha X}(\beta \xi).$$

This gives

$$A_{\beta\xi}(\alpha X) = \alpha\beta A_\xi X,$$

and

$$\nabla^\perp_{\alpha X}(\beta \xi) = \alpha(X\beta)\xi + \alpha\beta\nabla^\perp_X \xi.$$

The additivity property is trivially established and hence we have shown that $A_\xi(X)$ is bilinear. We note incidentally that we have also shown that $\nabla^\perp$ is a connexion on the normal bundle, which we shall call the *normal connexion*. The relation between $A_\xi(X)$ and $h(X, Y)$ is given by the next result.

Proposition (4.1.4) For each normal vector field ξ on M we have

$$g(A_\xi X, Y) = \tilde{g}(h(X, Y), \xi). \qquad (4.8)$$

To prove this we differentiate the equation $\tilde{g}(Y, \xi) = 0$ to get

$$0 = \tilde{g}(\tilde{\nabla}_X Y, \xi) + \tilde{g}(Y, \tilde{\nabla}_X \xi)$$
$$= \tilde{g}(\nabla_X Y, \xi) + \tilde{g}(h(X, Y), \xi) + \tilde{g}(Y, -A_\xi X + \nabla^\perp_X \xi)$$
$$= 0 + \tilde{g}(h(X, Y), \xi) - \tilde{g}(Y, A_\xi X) + 0$$

$$= \tilde{g}(h(X, Y), \xi) - \tilde{g}(Y, A_\xi X),$$

and the proposition is proved.

As a consequence we see that for fixed ξ, A_ξ is a self-adjoint endomorphism of the tangent space to M at p. This is called the *second fundamental tensor* at p with respect to ξ.

Let $\{\xi_\alpha\}$ be an orthonormal frame of $TM^\perp$ at p. Then

$$\text{trace } h = \sum_{\alpha, i} \tilde{g}(h(e_i, e_i), \xi_\alpha) = \sum_{\alpha, i} \tilde{g}(A_{\xi_\alpha}(e_i), e_i)$$

$$= \sum_\alpha \text{trace}(A_{\xi_\alpha}).$$

So an alternative definition of the mean curvature vector would be given by

$$H = \frac{1}{m} \sum_\alpha \text{trace}(A_{\xi_\alpha}).$$

If we write

$$h(e_i, e_j) = \sum_\alpha h_{ij}^\alpha \xi_\alpha$$

we see that the square of the length of the second fundamental form is given by

$$l^2 = \sum_{i,j} \tilde{g}(h(e_i, e_j), h(e_i, e_j)) = \sum (h_{ij}^\alpha)^2.$$

We now show that l^2 is *independent of the choice of orthonormal frame* (e_i). For, if we write

$$\tilde{e}_i = e_l b_i^l,$$

where (b_i^l) is an orthogonal $m \times m$ matrix, we have

$$\sum_{i,j} \tilde{g}(h(\tilde{e}_i, \tilde{e}_j), h(\tilde{e}_i, \tilde{e}_j)) = \sum_{i,j,l,k,r,s} \tilde{g}(h(e_l b_i^l, e_k b_j^k), h(e_r b_i^r, e_s b_j^s))$$

$$= \sum \tilde{g}(h(e_l, e_r), h(e_r, e_s)) b_i^l b_i^r b_j^k b_j^s$$

$$= \sum \tilde{g}(h(e_l, e_k), h(e_l, e_k))$$

where we have used the property $\Sigma_i b_i^l b_i^r = \delta_{lr}$. Moreover

$$h^\alpha(X, Y) = \tilde{g}(h(X, Y), \xi_\alpha) = g(A_{\xi_\alpha} X, Y), \quad \alpha = m+1, \ldots, n$$

are a set of real-valued symmetric bilinear forms.

We now have the following result.

Proposition (4.1.5) *The normal connexion $\nabla^\perp$ is a metric connexion in the normal bundle $TM^\perp$ with respect to the induced metric on $TM^\perp$.*

To prove this we note that for any normal vector fields ξ and η, and for any vector field X tangent to M we have

$$\tilde{\nabla}_X \xi = -A_\xi X + \nabla^\perp_X \xi, \qquad \tilde{\nabla}_X \eta = -A_\eta X + \nabla^\perp_X \eta;$$

hence

$$\tilde{g}(\nabla^\perp_X \xi, \eta) + \tilde{g}(\xi, \nabla^\perp_X \eta) = \tilde{g}(\tilde{\nabla}_X \xi, \eta) + \tilde{g}(\xi, \tilde{\nabla}_X \eta)$$
$$= \tilde{\nabla}_X \tilde{g}(\xi, \eta)$$
$$= \nabla^\perp_X g(\xi, \eta)$$

and we are through.

For a unit normal vector ξ of M at p, the second fundamental tensor A_ξ is self-adjoint and therefore we may choose an orthonormal basis $e_1, \ldots, e_m$ of M which are eigenvectors of A_ξ.
Then we have

$$A_\xi(e_i) = k_i e_i \qquad \text{(not summed for } i\text{)}$$

for some real numbers k_i. These numbers are called the *principal curvatures* of the normal direction ξ, and the eigenvectors are called the *principal directions*.

4.2 THE EQUATIONS OF GAUSS, CODAZZI, AND RICCI

The curvature operator of the Riemannian manifold N is given by

$$\tilde{R}(\tilde{X}, \tilde{Y})\tilde{Z} = (\tilde{\nabla}_{\tilde{X}} \tilde{\nabla}_{\tilde{Y}} - \tilde{\nabla}_{\tilde{Y}} \tilde{\nabla}_{\tilde{X}})\tilde{Z} - \tilde{\nabla}_{[\tilde{X}, \tilde{Y}]} \tilde{Z}$$

for vectors fields $\tilde{X}, \tilde{Y}, \tilde{Z}$ tangent to N. Let X, Y, Z be vector fields tangent to the submanifold M. Then

$$\tilde{R}(X, Y)Z = (\tilde{\nabla}_X \tilde{\nabla}_Y - \tilde{\nabla}_Y \tilde{\nabla}_X)Z - \tilde{\nabla}_{[X, Y]} Z.$$

We substitute for $\tilde{\nabla}$ in terms of ∇ from

$$\tilde{\nabla}_X Y = \nabla_X Y + h(X, Y)$$

to get

$$\tilde{R}(X, Y)Z = \tilde{\nabla}_X(\nabla_Y Z + h(Y, Z)) - \tilde{\nabla}_Y(\nabla_X Z + h(X, Z))$$
$$- \nabla_{[X, Y]} Z - h([X, Y], Z)$$
$$= \nabla_X \nabla_Y Z + h(X, \nabla_Y Z) + \tilde{\nabla}_X h(Y, Z) - \nabla_Y \nabla_X Z$$
$$- h(Y, \nabla_X Z) - \tilde{\nabla}_Y h(X, Z) - \nabla_{[X, Y]} Z - h([X, Y], Z)$$

$$= R(X, Y)Z - h([X, Y], Z) + h(X, \nabla_Y Z) - h(Y, \nabla_X Z)$$
$$+ \tilde{\nabla}_X h(Y, Z) - \tilde{\nabla}_Y h(X, Z).$$

From Weingarten's equation we have
$$\tilde{\nabla}_X h(Y, Z) = -A_{h(Y,Z)} X + \nabla^\perp_X h(Y, Z).$$

So we get
$$\tilde{R}(X, Y)Z = (R(X, Y)Z - A_{h(Y,Z)} X + A_{h(X,Z)} Y) - h([X, Y], Z)$$
$$+ h(X, \nabla_Y Z) - h(Y, \nabla_X Z) + \nabla^\perp_X h(Y, Z)$$
$$- \nabla^\perp_Y h(X, Z).$$

The first three terms of the right-hand member are tangential, whilst the remaining terms are normal. We now take the inner product of both sides of the equation with the vector field W, tangential to M, to get
$$\tilde{g}(\tilde{R}(X, Y)Z, W) = g(R(X, Y)Z, W) - g(A_{h(Y,Z)} X, W)$$
$$+ g(A_{h(X,Z)} Y, W).$$

Using Proposition (4.1.4) the right-hand member becomes
$$g(R(X, Y)Z, W) - \tilde{g}(h(Y, Z), h(X, W)) + \tilde{g}(h(X, Z), h(Y, W)).$$

Thus we derive the equation of Gauss in the form
$$\tilde{g}(\tilde{R}(X, Y)Z, W) - g(R(X, Y)Z, W) = \tilde{g}(h(X, Z), h(Y, W))$$
$$- \tilde{g}(h(X, W), h(Y, Z))$$

or
$$\tilde{R}(X, Y, W, Z) - R(X, Y, W, Z) = \tilde{g}(h(X, Z), h(Y, W))$$
$$- \tilde{g}(h(X, W), h(Y, Z)). \qquad (4.9)$$

If we denote the normal part of $\tilde{R}(X, Y)Z$ by $(R(X, Y)Z)^\perp$ we have
$$(\tilde{R}(X, Y)Z)^\perp = \nabla^\perp_X h(Y, Z) - h(Y, \nabla_X Z) - \nabla^\perp_Y h(X, Z)$$
$$+ h(X, \nabla_Y Z) - h([X, Y], Z).$$

We recall that if we have connexions ∇^E, ∇^F on vector bundles E and F over M, we can extend these to the vector bundle of r-multilinear maps
$$(E \times \cdots \times E, F) \xrightarrow{\pi} M$$

by
$$(\nabla_X T)(s_1, \ldots, s_r) = \nabla^F_X(T(s_1, \ldots, s_r)) - \sum_{k=1}^{r} T(s_1, \ldots, \nabla^E_X s_k, \ldots, s_r),$$

4.2 The equations of Gauss, Codazzi, and Ricci

where $s_k: M \to E$ are sections of E and X is a vector field on M. In the particular case above, we define

$$(\nabla_X^\perp h)(Y, Z) = \nabla_X^\perp (h(Y, Z)) - h(\nabla_X Y, Z) - h(Y, \nabla_X Z).$$

Using this relation we obtain the condition

$$(\tilde{R}(X, Y)Z)^\perp = (\tilde{\nabla}_X h)(Y, Z) - (\tilde{\nabla}_Y h)(X, Z), \tag{4.10}$$

which is the *Codazzi equation*.

If we write $h = h^\alpha \xi_\alpha$, then $\tilde{\nabla}_X h = (\nabla_X h^\alpha)\xi_\alpha + h^\alpha \nabla_X^\perp \xi_\alpha$ and therefore

$$(\tilde{R}(X, Y)Z)^\perp = ((\nabla_X h^\alpha)(Y, Z) - (\nabla_Y h^\alpha)(X, Z))\xi_\alpha$$
$$+ h^\alpha(Y, Z)\nabla_X^\perp \xi_\alpha - h^\alpha(X, Z)\nabla_Y^\perp \xi_\alpha \tag{4.11}$$

which is another form of the Codazzi equations.

Before proceeding further we examine the particular form of the Gauss and Codazzi equations when N is a space of constant curvature K. The Gauss equation becomes

$$R(X, Y, Z, W) = K(g(X, W)g(Y, Z) - g(X, Z)g(Y, W))$$
$$+ \tilde{g}(h(X, W), h(Y, Z)) - \tilde{g}(h(X, Z), h(Y, W)). \tag{4.12}$$

The scalar curvature is given in terms of an orthonormal basis by

$$u = \sum_{i,j} R(e_i, e_j, e_i, e_j)$$

$$= \sum_{i,j} K(g_{ii}g_{jj} - g_{ij}g_{ij}) + \sum_{i,j,\alpha,\beta} \tilde{g}(h_{ii}^\alpha \xi_\alpha, h_{jj}^\beta \xi_\beta) - \sum_{i,j,\alpha,\beta} \tilde{g}(h_{ij}^\alpha \xi_\alpha, h_{ij}^\beta \xi_\beta),$$

$$= Km(m-1) - \sum_{i,j,\alpha} h_{ij}^\alpha h_{ij}^\alpha + \sum_{i,j,\alpha} h_{ii}^\alpha h_{jj}^\alpha.$$

Thus

$$u = Km(m-1) - l^2 + m^2 |H|^2. \tag{4.13}$$

The Codazzi equations simplify considerably in this special case because the normal component of $\tilde{R}$ vanishes, and the equation reduces to

$$(\nabla_X h)(Y, Z) = (\nabla_Y h)(X, Z). \tag{4.14}$$

Thus, in this case, the trilinear map $(X, Y, Z) \to (\nabla_X h)(Y, Z)$ is symmetric in all three arguments. It frequently happens that relation (4.14) appears as a natural restriction on many symmetric covariant tensors of order 2 in geometric problems, and such tensors are called *Codazzi tensors*.

We now obtain the equation of Ricci for a general submanifold. Let ξ, η be two normal vector fields on M. Then
$$R(X, Y)\xi = \tilde{\nabla}_X\tilde{\nabla}_Y\xi - \tilde{\nabla}_Y\tilde{\nabla}_X\xi - \tilde{\nabla}_{[X,Y]}\xi.$$
Using the Weingarten equation the right-hand side of this becomes
$$-\tilde{\nabla}_X(A_\xi(Y)) + \tilde{\nabla}_X\tilde{\nabla}_Y^\perp\xi + \tilde{\nabla}_Y(A_\xi(X))$$
$$- \tilde{\nabla}_Y\tilde{\nabla}_X^\perp\xi + A_\xi([X, Y]) - \nabla^\perp_{[X,Y]}\xi.$$
We wish to take the inner product of this with the normal vector η, so in simplifying this expression further we will omit terms which are tangential. This leads to the expression
$$-h(X, A_\xi(Y)) + h(Y, A_\xi(X)) + \nabla_X^\perp\nabla_Y^\perp\xi - \nabla_Y^\perp\nabla_X^\perp\xi - \nabla^\perp_{[X,Y]}\xi.$$
If we write
$$R^\perp(X, Y)\xi = (\nabla_X^\perp\nabla_Y^\perp - \nabla_Y^\perp\nabla_X^\perp - \nabla^\perp_{[X,Y]})\xi$$
we have
$$\tilde{g}(\tilde{R}(X, Y)\xi, \eta) = \tilde{g}(R^\perp(X, Y)\xi, \eta) + \tilde{g}(h(Y, A_\xi X), \eta)$$
$$- \tilde{g}(h(X, A_\xi Y), \eta)$$
$$= \tilde{g}(R^\perp(X, Y)\xi, \eta) + g(A_\eta A_\xi(X), Y)$$
$$- g(A_\xi A_\eta(X), Y)$$
$$= \tilde{g}(R^\perp(X, Y)\xi, \eta) - g([A_\xi, A_\eta](X), Y),$$
where we have written
$$[A_\xi, A_\eta] = A_\xi A_\eta - A_\eta A_\xi \qquad (4.15)$$
Thus we get
$$\tilde{R}(X, Y, \eta, \xi) = R^\perp(X, Y, \eta, \xi) - g([A_\xi, A_\eta]X, Y) \qquad (4.16)$$
This is the equation of Ricci. In the particular case when N is of constant curvature, this reduces to
$$R^\perp(X, Y, \eta, \xi) = g([A_\xi, A_\eta]X, Y). \qquad (4.17)$$
We close this section with a statement of the *fundamental theorem of submanifolds*. We will not prove this theorem, but refer the reader to the book by Bishop and Crittenden (1964).

Theorem (4.2.1) (Fundamental theorem) *Let M be a simply connected m-dimensional Riemannian manifold. Assume that there is a Riemannian q-plane bundle E over M with a second fundamental form h and an associated second fundamental tensor A. If these quantities satisfy equations (4.12), (4.14), and (4.17) namely the special equations*

4.3 The method of moving frames

of Gauss, Codazzi, and Ricci, then M can be isometrically immersed in an $(m + q)$-dimensional space of constant curvature K with normal bundle E.

4.3 THE METHOD OF MOVING FRAMES

In this section we use the method of moving frames to obtain the equations of Gauss, Codazzi, and Ricci. We shall find that these are consequences of the Cartan structural equations for a submanifold.

Let N be a Riemannian manifold of dimension n and metric $\tilde{g}$. We can assume that M is a submanifold of N, of dimension m, locally imbedded in N. We denote by $F(N, M)$ the submanifold of the orthonormal frame bundle over N which consists of *adapted frames*, that is, frames (e_A), $A = 1, \ldots, n$ of which (e_i), $i = 1, \ldots, m$ are tangent to M and (e_α), $\alpha = m + 1, \ldots, n$ are normal to M. Two adapted frames $\tilde{e}$ and e are related by the matrix equation

$$\tilde{e} = eK \tag{4.18}$$

where K is the matrix

$$\begin{pmatrix} A & O \\ \hline O & B \end{pmatrix} \tag{4.19}$$

where $A \in O(m)$ and $B \in O(n - m)$.

The Levi-Civita connexion for N is given by the matrix of 1-forms (ω_B^A) where

$$d\omega^A + \omega_B^A \wedge \omega^B = 0,$$

$$\omega_B^A + \omega_A^B = 0.$$

We note that the restriction of (ω_B^A) to (ω_j^i) over M gives a connexion over M. This follows because the equation

$$d\omega^A = -\omega_B^A \wedge \omega^B$$

when restricted to M gives

$$d\omega^i = -\omega_B^i \wedge \omega^B = -\omega_j^i \wedge \omega^j$$

because $\omega^\alpha = 0$ on M. Moreover $\omega_j^i = -\omega_i^j$, so (ω_j^i) is the uniquely defined Levi-Civita connexion corresponding to the metric on M induced from that on N.

For tangent vectors on M we have $\omega^\alpha = 0$. We take the exterior derivative of this equation to get

$$0 = d\omega^\alpha = -\omega_A^\alpha \omega^A = -\omega_i^\alpha \wedge \alpha^i.$$

From Cartan's lemma it follows that

$$\omega_i^\alpha = h_{ij}^\alpha \omega^j$$

for smooth functions $h_{ij}^\alpha = h_{ji}^\alpha$.

We have

$$\nabla e_\alpha = \omega_\alpha^j e_j + \omega_\alpha^\beta e_\beta.$$

The scalar product of this with

$$-dp = -\omega^i e_i$$

gives

$$\tilde{g}(\nabla e_\alpha, -\omega^i e_i) = -\omega_\alpha^i \omega^i = \omega_i^\alpha \omega^i = h_{ij}^\alpha \omega^i \omega^j,$$

so that h_{ij}^α are none other than the components of the second fundamental form in the direction e_α. In fact, from the relation $(\tilde{\nabla}_X \xi)^T = -A_\xi X$ we write $\xi = e_\alpha$ and find

$$(\tilde{\nabla}_{e_i} e_\alpha)^T = (\omega_\alpha^A(e_i) e_A)^T = \omega_\alpha^j(e_i) e_j = -A_{e_\alpha}(e_i).$$

We write $A^\alpha = A_{e_\alpha}$ and obtain

$$A^\alpha(e_i) = -\omega_\alpha^j(e_i) e_j$$
$$= \omega_j^\alpha(e_i) e_j$$
$$= h_{jk}^\alpha \omega^k(e_i) e_j$$

from which

$$A_{(e_i)}^\alpha = h_{ij}^\alpha e_j. \qquad (4.20)$$

Similarly from the equation

$$d\omega^A + \omega_B^A \wedge \omega^B = 0$$

we may restrict our attention to normal vectors, that is, we consider the above equation on the normal bundle to get

$$d\omega^\alpha + \omega_B^\alpha \wedge \omega^B = d\omega^\alpha + \omega_\beta^\alpha \wedge \omega^\beta = 0$$

since $\omega^i = 0$ on the normal bundle. Moreover $\omega_\beta^\alpha = -\omega_\alpha^\beta$. Thus (ω_β^α) are the connexion forms of the metric connexion $\nabla^\perp$.

From the second structural equation for N we have

$$d\omega_B^A + \omega_C^A \wedge \omega_B^C = \tilde{\Omega}_B^A.$$

Restricting the indices range gives

$$d\omega_j^i + \omega_k^i \wedge \omega_j^k + \omega_\alpha^i \wedge \omega_j^\alpha = \tilde{\Omega}_j^i. \qquad (4.21)$$

The second structural equation for the connexion on M gives

$$d\omega_j^i + \omega_k^i \wedge \omega_j^k = \Omega_j^i. \qquad (4.22)$$

4.3 The method of moving frames

Moreover we have
$$\omega_\alpha^i \wedge \omega_j^\alpha = -h_{ik}^\alpha \omega^k \wedge h_{jl}^\alpha \omega^l = -h_{ik}^\alpha h_{jl}^\alpha \omega^k \wedge \omega^l.$$

Thus we get
$$\tilde{\Omega}_j^i - \Omega_j^i = -h_{ik}^\alpha h_{jl}^\alpha \omega^k \wedge \omega^l. \tag{4.23}$$

We evaluate this relation on the vectors e_r, e_s to get
$$\tilde{R}_{ijrs} - R_{ijrs} = -h_{ik}^\alpha h_{jl}^\alpha (\delta_r^k \delta_s^l - \delta_s^k \delta_r^l)$$
$$= (h_{is}^\alpha h_{jr}^\alpha - h_{ir}^\alpha h_{js}^\alpha). \tag{4.24}$$

Here we have used
$$\Omega_j^i(e_r, e_s) = R_{jrs}^i = R_{ijrs}$$
since the frames are orthonormal, and also the relation
$$(\omega^k \wedge \omega^l)(X, Y) = \omega^k(X)\omega^l(Y) - \omega^l(X)\omega^k(Y).$$

Equation (4.24) is the equation of Gauss and is precisely the same as equation (4.9) when we write
$$X = e_r, \quad Y = e_s, \quad Z = e_i, \quad W = e_j.$$

We now obtain the Codazzi equations by the moving frame method. From the second structural equation for N we have
$$d\omega_B^A + \omega_C^A \wedge \omega_B^C = \tilde{\Omega}_B^A.$$

Restrict the indices range to $A = \alpha$, $B = i$ to get
$$d\omega_i^\alpha + \omega_j^\alpha \wedge \omega_i^j + \omega_\beta^\alpha \wedge \omega_i^\beta = \tilde{\Omega}_i^\alpha.$$

Now
$$d\omega_i^\alpha = d(h_{ik}^\alpha \omega^k) = dh_{ik}^\alpha \wedge \omega^k - h_{ij}^\alpha \omega_k^j \wedge \omega^k$$
where in the last term we have used the structural equation for M. Moreover we have
$$\omega_j^\alpha \wedge \omega_i^j = h_{jk}^\alpha \omega^k \wedge \omega_i^j = -h_{jk}^\alpha \omega_i^j \wedge \omega^k$$
$$\omega_\beta^\alpha \wedge \omega_i^\beta = \omega_\beta^\alpha \wedge (h_{ik}^\beta \omega^k) = h_{ik}^\beta \omega_\beta^\alpha \wedge \omega^k.$$

So we have proved that
$$(dh_{ik}^\alpha - h_{ij}^\alpha \omega_k^j - h_{jk}^\alpha \omega_i^j + h_{ik}^\beta \omega_\beta^\alpha) \wedge \omega^k = \tilde{\Omega}_i^\alpha.$$

We evaluate this relation on e_r, e_s to get
$$dh_{is}^\alpha(e_r) - h_{ij}^\alpha \omega_s^j(e_r) - h_{js}^\alpha \omega_i^j(e_r) + h_{is}^\beta \omega_\beta^\alpha(e_r) - dh_{ir}^\alpha(e_s)$$
$$+ h_{ij}^\alpha \omega_r^j(e_s) + h_{jr}^\alpha \omega_i^j(e_s) - h_i^\beta \omega_\beta^\alpha(e_s) = \tilde{R}_{\alpha irs}.$$

This can be rewritten as

$$\tilde{R}_{\alpha irs} = (\nabla_{e_r} h^\alpha)(e_s, e_i) - (\nabla_{e_s} h^\alpha)(e_r, e_i) + h^\beta(e_s, e_i)\tilde{g}(\nabla^\perp_{e_r} e_\beta, e_\alpha)$$
$$- h^\beta(e_r, e_i)\tilde{g}(\nabla^\perp_{e_s} e_\beta, e_\alpha). \tag{4.25}$$

However, this is none other than the Codazzi equation (4.11) which we can see if we put

$$X = e_s, \quad Y = e_r, \quad Z = e_i$$

and take the inner product with $\xi_\alpha = e_\alpha$.

We now obtain the Ricci equations. From the second structural equation for N we have

$$d\omega^A_B + \omega^A_C \wedge \omega^C_B = \tilde{\Omega}^A_B.$$

We now restrict the indices to the range $m+1, \ldots, n$, to get

$$d\omega^\alpha_\beta + \omega^\alpha_i \wedge \omega^i_\beta + \omega^\alpha_\gamma \wedge \omega^\gamma_\beta = \tilde{\Omega}^\alpha_\beta.$$

We have

$$\omega^\alpha_i \wedge \omega^i_\beta = -h^\alpha_{ij}\omega^j \wedge h^\beta_{ik}\omega^k = -h^\alpha_{ij}h^\beta_{ik}\omega^j \wedge \omega^k,$$

and

$$d\omega^\alpha_\beta + \omega^\alpha_\gamma \wedge \omega^\gamma_\beta = \Omega^{\perp\alpha}_\beta$$

so

$$\tilde{\Omega}^\alpha_\beta - \Omega^{\perp\alpha}_\beta = -h^\alpha_{ij}h^\beta_{ik}\omega^j \wedge \omega^k.$$

We evaluate on e_r, e_s to get

$$\tilde{R}_{\alpha\beta rs} - R^\perp_{\alpha\beta rs} = -h^\alpha_{ij}h^\beta_{ik}(\delta^j_r\delta^k_s - \delta^j_s\delta^k_r),$$

giving

$$\tilde{R}_{rs\alpha\beta} - R^\perp_{rs\alpha\beta} = h^\alpha_{is}h^\beta_{ir} - h^\alpha_{ir}h^\beta_{is}. \tag{4.26}$$

We claim that (4.26) is one form of the Ricci equations.

We put

$$X = e_r, \quad Y = e_s, \quad \eta = e_\alpha, \quad \xi = e_\beta$$

in the Ricci equation (4.16) previously obtained to get

$$\tilde{R}_{rs\alpha\beta} - \tilde{R}^\perp_{rs\alpha\beta} = -\tilde{g}([A_{e_\beta}, A_{e_\alpha}]e_r, e_s)$$
$$= -\tilde{g}([A^\beta, A^\alpha]e_r, e_s)$$
$$= \tilde{g}(A^\alpha A^\beta e_r, e_s) - \tilde{g}(A^\beta A^\alpha e_r, e_s)$$

Now $A^\alpha = -e_j\omega^j_\alpha = e_j\omega^\alpha_j = e_j h^\alpha_{ij}\omega^i$.

Substitute in the right-hand member to get

4.4 The theory of curves

$$\tilde{g}(h^\beta_{rt}h^\alpha_{tu}e_u, e_s) - \tilde{g}(h^\alpha_{rt}h^\beta_{tu}e_u, e_s) = h^\alpha_{is}h^\beta_{ir} - h^\alpha_{ir}h^\beta_{is}.$$

Thus we obtain as an equivalent form of (4.16) the equation

$$\tilde{R}_{rs\alpha\beta} - R^\perp_{rs\alpha\beta} = h^\alpha_{is}h^\beta_{ir} - h^\alpha_{ir}h^\beta_{is}$$

which is none other than equation (4.26).

It is interesting to see how the equations of Gauss, Codazzi, and Ricci are all obtained from the Cartan structural equations by restricting the range of indices appropriately.

4.4 THE THEORY OF CURVES

This is a very special but very important case of the previous theory when $M = \mathbb{R}^1$. Indeed, the theory of the differential geometry of curves in Euclidean 3-space is as old as the calculus itself. The Frenet–Serret formulas are included in every mathematics undergraduate course. We remind the reader that in terms of vector calculus these equations are

$$\frac{d\mathbf{t}}{ds} = \kappa\mathbf{n}, \qquad \frac{d\mathbf{b}}{ds} = -\tau\mathbf{n}, \qquad \frac{d\mathbf{n}}{ds} = \tau\mathbf{b} - \kappa\mathbf{t},$$

where s is arc length, $\mathbf{t}$ is the unit tangent vector in the direction of increasing s, $\mathbf{n}$ is the principal normal, κ is the curvature, $\mathbf{b} = \mathbf{t} \times \mathbf{n}$ is the binormal, and τ is the torsion.

We assume that the curve is given by the vector-valued function $t \to x(t)$. The arc length s is given by

$$s = \int_0^t \langle \dot{x}(t), \dot{x}(t) \rangle^{1/2} dt$$

where the inner product is that induced from $N = \mathbb{R}^3$. Since $\langle \mathbf{t}, \mathbf{t} \rangle = 1$ we have $\langle \nabla_{d/ds}\mathbf{t}, \mathbf{t} \rangle = 0$. Hence we can write

$$\nabla_{\frac{d}{ds}}\mathbf{t} = \kappa\mathbf{n}$$

where $\mathbf{n}$ is defined to be the unit principal normal. This definition makes sense provided that $\kappa \neq 0$. If $\kappa = 0$ for values of s in an interval, then the curve is a straight line, and the corresponding principal normal is indeterminate. If $\kappa = 0$ at an isolated point, we can define $\mathbf{n}$ at that point by a continuity argument. We choose a frame field (e_1, e_2, e_3) over $\mathbb{R}^3$ whose restriction to the curve satisfies $e_1 = \mathbf{t}$, $e_2 = \mathbf{n}$, $e_3 = \mathbf{b}$. We have

$$\nabla e_1 = \omega_1^2 e_2 + \omega_1^3 e_3$$

from which we deduce that $\omega_1^3 = 0$, $\omega_1^2 = \kappa\, ds$. Then

$$\nabla e_2 = \omega_2^1 e_1 + \omega_2^3 e_3,$$
$$\nabla e_3 = \omega_3^1 e_1 + \omega_3^2 e_2 = -\omega_2^3 e_2.$$

We write $\alpha_2^3 = \tau \, ds$ and the Frenet–Serret equations follow immediately from the standard equations

$$\nabla e_i = \omega_i^j e_j, \qquad \omega_j^i + \omega_i^j = 0.$$

The general existence theorem mentioned at the end of Section 4.2 becomes the following.

Theorem (4.4.1) Let $\kappa(s)$, $\tau(s)$ be given smooth functions of the real variable s where $0 \leqslant s \leqslant c$. Then there exists a curve in $\mathbb{R}^3$ for which s is the arc length, κ is the curvature, and τ is the torsion. Moreover such a curve is uniquely determined to within a Euclidean motion of $\mathbb{R}^3$.

The existence of such a curve depends upon Picard's theorem for the existence of solutions of a system of differential equations of the Frenet–Serret type. The uniqueness is easily proved as follows. Suppose that C_1, C_2 are two curves in $\mathbb{R}^3$ determined by the data s, $\kappa(s)$, and $\tau(s)$. By means of a Euclidean motion we can arrange that C_1 intersects C_2 at some point p and by a suitable orthogonal transformation we can assume that, at p, we have

$$\mathbf{t}_1 = \mathbf{t}_2, \qquad \mathbf{n}_1 = \mathbf{n}_2, \qquad \mathbf{b}_1 = \mathbf{b}_2.$$

Making use of the hypothesis that $\kappa_1(s) = \kappa_2(s)$, $\tau_1(s) = \tau_2(s)$ we have

$$\frac{d}{ds}(\langle t_1, t_2\rangle + \langle n_1, n_2\rangle + \langle b_1, b_2\rangle)$$
$$= \kappa_1\langle n_1, t_2\rangle + \kappa_1\langle t_1, n_2\rangle + \tau_1\langle b_1, n_2\rangle - \kappa_1\langle t_1, n_2\rangle + \tau_1\langle n_1, b_2\rangle$$
$$- \kappa_1\langle n_1, t_2\rangle - \tau_1\langle n_1, b_2\rangle - \tau_1\langle n_2, b_1\rangle = 0,$$

and hence the expression has the constant value 3 along the curve. Since $\langle t_1, t_2\rangle \leqslant 1$, this implies that $t_1 = t_2$, $n_1 = n_2$, and $b_1 = b_2$ for all values of s and hence the two curves coincide. The theorem implies that there are essentially only two independent differential invariants of curves in $\mathbb{R}^3$, namely the curvature and the torsion.

The reader should not dismiss as trivial the study of curves in $\mathbb{R}^3$. For example we have the following theorem due to Fenchel (1929) and to Fary (1949).

Theorem (4.4.2) For any closed space curve C we have

$$\int_C |\kappa| \, ds \geqslant 2\pi,$$

4.4 The theory of curves

equality holding only when C is the boundary of a convex set in $\mathbb{R}^2$.

A really fascinating theorem was proved by J. Milnor (1950) when he was an undergraduate.

Theorem (4.4.3) Let C be a closed space curve in $\mathbb{R}^3$ such that

$$\int_C |\kappa|\, ds < 4\pi.$$

Then the curve C is unknotted.

We now mention an unsolved problem concerning closed curves in $\mathbb{R}^3$. Clearly if C is closed and of length l, then $\kappa(s)$ and $\tau(s)$ are periodic functions of period l. The problem is to determine necessary and sufficient conditions on the functions $\kappa(s)$ and $\tau(s)$ which will ensure that the corresponding curve C is closed. That periodicity is not sufficient is shown by the example of a circular helix.

The following theorem due to Totaro (1990) shows that properties of curves in E^3 still attract research workers.

Theorem (4.4.4) Let C be a closed curve in E^3 with nowhere vanishing curvature κ and torsion τ. Then

$$\int_C \sqrt{(\kappa^2 + \tau^2)}\, ds > 4\pi.$$

Let us now consider the case when N is a general Riemannian manifold. Let $t \to x^i(t)$ be a representation of the curve in some local coordinate system. Then the arc length s is given by

$$s = \int \langle x(t), x(t) \rangle^{1/2}\, dt = \int (g_{ij}\dot{x}^i\dot{x}^j)^{1/2}\, dt$$

where $g_{ij}(t)$ is the induced metric on the curve. We see that the tangent vector $t = d/ds$ is a unit vector along the curve and hence $\nabla_{d/ds} t$ is perpendicular to t. We set

$$\nabla_{d/ds} t = \kappa_g n_g$$

where $\mathbf{n}_g$ is a unit normal vector defined by this equation except when $\kappa_g = 0$. In that case the curve is a geodesic since this is the condition for self-parallelism. We see that κ_g measures the deviation of a curve from being a geodesic. The function κ_g is called the *geodesic curvature*. In Section (4.7) we shall use the concept of geodesic curvature to prove one of the most important theorems in differential geometry, namely the Gauss-Bonnet theorem.

4.5 THE THEORY OF SURFACES

In this section we illustrate the previous general theory in the special case when M is a two-dimensional manifold and N is $\mathbb{R}^3$ equipped with a Euclidean metric. We consider adapted orthonormal frames (e_1, e_2, e_3) on M with e_1, e_2 tangent to M and e_3 along the outer surface normal so that (e_1, e_2, e_3) is a right-handed trihedral. We have the Cartan equations

$$d\omega^1 + \omega^1_2 \wedge \omega^2 + \omega^1_3 \wedge \omega^3 = 0, \qquad \nabla e_1 = \omega^2_1 e_2 + \omega^3_1 e_3,$$
$$d\omega^2 + \omega^2_1 \wedge \omega^1 + \omega^2_3 \wedge \omega^3 = 0, \qquad \nabla e_2 = \omega^1_2 e_1 + \omega^3_2 e_3,$$
$$d\omega^3 + \omega^3_1 \wedge \omega^1 + \omega^3_2 \wedge \omega^2 = 0, \qquad \nabla e_3 = \omega^1_3 e_1 + \omega^2_3 e_2,$$

which simplify when restricted to M since $\omega^3 = 0$. Then

$$d\omega^1 = -\omega^1_2 \wedge \omega^2,$$
$$d\omega^2 = -\omega^2_1 \wedge \omega^1.$$

Writing

$$\omega^3_1 = h^3_{11}\omega^1 + h^3_{12}\omega^2,$$
$$\omega^3_2 = h^3_{21}\omega^1 + h^3_{22}\omega^2,$$

and using $d\omega^3 = 0$ we get

$$h^3_{12}\omega^2 \wedge \omega^1 + h^3_{21}\omega^1 \wedge \omega^2 = 0,$$

giving

$$h^3_{12} = h^3_{21}.$$

The second fundamental tensor $A^3 = A_{e_3}$ is given by

$$-A_{e_3}X = \text{tangential part of } \nabla_X e_3$$
$$= \omega^1_3(X)e_1 + \omega^2_3(X)e_2$$
$$= -\omega^3_1(X)e_1 - \omega^3_2(X)e_2$$
$$= -(h^3_{11}\omega^1 + h^3_{12}\omega_2)(X)e_1 - (h^3_{21}\omega^1 + h^3_{22}\omega_1)(X)e_2.$$

The matrix representation of A_{e_3} is

$$\begin{bmatrix} h^3_{11} & h^3_{12} \\ h^3_{21} & h^3_{22} \end{bmatrix}.$$

By a change of adapted frame we may diagonalize the above matrix so that it assumes the form

$$\begin{bmatrix} r_1 & 0 \\ 0 & r_2 \end{bmatrix}.$$

4.5 The theory of surfaces

Then r_1, r_2 are called the *principal curvatures* and the corresponding directions of the basis vectors are called the *principal directions*. With this choice of moving frames we have

$$\omega_1^3 = r_1 \omega^1, \qquad \omega_2^3 = r_2 \omega^2, \tag{4.27}$$

since we can now take $h_{12}^3 = h_{21}^3 = 0$.

We take the exterior derivative of the equation

$$\omega_1^3 = r_1 \omega^1$$

to get

$$d\omega_1^3 = dr_1 \wedge \omega^1 + r_1 d\omega^1.$$

Now use the structure equations to substitute for $d\omega_1^3$ and $d\omega^1$, to get

$$-\omega_2^3 \wedge \omega_1^2 = dr_1 \wedge \omega^1 + r_1 \omega_1^2 \wedge \omega^2,$$

that is,

$$dr_1 \wedge \omega^1 = (r_2 - r_1)\omega_1^2 \wedge \omega^2. \tag{4.28}$$

Similarly we may obtain

$$dr_2 \wedge \omega^2 = (r_1 - r_2)\omega_1 \wedge \omega_1^2. \tag{4.29}$$

Equations (4.28) and (4.29) are none other than the Codazzi equations and could have been written down immediately by substituting in the general formula (4.11).

We now write

$$\omega_1^2 = h\omega^1 + k\omega^2,$$

$$dh = h_1 \omega^1 + h_2 \omega^2,$$

$$dk = k_1 \omega^1 + k_2 \omega^2,$$

and consider the structural equation for $\mathbb{R}^3$ with Euclidean metric,

$$d\omega_1^2 + \omega_A^2 \wedge \omega_1^A = \tilde{\Omega}_1^2 = 0, \qquad A = 1, 2, 3.$$

This gives

$$(h_1 \omega^1 + h_2 \omega^2) \wedge \omega^1 + h d\omega^1 + (k_1 \omega^1 + k_2 \omega^2) \wedge \omega^2 + k d\omega^2$$
$$- r_2 \omega^2 \wedge r_1 \omega^1 = 0.$$

That is

$$h_2 - k_1 = r_1 r_2 + h^2 + k^2. \tag{4.30}$$

The Codazzi equations can now be expressed in terms of h and k. Altogether we obtain the following equations which play a fundamental part in the theory of surfaces in Euclidean space.

$$d\omega^1 = h\omega^1 \wedge \omega^2, \quad d\omega^2 = k\omega^1 \wedge \omega^2,$$
$$dr_1 \wedge \omega^1 = h(r_2 - r_1)\omega^1 \wedge \omega^2,$$
$$dr_2 \wedge \omega^2 = k(r_1 - r_2)\omega^1 \wedge \omega^2, \tag{4.31}$$
$$r_1 r_2 = h_2 - k_1 - h^2 - k^2.$$

The product of the principal curvatures
$$K = r_1 r_2 \tag{4.32}$$
is called the *Gaussian curvature* of the surface. The arithmetic mean of the principal curvatures
$$H = \frac{1}{2}(r_1 + r_2) \tag{4.33}$$
is called the *mean curvature* of the surface. It is the magnitude of the mean curvature vector which in this case is He_3.

The following theorem is a profound discovery of Gauss which had considerable influence on the development of differential geometry. Gauss referred to it as 'Theorema egregium', meaning a most excellent theorem.

Theorem (4.5.1) Let M_1, M_2 be two surfaces in Euclidean 3-space which are both diffeomorphic and isometric. Then the Gauss curvature K is the same at corresponding points.

This means that although K was defined extrinsically, that is, in terms of the imbedding in $\mathbb{R}^3$, in fact it involves only the metric properties of the surface and therefore is an *intrinsic invariant*.

To prove the theorem we evaluate the curvature 2-form Ω_1^2 on the pair (e_1, e_2) and obtain an expression which must depend only on the Riemannian structure. We have

$$\Omega_1^2 = d\omega_1^2 = d(h\omega^1 + k\omega^2)$$
$$= dh \wedge \omega^1 + h d\omega^1 + dk \wedge \omega^2 + k d\omega^2$$
$$= h_2 \omega^2 \wedge \omega^1 + h\omega_1^2 \wedge \omega^2 + k_1 \omega^1 \wedge \omega^2 - k\omega_1^2 \wedge \omega^1$$
$$= (-h_2 + h^2 + k_1 + k^2)\omega^1 \wedge \omega^2.$$

Then
$$\Omega_1^2(e_1, e_2) = h^2 + k^2 + k_1 - h_2.$$

From the last equation of (4.31) we see that
$$\Omega_1^2 = -K,$$
showing that K is a Riemannian invariant and thus proving the theorem.

We now consider the 'spherical mapping' of M, sometimes called the

4.5 The theory of surfaces

normal Gauss map, which is defined as follows. To each point $p \in M$ we make correspond the unit normal vector e_3. This maps the surface M into the unit sphere $S^2(1)$, the point of $S^2(1)$ which corresponds to $p \in M$ being the 'end point' of the unit vector obtained by transporting e_3 by parallelism in $\mathbb{R}^3$ to some convenient origin.

For our specially adapted frames with e_1, e_2 along principal curvature directions we have

$$\nabla e_3 = \omega_3^1 e_1 + \omega_3^2 e_2$$
$$= -\omega_1^3 e_1 - \omega_2^3 e_2$$
$$= -r_1 \omega^1 e_1 - r_2 \omega^2 e_2.$$

The volume element at p is $\omega^1 \wedge \omega^2$ whilst the corresponding volume element on $S^2(1)$ is $r_1 \omega^1 \wedge r_2 \omega^2 = K \omega^1 \wedge \omega^2$.

Thus we have shown the following result.

Theorem (4.5.2) At any point p of a surface M we have

$$d\sigma = |K|\, dS \tag{4.34}$$

where dS, $d\sigma$ are, respectively, elements of area of M and its spherical image under the Gauss map, and K is the Gaussian curvature of M at p.

To illustrate the fact that two surfaces of quite different appearance in $\mathbb{R}^3$ can be locally isometric we consider the helicoid and the catenoid. The helicoid is defined in terms of Cartesian coordinates in $\mathbb{R}^3$ by

$$(u, r) \to (u \cos r, u \sin r, v) \qquad u > 0, -\infty < v < \infty.$$

Its shape resembles that of a spiral staircase.
The catenoid is given by

$$(z, \theta) \to (\cosh z \cos \theta, \cosh z \sin \theta, z), \qquad -\infty < z < \infty, 0 < \theta < 2\pi,$$

and has a shape resembling that of a cooling tower.
The metric of the helicoid is

$$du^2 + (1 + u^2)\, dr^2$$

while that of the catenoid is

$$\cosh^2 z\, dz^2 + \sinh^2 z\, d\theta^2.$$

The *local* isometry between the surfaces is given by

$$v = \theta, \qquad u = \sinh z$$

from which it can be seen that the whole catenoid is isometric to just one complete spiral of the helicoid.

The curvatures r_1 and r_2 are the extremal values of the curvature of normal sections at p. We can choose frames such that $r_1 \geq r_2$ with r_1

the curvature in the direction e_1. If at a point p the two principal curvatures are equal, the point is called an *umbilic*. At such a point the principal directions are indeterminate. We now prove a useful lemma.

Lemma (4.5.3) *Let M have the property that every point is an umbilic. Then M is part of a plane or a sphere.*

To prove the lemma, we note from Codazzi's equations, on putting $r_1 = r_2 = r$, that we have

$$dr \wedge \omega^1 = 0, \qquad dr \wedge \omega^2 = 0.$$

Writing $dr = a\omega^1 + b\omega^2$ we see that the previous equations force $a = 0$ and $b = 0$ and hence r takes the same value at all points. Then

$$\nabla e_3 = \omega_3^1 e_1 + \omega_3^2 e_2 = -r(\omega_1 e_1 + \omega_2 e_2) = -r\, dp.$$

If $r = 0$, then e_3 is constant and the surface is part of a plane. If $r =$ constant k and $k \neq 0$, then $dp = -\frac{1}{k}\nabla e_3$ from which

$$p - c = -\frac{e_3}{k}$$

for some constant vector c. Then

$$(p - c) \cdot (p - c) = \frac{1}{k^2}$$

and M is part of a sphere. Thus the lemma is proved.

The surface M is called *convex* if $K = r_1 r_2$ is everywhere positive. It is called a *Weingarten surface* (W-surface) if dr_1 and dr_2 are linearly dependent, that is, if functions λ_1, λ_2 exist, not both zero, such that

$$\lambda_1 dr_1 + \lambda_2 dr_2 = 0. \tag{4.35}$$

This implies either that both r_1 and r_2 are constants, or that they are connected by a functional relation of the type

$$F(r_1, r_2) = 0. \tag{4.36}$$

We shall say that the W-surface is *special* if the functions λ_i in (4.35) can be chosen to be positive: $\lambda_i > 0$, $i = 1, 2$. Unless r_1, r_2 are constant, this implies that one of the principal curvatures is a strictly monotone increasing function of the other. Examples of such special W-surfaces are those for which $K =$ constant $= r_1 r_2 > 0$, those for which $H = \frac{1}{2}(r_1 + r_2) =$ constant, and so on. We now prove the following.

Theorem (4.5.4) *A convex special W-surface is a sphere.*

As corollaries we have two results.

4.5 The theory of surfaces

Corollary (4.5.5) A closed surface of constant Gauss curvature is a sphere.

Corollary (4.5.6) A convex surface of constant mean curvature is a sphere.

To prove Theorem (4.5.4) we proceed as follows. Let O be a point on M at which the continuous function r_1 attains its maximum value. Suppose that O is an umbilic and denote the normal curvature at O by r_0. Let r_1 and r_2 be the principal curvatures at any other point of M. Then $r_0 \geq r_1 \geq r_2$. However, since r_2 is either a monotone decreasing function of r_1 or else a constant, we must have $r_0 \leq r_2$. Combining the two sets of inequalities we get $r_0 = r_1 = r_2$. Hence every point on M is an umbilic and hence M must be a sphere.

Suppose now that O is not an umbilic, and consider some neighbourhood of O which is free from umbilics. Then we have

$$dr_1 = \lambda \, dr_2, \qquad \lambda < 0.$$

From the Codazzi equations (4.28) and (4.29) we get

$$dr_1 \wedge \omega^1 = h(r_2 - r_1)\omega^1 \wedge \omega^2,$$
$$dr_1 \wedge \omega^2 = \lambda \, dr_2 \wedge \omega^2 = \lambda k(r_1 - r_2)\omega^1 \wedge \omega^2.$$

Writing

$$dr_1 = (r_1 - r_2)(a\omega^1 + b\omega^2)$$

we get

$$-(r_1 - r_2)b = h(r_2 - r_1), \qquad \text{so} \qquad b = h,$$
$$(r_1 - r_2)a = \lambda k(r_1 - r_2), \qquad \text{so} \qquad a = \lambda k.$$

Thus

$$dr_1 = (r_1 - r_2)(\lambda k \omega^1 + h\omega^2).$$

Since r_1 attains its maximum at O we have, at O, $h = k = 0$. At O the second differential is

$$d^2r_1 = (r_1 - r_2)(\lambda \, dk \, \omega^1 + dh \, \omega^2)|_0.$$

Write

$$dk = k_1\omega^1 + k_2\omega^2,$$
$$dh = h_1\omega^1 + h_2\omega^2.$$

The condition that r_1 attains its maximum at O gives, at O,

$$\lambda k_1 \leq 0, \qquad h_2 \leq 0,$$

or, since $\lambda < 0$ we have $k_1 \geq 0$, $h_2 \leq 0$.

Now use these results in the last equation of (4.31) to get $r_1 r_2 \leq 0$, giving a contradiction to the convexity hypothesis. Thus the previous case must hold and M must be a sphere.

For many years it seemed likely that a closed analytic surface of constant mean curvature must be a sphere. However recently it was proved by Wente (1986) that there exists a torus in $\mathbb{R}^3$ of constant mean curvature. However, this torus is immersed but not embedded in $\mathbb{R}^3$.

Let C be a curve on M parametrized by arc length s, the surface M being a submanifold of $\mathbb{R}^3$. Apply the Gauss equation with tangent vectors $X = Y = d/ds$ to get

$$\tilde{\nabla}_{\frac{d}{ds}} \left(\frac{d}{ds} \right) = \nabla_{\frac{d}{ds}} \left(\frac{d}{ds} \right) + h\left(\frac{d}{ds}, \frac{d}{ds} \right)$$

which is equivalent to

$$\kappa n = \kappa_g n_g + h\left(\frac{d}{ds}, \frac{d}{ds} \right).$$

This follows because the right-hand side is just κn from the Frenet–Serret formula. The tangential component is just the geodesic curvature of C, determined by C and the induced Riemannian metric on M. Thus the geometrical interpretation of the geodesic curvature vector at a point p of a curve on a surface M in $\mathbb{R}^3$ is none other than the orthogonal projection of the normal curvature vector of C at p, considered as a space curve of $\mathbb{R}^3$. In the case of a geodesic, the principal normal to C must lie along the surface normal at all points on C.

Let us denote the Gauss map of a surface S in E by $g: S \to \Sigma$ where Σ is the unit sphere, given by $g: p \to N(p)$ where $N(p)$ is the unit normal at p chosen to determine the orientation. Then equation (4.34) may be written

$$|K| = |\det(dg_p)| = \lim_{U \to p} \frac{\text{Area } g(U)}{\text{Area } U}.$$

This approach to curvature is the one taken by Gauss in his fundamental paper on surfaces (see Dombrowski (1979)). In fact, it appears that the Gauss map had been considered by Rodrigues (1814–16) many years before Gauss. Rodrigues observed that the differential of the Gauss map is diagonalized by the principal curvatures. More precisely, if $x(t)$ defines a curve C on the surface and $N(t)$ is the unit normal along C, then $N'(t) = \lambda x'(t)$ for some real λ if and only if the plane spanned by $x'(t)$ and $N(t)$ intersects S in a curve whose curvature is κ_a or κ_b.

Similarly for a hypersurface in E^n, we have orthonormal vectors $e_1, \ldots, e_{n-1}$ as principal curvature directions at p with corresponding

4.6 Surface theory in classical notation

curvatures $\kappa_1, \ldots, \kappa_{n-1}$. Just as in E^3, we have the Gauss map $g: p \to N(p)$ whose differential

$$dg_p : x'(t) - N'(t), \qquad (x(t) = p)$$

again satisfies Rodrigues' equation

$$dg_p(e_i) = \kappa_i e_i \qquad \text{(not summed)}.$$

We have the Gauss–Kronecker curvature

$$\kappa_1 \kappa_2 \cdots \kappa_{n-1} = (-1)^{n-1} \det(d\gamma_p),$$

and the mean curvature

$$H - \frac{1}{n-1}(\kappa_1 + \cdots + \kappa_{n-1}) - -\frac{1}{(n-1)} \operatorname{tr}(dg_p).$$

Exercise (4.5.7) If we denote by H_j the jth-order mean curvature of a hypersurface in E^n, normalized so that

$$\prod_{j=1}^{n-1}(1 + tk_j) = \sum_{j=0}^{n-1} \binom{n-1}{j} H_j t^j$$

prove that H_2 and the scalar curvature τ of the hypersurface considered as a Riemannian manifold with metric induced from that of E^n are related by

$$\tau = (n-1)(n-2)H_2.$$

4.6 SURFACE THEORY IN CLASSICAL NOTATION

So far in this chapter we have used tensor calculus and the Cartan calculus for dealing with submanifolds, including surfaces in E^3. However, for surfaces in E^3 we could have used vector methods which are not available in E^n with $n > 3$. In fact if we were only interested in curves and surfaces living in E^3, it could be argued that we have been using a steam-roller to crack a nut. We now give a brief account of surface theory in E^3 using vector methods.

Locally a surface in E^3 is given by a vector-valued map $\mathbb{R}^2 \to \mathbb{R}^3$: $(u, v) \to \mathbf{r}(u, v)$, and we regard $\mathbf{r}(u, v)$ as the position vector of a point P on the surface. We denote $\partial \mathbf{r}/\partial u$ by $\mathbf{r}_1$, $\partial \mathbf{r}/\partial v$ by $\mathbf{r}_2$, $\partial^2 \mathbf{r}/\partial u \partial v$ by $\mathbf{r}_{12}, \ldots$. Clearly $\mathbf{r}_1$ and $\mathbf{r}_2$ are tangent vectors to the surface at P. We assume that $\mathbf{r}_1$ and $\mathbf{r}_2$ are linearly independent at P, that is, we assume that the surface is *regular* at P. We deal only with *regular surfaces*, that is, surfaces regular at all points.

The tangent plane to S at P is spanned by $\mathbf{r}_1$ and $\mathbf{r}_2$. The inner product induced from E^3 is given by

$$(d\mathbf{r})^2 = (\mathbf{r}_1\,du + \mathbf{r}_2\,dv)^2 = E\,du^2 + 2F\,du\,dv + G\,dv^2,$$

where $E = \mathbf{r}_1^2$, $F = \mathbf{r}_1 \cdot \mathbf{r}_2$, $G = \mathbf{r}_2^2$.

The unit normal $\mathbf{N}$ at P is given by

$$\mathbf{N} = \frac{\mathbf{r}_1 \times \mathbf{r}_2}{|\mathbf{r}_1 \times \mathbf{r}_2|}.$$

By using the classical identity

$$(\mathbf{r}_1 \times \mathbf{r}_2) \cdot (\mathbf{r}_1 \times \mathbf{r}_2) = (\mathbf{r}_1 \cdot \mathbf{r}_1)(\mathbf{r}_2 \cdot \mathbf{r}_2) - (\mathbf{r}_1 \cdot \mathbf{r}_2)(\mathbf{r}_1 \cdot \mathbf{r}_2)$$

$$= EG - F^2,$$

and writing $H = \sqrt{(EG - F^2)}$ we see that

$$\mathbf{r}_1 \times \mathbf{r}_2 = H\mathbf{N}.$$

The normal curvature of S at P in the direction of the unit vector $l\mathbf{r}_1 + m\mathbf{r}_2$ (where $El^2 + 2Flm + Gm^2 = 1$) is obtained as follows. We take the curve $s \to \mathbf{r}(s)$ through P parametrized by arc length s, such that at P we have

$$\mathbf{r}' = l\mathbf{r}_1 + m\mathbf{r}_2, \qquad l = u', \; m = v'.$$

Then

$$\mathbf{r}'' = l'\mathbf{r}_1 + m'\mathbf{r}_2 + l(\mathbf{r}_{11}l + \mathbf{r}_{12}m) + m(\mathbf{r}_{21}l + \mathbf{r}_{22}m)$$

so that

$$\kappa\mathbf{n} = (l'\mathbf{r}_1 + m'\mathbf{r}_2) + \mathbf{r}_{11}l^2 + 2\mathbf{r}_{12}lm + \mathbf{r}_{22}m^2.$$

The projection along the normal $\mathbf{N}$ is given by

$$\kappa_n = (\mathbf{r}_{11} \cdot \mathbf{N})l^2 + 2(\mathbf{r}_{12} \cdot \mathbf{N})lm + (\mathbf{r}_{22} \cdot \mathbf{N})m^2.$$

Writing $(\mathbf{r}_{11} \cdot \mathbf{N}) = L$, $(\mathbf{r}_{12} \cdot \mathbf{N}) = M$, $(\mathbf{r}_{22} \cdot \mathbf{N}) = N$ we get

$$\kappa_n = Ll^2 + 2Mlm + Nm^2.$$

The matrix

$$\begin{pmatrix} L & M \\ M & N \end{pmatrix}$$

is that of the second fundamental form. As l, m vary subject to $El^2 + 2Flm + Gm^2 = 1$, the normal curvature will vary. Its extreme values may be found by making use of Lagrange's multipliers. Write

$$\kappa = Ll^2 + 2Mlm + Nm^2 - \lambda(El^2 + 2Flm + Gm^2 - 1).$$

Then when κ is stationary we have

4.6 Surface theory in classical notation

$$\frac{1}{2}\frac{\partial \kappa}{\partial l} = Ll + Mm - \lambda El - \lambda Fm = 0,$$

$$\frac{1}{2}\frac{\partial \kappa}{\partial m} = Ml + Nm - \lambda Fl - \lambda Gm = 0.$$

Multiply the first equation by l, the second by m, and add. This gives $\lambda = \kappa$. Now eliminate l and m from the two equations to get

$$\kappa^2(EG - F) - \kappa(EN + GL - 2FM) + (LN - M^2) = 0.$$

The roots κ_1, κ_2 of this equation are the principal curvatures. The mean curvature is defined by

$$\mu - \frac{1}{2}(\kappa_1 + \kappa_2) = \frac{EN + GL - 2FM}{2(EG - F^2)},$$

and the Gauss curvature K given by

$$K = \kappa_1\kappa_2 = \frac{LN - M^2}{EG - F^2}.$$

The principal directions corresponding to the principal curvatures are obtained by eliminating λ from the two previous equations to give

$$(EM - FL)l^2 + (EN - GL)lm + (FN - GM)m^2 = 0.$$

The discriminant of this quadratic form can be written

$$4\left[\frac{EG - F^2}{E^2}\right](EM - FL)^2 + \left\{(EN - GL) - \frac{2F}{E}(EM - FL)\right\}^2.$$

It follows that the roots of the equation are real and different provided that we do *not* have the relations

$$\frac{L}{E} = \frac{M}{F} = \frac{N}{G}.$$

Where this relation holds the principal directions are indeterminate and the normal curvature is the same in all directions. Such a point is called an *umbilic*.

Two directions $l\mathbf{r}_1 + m\mathbf{r}_2$, $l'\mathbf{r}_1 + m'\mathbf{r}_2$ will be orthogonal if and only if

$$Ell' + F(lm' + l'm) + Gmm' = 0.$$

It follows that the two principal directions will be orthogonal if and only if

$$E(FN - GL) - F(EN - GL) + G(EM - FL) = 0$$

which is clearly satisfied. By choosing lines of curvature as parametric curves computations become simpler since we have $F = 0$, $M = 0$.

The equations of Weingarten when expressed in terms of E, F, G, L, M, N are

$$H^2\mathbf{N}_1 = (FM - GL)\mathbf{r}_1 + (FL - EM)\mathbf{r}_2,$$
$$H^2\mathbf{N}_2 = (FN - GM)\mathbf{r}_1 + (FM - EN)\mathbf{r}_2.$$

To obtain these equations *ab initio* we write

$$\mathbf{N}_1 = a\mathbf{r}_1 + b\mathbf{r}_2, \qquad \mathbf{N}_2 = c\mathbf{r}_1 + d\mathbf{r}_2$$

since both vectors lie in the tangent plane. Multiplying both equations by $\mathbf{r}_1, \mathbf{r}_2$ and using $L = \mathbf{r}_{11} \cdot \mathbf{N} = -\mathbf{r}_1 \cdot \mathbf{N}_1$, and so on, we get four equations for a, b, c, d and these lead to the given equations.

We note that

$$H^4\mathbf{N}_1 \times \mathbf{N}_2 = [(FM - GL)(FM - EN) - (FL - EM)(FN - GM)]H\mathbf{N}$$

and this simplifies to

$$H\mathbf{N}_1 \times \mathbf{N}_2 = (LN - M^2)\mathbf{N}.$$

4.7 THE GAUSS–BONNET THEOREM

In this section we deal with one of the deepest theorems in the differential geometry of two-dimensional Riemannian manifolds. The theorem itself is intrinsic in the sense that it depends only on the Riemannian metric and is independent of considerations of whether the manifold is a submanifold of another Riemannian manifold. However, the theorem involves one or more curves on the manifold and these can be regarded as one-dimensional submanifolds of the two-dimensional manifold, so we find it convenient to deal with it in the present chapter.

The geometrical arguments involved in the theorem and its proof are straightforward, and the difficulties arise from topological questions which we shall describe but refer the reader elsewhere for a more detailed treatment including rigorous proofs.

The first statement of the theorem, due to Gauss, deals with geodesic triangles on surfaces (that is, triangles whose sides are geodesics).

Theorem (4.7.1) (Local Gauss–Bonnet: version 1) *Let T be a geodesic triangle on a two-dimensional Riemannian manifold, with interior angles $\alpha_1, \alpha_2, \alpha_3$. Then*

$$\alpha_1 + \alpha_2 + \alpha_3 - \pi = \int_T K \, dS, \tag{4.37}$$

4.7 The Gauss–Bonnet theorem

where K is the Gaussian curvature and the integral is taken over the interior of the triangle T.

Thus the theorem states that the excess of the sum of the interior angles over π is equal to the integral of the Gaussian curvature over the triangle.

To prove the theorem we assume that the triangle T lies inside a domain that can be covered with an orthonormal frame (e_1, e_2). We consider the angle between the unit tangent vector at points different from a vertex, and the vector e_1, and see how this changes along the geodesic arc. We have

$$\nabla_{\frac{d}{ds}} \left\langle e_1, \frac{d}{ds} \right\rangle = \left\langle \nabla_{\frac{d}{ds}} e_1, \frac{d}{ds} \right\rangle + \left\langle e_1, \nabla_{\frac{d}{ds}} \frac{d}{ds} \right\rangle$$

$$= \left\langle \omega_1^2 \left(\frac{d}{ds} \right) e_2, \frac{d}{ds} \right\rangle + 0,$$

the last term vanishing since the arc is a geodesic. Let ϕ denote the angle between d/ds and e_1 at a non-vertex. Then

$$\frac{d}{ds}(\cos\phi) = \omega_1^2 \left(\frac{d}{ds} \right) \cdot \sin\phi$$

giving

$$d\phi = -\omega_1^2 \qquad (4.38)$$

along the geodesic arc. It seems intuitively obvious that as the point p traverses the whole triangle, the total angle turned through must be precisely 2π. Assuming this to be the case we have

$$-\int_{\partial T} \omega_1^2 + \sum_{i=1}^{3} (\pi - \alpha_i) = 2\pi$$

giving

$$-\int_{\partial T} \omega_1^2 = \alpha_1 + \alpha_2 + \alpha_3 - \pi.$$

However, from Stokes' theorem we have

$$\int_{\partial T} \omega_1^2 = \int_T d\omega_1^2 = -\int_T K\omega^1 \wedge \omega^2.$$

Thus we get

$$\alpha_1 + \alpha_2 + \alpha_3 - \pi = \int_T K\omega^1 \wedge \omega^2 = \int_T K\,dS,$$

and the theorem has been proved, modulo certain assumptions.

4 Submanifold theory

We now pass on to a more general result which is Bonnet's modification of the original version.

Theorem (4.7.2) (Local Gauss–Bonnet: version 2) *Let T be a triangle on a two-dimensional Riemannian manifold with interior angles $\alpha_1, \alpha_2, \alpha_3$. Then*

$$\alpha_1 + \alpha_2 + \alpha_3 - \pi = \int_{\partial T} \kappa_g \, ds + \int_T K \, dS \qquad (4.39)$$

where κ_s is the geodesic curvature (this is well defined at all points on the sides of the triangle except at the vertices).

We see that in version 2, the fact that the arcs are no longer required to be geodesics produces the extra term on the right-hand member. The proof is similar to that above except that we now have

$$\frac{d}{ds}(\cos\phi) = \left\langle \omega_1^2 \left(\frac{d}{ds}\right) e_2, \frac{d}{ds} \right\rangle + \langle e_1, \kappa_g n_g \rangle$$

$$= \omega_1^2 \left(\frac{d}{ds}\right) \cdot \sin\phi - \kappa_g \sin\phi,$$

and hence

$$d\phi = \kappa_g \, ds - \omega_1^2. \qquad (4.40)$$

The same argument as before now gives (4.39) and the theorem is proved.

The weak point in the previous argument is the assertion that the total angle turned through is 2π. This assertion is indeed correct, but its proof is quite sophisticated and depends on a number of non-trivial theorems which we now mention. The first is known as *the theorem of turning tangents*.

Let C be a curve in the Euclidean plane $\mathbb{R}^2$ given by

$$s \to X(s)$$

where s is the arc length and X is an $\mathbb{R}^2$-valued smooth function. The curve is *closed* if $X(s)$ is periodic of period L, where L is the length of C. The curve C is *simple* if $X(s_1) \neq X(s_2)$ when $0 < |s_1 - s_2| < L$. We assume now that C is an oriented closed curve of length L. Let O be a fixed point in the plane. Denote by Γ the unit circle about O. We define the *tangential map* $T: C \to \Gamma$ as that which maps a point p of C to the end point of the unit vector through O parallel to the tangent vector to C at p. As the point p moves round C once, its image goes round Γ an integral number of times. The number is called the *rotation index* of C. Then the result in question can be stated as follows.

4.7 The Gauss–Bonnet theorem

Theorem (4.7.3) *The rotation index of a simple closed curve is* ± 1.

For a rigorous proof of this theorem the reader is referred to Chern (1967).

We now consider *sectionally smooth curves*. Such a curve consists of a finite number of smooth arcs $A_0A_1, A_1A_2, \ldots, A_{m-1}A_m$ where the tangent vectors of the two arcs at the common vertex A_i, $i = 1, \ldots, m-1$, are different in general. The curve is closed if $A_m = A_0$. An example of a sectionally smooth curve is a closed rectilinear polygon. The concept of rotation index and the theorem of turning tangents can be extended to apply to this more general class of closed curves.

We now introduce a concept which we shall describe in greater detail in later chapters, namely, conformally related Riemannian metrics. Let the two-dimensional manifold M possess two Riemannian metrics g and g' which are related by

$$g' = e^{2\lambda} g \tag{4.41}$$

where λ is a smooth real-valued function on M. Then the metrics are said to be *conformally related*.

Let $t \to C_1(t)$, $t \to C_2(t)$ be two smooth curves intersecting at $t = t_0$. Let α, α' be the angles between the curves at t_0 as measured by the metrics g, g', respectively. Then

$$\cos \alpha = \frac{\langle \dot{C}_1(t_0), \dot{C}_2(t_0) \rangle_g}{|\dot{C}_1(t_0)|_g |\dot{C}_2(t_0)|_g},$$

$$\cos \alpha = \frac{\langle \dot{C}_1(t_0), \dot{C}_2(t_0) \rangle_{g'}}{|\dot{C}_1(t_0)|_{g'} |\dot{C}_2(t_0)|_{g'}}$$

Thus

$$\cos \alpha' = e^{2\lambda(t_0)} \cdot \cos \alpha \cdot e^{-\lambda(t_0)} \cdot e^{-\lambda(t_0)}$$

$$= \cos \alpha$$

so that the angle between the curves is preserved at least up to sign. On the other hand, distances are in general not preserved.

We now appeal to an existence theorem for whose delicate proof the reader is referred to a research paper by Chern (1955).

Theorem (4.7.4) *For any smooth two-dimensional Riemannian manifold it is always possible to choose local coordinates* (u, v) *relative to which the metric takes the form*

$$ds^2 = e^{2\lambda}(du^2 + dv^2). \tag{4.42}$$

Such coordinates are called *isothermal*. In fact Chern's proof holds under differentiability conditions much less stringent than the assumption of smoothness. From this theorem it follows that any two surfaces are locally conformal. In particular, any surface is locally conformal to the Euclidean plane.

It follows that the theorem of turning tangents is valid for sectionally smooth simple closed curves on a two-dimensional Riemannian manifold. This justifies the assumptions made in the proofs of the previous two theorems. Moreover, it enables us to prove the next theorem.

Theorem (4.7.5) (Local Gauss–Bonnet: version 3) *Let U be a coordinate neighbourhood of a two-dimensional Riemannian manifold. Let C be a simple sectionally smooth closed curve contained in U, positively oriented, parametrized by arc length s, and let $\alpha_1, \alpha_2, \ldots, \alpha_k$ be the internal angles at the vertices. Let R be the simply connected region bounded by C. Then*

$$\int_C \kappa_g \, ds + \int_R K \, dS + \sum_{i=0}^{k} (\pi - \alpha_i) = 2\pi. \qquad (4.43)$$

It is easy to see that virtually the same proof as that of theorem (4.72) goes over in this case.

We now pass on to consider two global versions of the Gauss–Bonnet theorem. We restrict ourselves to compact orientable two-dimensional Riemannian manifolds. Let R be a closed region of M with the property that its boundary is a closed simple curve consisting of a finite number of sectionally continuous arcs which do not intersect. Such a region will be called *regular*.

A *triangulation* of R is a finite family $\mathfrak{J}$ of triangles T_i, $i = 1, \ldots, n$ such that:

(i) $\cup T_i = R$;
(ii) if $T_i \cap T_j \neq \varnothing$, then $T_i \cap T_j$ is either a common edge of T_i and T_j or else a common vertex of T_i and T_j.

For a given triangulation we denote the number of triangles by f, the number of sides by e, and the number of vertices by v. The number

$$f - e + v = \chi$$

is called the *Euler–Poincaré characteristic* of the triangulation.

The following facts are presented without proof – the reader is referred, for example, to Ahlfors and Sario (1960) Chapter 1, for detailed proofs.

Theorem (4.7.6) *Every regular region admits a triangulation.*

4.7 The Gauss-Bonnet theorem

Theorem (4.7.7) *Let M be an orientable compact two-dimensional Riemannian manifold, and let $R \subset M$ be a regular region. Then there exists a triangulation of R such that every triangle is contained in some coordinate neighbourhood of M. Moreover, if the boundary of every triangle is positively oriented, adjacent triangles determine opposite orientations along the common edge.*

Theorem (4.7.8) *The Euler-Poincaré characteristic of R does not depend on the particular triangulation of R and can therefore be denoted by $\chi(R)$.*

We can now state the first version of the global Gauss-Bonnet theorem.

Theorem (4.7.9) (Global Gauss-Bonnet: version 1) *Let $R \subset M$ be a regular region of a compact orientable two-dimensional Riemannian manifold. Let $C_1, \ldots, C_n$ be closed, simple, sectionally differentiable curves forming the boundary ∂R. We assume that each C_i is positively oriented. Let α_i, $i = 1, \ldots, p$, be the interior angles at the vertices of ∂R. Then we have*

$$\int_{\partial R} \kappa_g \, ds + \int_R K \, dS + \sum_{i=1}^{p} (\pi - \alpha_i) = 2\pi\chi(R). \quad (4.44)$$

The proof is obtained by applying the local Gauss-Bonnet theorem: version 2 to each of the triangles of a triangulation of R and then summing over all triangles.

To prove theorem (4.7.9) let $\mathfrak{I}$ be a triangulation of R such that every triangle T_i of $\mathfrak{I}$ is contained in a family of neighbourhoods with orthonormal frames compatible with the orientation of M. Moreover, if the boundary of each triangle is positively oriented, edges common to adjacent triangles will have opposite orientations and the line integrals along such edges will cancel. Hence we have

$$\int_{\partial R} \kappa_g \, ds + \int_R K \, dS + \sum_{j,k}^{f,3} (\pi - \phi_{jk}) = 2\pi f,$$

where f denotes the number of triangles in the triangulation and $\alpha_{j1}, \alpha_{j2}, \alpha_{j3}$ are the interior angles of triangle T_j.

Let e_e be the number of external edges of $\mathfrak{I}$, e_i the number of internal edges, v_e the number of external vertices, and v_i the number of internal vertices. Since the curves C_i are closed we have $e_e = v_e$. Moreover $3f = 2e_i + e_e$. Hence

$$\sum_{j,k} (\pi - \phi_{jk}) = 3\pi f - \sum_{j,k} \phi_{jk}$$

$$= 2\pi e_i + \pi e_e - \sum_{j,k} \phi_{jk}.$$

We note that the external vertices may be vertices of some C_i or else vertices introduced by the triangulation $\mathfrak{J}$. We denote by v_{ec} the number of external vertices arising from the C_i and by v_{et} the number arising from the triangulation. Thus $v_e = v_{ec} + v_{et}$. Then we have

$$\sum_{j,k} (\pi - \phi_{jk}) = 2\pi e_i + \pi e_e - 2\pi v_i - \sum_{i=1}^{p} \alpha_i$$

$$= 2\pi(e_i + e_e) - 2\pi v_i - \pi v_e - \pi v_{et} - \sum_{i=1}^{p} \alpha_i$$

$$= 2\pi e - 2\pi v_i - 2\pi v_e + \pi v_{ec} - \sum_{i=1}^{p} \alpha_i$$

$$= 2\pi e - 2\pi v + \sum_{i=1}^{p} (\pi - \alpha_i).$$

Hence

$$\int_{\partial R} \kappa_g \, ds + \int_R K \, dS + 2\pi e - 2\pi v + \sum_{i=1}^{p} (\pi - \alpha_i) = 2\pi f$$

from which (4.44) follows and the theorem is proved.

As a particular case when R is a simple connected region of M we have

$$\int_{\partial R} \kappa_g \, ds + \int_R K \, dS + \sum_{i=1}^{n} (\pi - \alpha_i) = 2\pi. \tag{4.45}$$

Finally we have the second version of the global Gauss-Bound theorem.

Theorem (4.7.10) (Global Gauss-Bonnet: version 2) *Let M be a compact two-dimensional orientable Riemannian manifold without boundary. Then*

$$\int_M K \, dS = 2\pi \chi(M). \tag{4.46}$$

This follows as a special case of theorem 4.7.9.

The Euler-Poincaré characteristic is known to give a complete classification of compact orientable two-dimensional manifolds. For a sphere, direct calculation gives $\chi = 2$. For a torus $\chi = 0$. For a double torus $\chi = -2$. Another topological invariant which often appears in algebraic geometry is the *genus g* of a two-dimensional manifold. In the case of orientable manifolds this is related to χ by

$$2g = 2 - \chi. \tag{4.47}$$

This last version of the Gauss-Bonnet theorem has very striking consequences. For example, all possible shaped surfaces obtained from a

4.7 The Gauss–Bonnet theorem

round sphere in $\mathbb{R}^3$ by a diffeomorphism possess the same total curvature, that is, the integral of the Gaussian curvature over the surfaces has the same value.

A number of results follow immediately from Theorem (4.7.10). For example, suppose we have a compact surface in $\mathbb{R}^3$ which we know is not homeomorphic to a sphere. Then we must have

$$\int_M K \, dS \leq 0.$$

Since the surface is compact, it must have a point p where the height function attains its absolute maximum value. All the surface lies on one side of the tangent plane at p and hence $K > 0$ at p. Since K is a continuous function there must be some neighbourhood of p where K is strictly positive. However, since the total curvature is non-positive, there must be a neighbourhood where K is strictly negative. We can conclude therefore that there are points on the surface where K is strictly positive, points where K is strictly negative, and points where K is zero.

Another immediate consequence is the following. Let M be an oriented two-dimensional Riemannian manifold with strictly negative Gaussian curvature at all its points. Then it is impossible for two geodesic arcs to intersect so as to enclose a simply connected region. For, if they did so intersect, forming a lunar shaped region with interior angles α, β, we would have

$$\int_R K \, dS + (\pi - \alpha) + (\pi - \beta) = 2\pi$$

giving

$$\int_R K \, dS = \alpha + \beta > 0$$

and hence a contradiction.

The Gauss–Bonnet theorem was generalized by Hopf (1925). He showed that a generalized theorem was valid for a compact orientable closed $2p$-dimensional Riemannian manifold immersed as a hypersurface in $\mathbb{R}^{2p+1}$ with a metric induced from the Euclidean metric. This was later generalized by Allendoerfer and Weil (1943). However, the real breakthrough came with a paper by Chern (1944) in which he used the ideas of moving frames and fibre bundles to give an intrinsic proof of the theorem.

Let M be a $2p$-dimensional compact orientable Riemannian manifold without boundary. With each coordinate neighbourhood we associate an orthonormal frame $e_1, \ldots, e_{2p}$. Relative to this frame the curvature form has components $\Omega_{i_2}^{i_1}$. From this form we construct the $2p$-form

$$\Omega = \frac{(-1)^p \cdot 1}{2^{2p}\pi^p p!} \varepsilon_{i_1 \cdots i_{2p}} \Omega_{i_2}^{i_1} \cdots \Omega_{i_{2p}}^{i_{2p-1}} \qquad (4.48)$$

where $\varepsilon_{i_1 \cdots i_{2p}} = \pm 1$ according as $(i_1 \cdots i_{2p})$ is an even or odd permutation and is otherwise zero. It is readily checked that the form Ω remains invariant under a change of orthonormal frames and is therefore globally defined over M. Then Chern proved

$$\int_M \Omega = \chi(M) \qquad (4.49)$$

which is his generalization of the Gauss–Bonnet theorem.

Subsequent research papers concerning the Gauss–Bonnet theorem include Bishop and Goldberg (1964) and Chern (1956). In the particular case when $\dim M = 4$, Avez (1963) showed that the formula can be written as

$$\int_M (\|R\|^2 - 4\|\rho\|^2 + \tau^2) \, dV = 8\pi^2 \chi(M), \qquad (4.50)$$

where

$$\|R\|^2 = R_{hijk} R^{hijk},$$
$$\|\rho\|^2 = \rho_{ij} \rho^{ij},$$
$$\tau = g^{ij} \rho_{ij}.$$

An immediate corollary is the following theorem.

Theorem (4.7.11) *A four-dimensional torus with an Einstein metric is necessarily flat.*

It is known from algebraic topology that for two manifolds M and N the Euler–Poincaré characteristic satisfies the relation

$$\chi(M \times N) = \chi(M) \times \chi(N). \qquad (4.51)$$

Hence $\chi((S^1 \times S^1) \times (S^1 \times S^1)) = \chi(T^2) \times \chi(T^2) = 0$. Moreover, since the metric is Einstein, that is,

$$\rho_{ij} = \tfrac{1}{4} \tau g_{ij},$$

the formula becomes

$$\int_M \|R\|^2 \, dV = 8\pi^2 \chi = 0.$$

Hence $R = 0$ and the result follows.

It may be shown that when $\dim M = 6$, the Gauss–Bonnet formula may be written

4.8 Exercises

$$\chi(M) = \frac{1}{8\pi^2} \int_M \Omega_{123456}$$

where

$$\Omega_{123456} = \Omega_{12} \wedge \Omega_{3456} - \Omega_{13} \wedge \Omega_{2456} + \Omega_{14} \wedge \Omega_{2356} - \Omega_{15} \wedge \Omega_{2346} + \Omega_{16} \wedge \Omega_{2345}$$

and

$$\Omega_{ijkl} = \Omega_{ij} \wedge \Omega_{kl} - \Omega_{ik} \wedge \Omega_{jl} + \Omega_{il} \wedge \Omega_{jk}.$$

The fact that Ω_{123456} must vanish identically on a manifold of dimension 4 or 5 gives rise to a further identity satisfied by the curvature tensor on manifolds with these dimensions. However this formula is very complicated and it seems that further restrictions must be imposed on the metric before the expression assumes a manageable form.

4.8 EXERCISES

1. Let x, y, z be the standard coordinate system on Euclidean 3-space $\mathbb{R}^3$. Let u^1, u^2, u^3, u^4, u^5, be the standard coordinate system of $\mathbb{R}^5$. Let $\phi: \mathbb{R}^3 \to \mathbb{R}^5$ be defined by

$$u^1 = \frac{1}{\sqrt{3}} yz, \quad u^2 = \frac{1}{\sqrt{3}} zx, \quad u^3 = \frac{1}{\sqrt{3}} xy,$$

$$u^4 = \frac{1}{2\sqrt{3}} (x^2 - y^2), \quad u^5 = \tfrac{1}{6} (x^2 + y^2 - 2z^2).$$

Show that ϕ maps $S^2(\sqrt{3})$ isometrically into $S^4(1)$. By observing that (x, y, z) and $(-x, -y, -z)$ of $S^2(\sqrt{3})$ map into the same point of $S^4(1)$, show that ϕ is an embedding of the real projective plane into $S^4(1)$, and that it is embedded as a minimal surface.

2. Let M be an n-dimensional submanifold of Euclidean m-space $\mathbb{R}^m$. Let $\mathbf{r}$ be the position vector of a point of $\mathbb{R}^m$. Prove that at a point of M,

$$\Delta \mathbf{r} = n\mathbf{H}$$

where $\mathbf{H}$ is the mean curvature vector, and Δ is the Laplacian with respect to the metric of M.

5
Complex and almost-complex manifolds

5.1 COMPLEX STRUCTURES ON VECTOR SPACES

A complex structure on a real vector space V is a linear endomorphism J of V such that $J^2 = -1$, where 1 stands for the identity transformation of V. A real vector space with a complex structure can be given the structure of a complex vector space. We define scalar multiplication by complex numbers as

$$(a + ib)X = aX + bJX \quad \text{for } X \in V \text{ and } a, b \in \mathbb{R}.$$

Evidently the dimension m of V must be even, and we define its complex dimension to be $m/2$. Conversely if we are given a complex vector space V of complex dimension n, we can define a linear endomorphism J of V by

$$J(X) = iX,$$

for $X \in V$. Then V, considered as a real vector space of dimension $2n$, has J as its complex structure. We now choose a special class of bases related to the complex structure. More precisely we prove the following proposition.

Proposition (5.1.1) Let J be a complex structure associated with a real $2n$-dimensional vector space V. Then there exist vectors $X_1, \ldots, X_n$ of V such that $\{X_1, \ldots, X_n, JX_1, \ldots, JX_n\}$ is a basis for V.

To prove the proposition, we regard V as a complex vector space of n complex dimensions, for which we choose as a basis $X_1, \ldots, X_n$. Then $X_1, \ldots, X_n, JX_1, \ldots, JX_n$ form a basis for V as a real $2n$-dimensional space.

We denote by $\mathbb{C}^n$ the complex vector space of n-tuples of complex numbers $(z^1, \ldots, z^n)$. We set

$$z^k = x^k + iy^k, \; k = 1, \ldots, n.$$

Then $\mathbb{C}^n$ can be identified with the real vector space $\mathbb{R}^{2n}$ of $2n$-tuples of real numbers $(x^1, \ldots, x^n, y^1, \ldots, y^n)$. When we identify $\mathbb{C}^n$ with $\mathbb{R}^{2n}$ we shall use the correspondence $(z^1, \ldots z^n) \to (x^1, \ldots, x^n, y^1, \ldots, y^n)$. The complex structure J induced from that of $\mathbb{C}^n$, maps $(x^1, \ldots, x^n, y^1, \ldots, y^n)$ into $(y^1, \ldots, y^n, -x^1, \ldots, -x^n)$ and is called the *canonical complex structure* of $\mathbb{R}^{2n}$. In terms of the natural (coor-

5.1 Complex structures on vector spaces

dinate) basis of $\mathbb{R}^{2n}$, J is represented by the matrix $J_0 = \begin{pmatrix} O & I_n \\ -I_n & O \end{pmatrix}$, where I_n is the identity matrix of degree n.

Proposition (5.1.2) Let J, J' be complex structures on real vector spaces V and V' respectively. Let f be a real linear mapping of V into V'. Then, when V, V' are regarded as complex vector spaces, f is complex linear if and only if $J' \circ f = f \circ J$.

This follows trivially since J and J' are effectively multiplication by i when V and V' are regarded as complex vector spaces. Thus the complex linear group GL $(n; \mathbb{C})$ of degree n can be regarded as a subgroup of GL $(2n; \mathbb{R})$ consisting of matrices which commute with the matrix J_0. It follows that this representation of GL $(n; \mathbb{C})$ into GL $(2n; \mathbb{R})$, called the *real representation of* GL $(n; \mathbb{C})$, is given by

$$A + iB \to \begin{pmatrix} A & B \\ -B & A \end{pmatrix}$$

for $A + iB \in$ GL $(n; \mathbb{C})$, where A and B are real $n \times n$ matrices.

Proposition (5.1.3) Let J be a complex structure on the real vector space V and let V' be a (real) subspace of V. Then V' is invariant under the action of J if and only if V' is a complex subspace of V when V is considered as a complex vector space.

The proof follows because J means effectively multiplication by i.

Let V be a real vector space and let V^* denote its dual. Then a complex structure J on V induces a complex structure $\tilde{J}$ on V^*. We define $\tilde{J}$ by

$$X^*(JX) = \tilde{J}X^*(X)$$

for $X \in V$ and $X^* \in V^*$. We have

$$\tilde{J}(\tilde{J}X^*(X)) = \tilde{J}X^*(JX) = X^*(J^2X) = -1(X^*(X))$$

giving $\tilde{J}^2 = -1$ as required. It will be convenient from now on to use the notation J instead of $\tilde{J}$.

Complexification

Let V be a real vector space, and consider the tensor product $V \otimes \mathbb{C}$. This can be considered as the space of vectors with values in $\mathbb{C}$, and is denoted by V^c. Clearly V is a (real) subspace of V^c. Similarly we can form tensors of type (r, s) starting with V^c instead of V, and then the tensors of type (r, s) over V form a (real) subspace of tensors of type (r, s) over V^c.

The operation of complex conjugation in V^c is the real linear endomorphism defined by

$$Z = X + iY \to \bar{Z} = X - iY$$

for $X, Y \in V$. This endomorphism extends in a natural way to the space $T^r_s(V^c)$.

Assume now that V is a $2n$-dimensional real vector space with a complex structure J. We can extend the action of J canonically to V^c and we also denote this extension by J which fortunately retains the property $J^2 = -1$. Thus the eigenvalues of this extended J are i and $-i$. Set

$$V^{1,0} = \{Z \in V^c; JZ = iZ\}, \qquad V^{0,1} = \{Z \in V^c: JZ = -iZ\}.$$

Then

$$V^{1,0} = \{X - iJX : X \in V\}$$
$$V^{0,1} = \{X + iJX : X \in V\}$$
$$V^c = V^{1,0} + V^{0,1},$$

as a direct sum. Moreover complex conjugation defines a real linear isomorphism between $V^{1,0}$ and $V^{0,1}$.

It is clear that the complexification of the dual space V^*, denoted by $(V^*)^c$, is effectively the same space as $(V^c)^*$, the dual space of V^c. Again with respect to the eigenvalues i and $-i$ of the complex structure on V^* we have a direct sum decomposition

$$V^{*c} = V_{1,0} + V_{0,1}.$$

Here

$$V_{1,0} = \{X^* \in V^{*c} : X^*(X) = 0 \text{ for all } X \in V^{0,1}\}$$
$$V_{0,1} = \{X^* \in V^* : X^*(X) = 0 \text{ for all } X \in V^{1,0}\}.$$

The decomposition of V^c leads to a decomposition of the tensor space $T^r_s(V^c)$ into a direct sum of tensor products of vector spaces identical to one of the spaces $V^{1,0}$, $V^{0,1}$, $V_{1,0}$, and $V_{0,1}$.

In particular the exterior algebra ΛV^{*c} decomposes, and the exterior algebras $\Lambda V_{1,0}$ and $\Lambda V_{0,1}$ are subalgebras of ΛV^{*c}. If we denote by $\Lambda^{p,q} V^{*c}$ the subspace spanned by $\alpha \wedge \beta$ where $\alpha \in \Lambda^p V_{1,0}$ and $\beta \in \Lambda^q V_{0,1}$, then

$$\Lambda V^{*c} = \sum_{r=0}^{n} \Lambda^r V^{*c} \qquad \text{with} \qquad \Lambda^r V^{*c} = \sum_{p+q=r} \Lambda^{p,q} V^{*c}.$$

Moreover complex conjugation in V^{*c} extends in a natural way to ΛV^{*c}, and this gives a real linear isomorphism between $\Lambda^{p,q} V^{*c}$ and $\Lambda^{q,p} V^{*c}$.

5.2 Hermitian metrics

Let $\{\omega^1, \ldots, \omega^n\}$ be a basis for the complex vector space $V_{1,0}$. Then the complex conjugate vectors $\{\bar{\omega}^1, \ldots, \bar{\omega}^n\}$ form a basis for $V_{0,1}$. Moreover, the elements $\omega^{j_1} \wedge \cdots \wedge \omega^{j_p} \wedge \bar{\omega}^{k_1} \wedge \cdots \wedge \bar{\omega}^{k_q}$, where $1 \leq j_1 < j_2 < \cdots < j_p \leq n$, $1 \leq k_1 < k_2 < \cdots < k_q \leq n$ form a basis for $\Lambda^{p,q} V^{*c}$ over the complex numbers.

5.2 HERMITIAN METRICS

A Hermitian metric h is an inner product defined over a real vector space with complex structure J such that

$$h(JX, JY) = h(X, Y) \quad \text{for all vectors } X, Y \in V.$$

We note that $h(X, JX) = 0$ for every vector X. This follows since $h(X, JX) = h(JX, J^2X) = -h(JX, X) = -h(X, JX)$. This property gives rise to the following result.

Proposition (5.2.1) Let h be a Hermitian inner product in a $2n$-dimensional real vector space V with a complex structure J. Then there exist elements $X_1, \ldots, X_n$ of V such that $\{X_1, \ldots, X_n, JX_1, \ldots, JX_n\}$ is an orthonormal basis for V with respect to h.

The proof is by induction on the dimension of V. We know that the claim is valid for $n = 1$ by using X_1 and JX_1. Let W be the space spanned by X_1 and JX_1, and let $W^\perp$ be its orthogonal complement in V so that $V = W + W^\perp$, and $W^\perp$ is invariant under J. We now make use of the induction hypothesis that the proposition is true for $n - 1$, so that $W^\perp$ has an orthonormal basis $\{X_2, \ldots, X_n, JX_2, \ldots, JX_n\}$. Then $\{X_1, \ldots, X_n, JX_1, \ldots, JX_n\}$ is the required basis.

We now prove a further result.

Proposition (5.2.2) Let h be a Hermitian inner product on a real vector space V with complex structure J. Then h can be uniquely extended to a complex symmetric bilinear form, also denoted by h, of V^c such that

$$h(\bar{Z}, \bar{W}) = \overline{h(Z, W)},$$

$$h(Z, \bar{Z}) > 0 \quad \text{for all non-zero } Z,$$

$$h(Z, \bar{W}) = 0 \quad \text{for } Z \in V^{1,0} \text{ and } W \in V^{0,1}.$$

Conversely every complex symmetric bilinear form h on V^c satisfying these conditions is the natural extension of a Hermitian inner product of V.

The proof is quite straightforward. For example, writing $Z = X + iY$, $W = U + iV$, we have

156   5 Complex and almost-complex manifolds

$$h(\bar{Z}, \bar{W}) = h(X - iY, U - iV) = h(X, U) + h(Y, V)$$
$$- ih(X, V) - ih(Y, U),$$
$$h(Z, W) = h(X, U) + h(Y, V) + ih(X, V) + ih(Y, U),$$

giving $h(\bar{Z}, \bar{W}) = \overline{h(Z, W)}$. The converse is easily established.

To each Hermitian inner product h on a real vector space with complex structure J, there corresponds an element ϕ of $\Lambda^2 V^*$ such that $\phi(X, Y) = h(X, JY)$, for all $X, Y \in V$. The skew-symmetry follows since

$$\phi(X, Y) = h(JX, J^2Y) = -h(JX, Y) = -h(Y, JX) = -\phi(Y, X).$$

Moreover ϕ is invariant under J since

$$\phi(JX, JY) = h(JX, J^2Y) = -h(Y, JX) = -\phi(Y, X) = \phi(X, Y).$$

Since $\Lambda^2 V^*$ can be considered as a subspace of $\Lambda^2 V^{*c}$ it follows that ϕ uniquely determines a skew-symmetric bilinear form on V^c which we still denote by ϕ. Our previous results now imply that $\phi \in \Lambda^{1,1} V^{*c}$.

Let $Z_1, \ldots, Z_n$ be a basis for $V^{1,0}$ over $\mathbb{C}$ and let $\{\omega^1, \ldots, \omega^n\}$ be a dual basis of $V_{0,1}$. We write

$$h_{j\bar{k}} = h(Z_j, \bar{Z}_k) \quad \text{for } j, k = 1, \ldots, n.$$

Then

$$h_{j\bar{k}} = \bar{h}_{k\bar{j}} \quad \text{for } j, k = 1, \ldots, n;$$

$$\phi(Z, W) = -i \sum_{j,k=1}^{n} (h_{j\bar{k}}(\omega^j(Z)\bar{\omega}^k(W) - \omega^j(W)\bar{\omega}^k(Z))).$$

The verification of this claim is left to the reader.

We now show that *there is a natural 1-1 correspondence between the set of complex structures on $\mathbb{R}^{2n}$ and the homogeneous space $GL(2n; \mathbb{R})/GL(n; \mathbb{C})$; the coset represented by an element $T \in GL(2n; \mathbb{R})$ corresponds to the complex structure $TJ_0 T^{-1}$, where J_0 is the canonical complex structure.*

To prove this result we note that every element $T \in GL(2n; \mathbb{R})$ sends every complex structure J of $\mathbb{R}^{2n}$ into a complex structure TJT^{-1} of $\mathbb{R}^{2n}$. We consider $GL(2n; \mathbb{R})$ as a group of transformations acting on the set of complex structures of $\mathbb{R}^{2n}$. It will be sufficient to prove that this action is transitive and that the subgroup which leaves J_0 invariant is $GL(n; \mathbb{C})$. Let J and J' be two complex structures of $\mathbb{R}^{2n}$, and let $\{e_1, \ldots, e_n, Je_1, \ldots, Je_n\}$ and $\{e'_1, \ldots, e'_n, J'e'_1, \ldots, J'e'_n\}$ be bases of $\mathbb{R}^{2n}$. We define the element T of $GL(2n; \mathbb{R})$ by

$$Te_k = e'_k, \quad TJe_k = J'e'_k \quad \text{for } k = 1, \ldots, n.$$

Then $J' = TJT^{-1}$, proving that the group acts transitively. Moreover, an element T of $GL(2n;\mathbb{R})$ belongs to $GL(n;\mathbb{C})$ if and only if it commutes with J_0, that is, $J_0 = TJ_0T^{-1}$. This completes the proof.

A similar result states:

There is a natural 1-1 correspondence between the set of Hermitian inner products in $\mathbb{R}^{2n}$ with respect to the complex structure J_0 and the homogeneous space $GL(n;\mathbb{C})/U(n)$. The coset represented by an element $T \in GL(n;\mathbb{C})$ corresponds to the Hermitian inner product h defined by

$$h(X, Y) = h_0(TX, TY) \quad \text{for } X, Y \in \mathbb{R}^{2n},$$

where h_0 is the canonical Hermitian product in $\mathbb{R}^{2n}$.

The proof is along similar lines to that of the previous proposition. An element T of $GL(n;\mathbb{C})$ sends a Hermitian inner product h (with respect to J_0) into a Hermitian inner product h' according to

$$h'(X, Y) = h(TX, TY) \quad \text{for } X, Y \in \mathbb{R}^{2n}.$$

We consider $GL(n;\mathbb{C})$ as a group of transformations acting on the set of Hermitian inner products in $\mathbb{R}^{2n}$ with respect to J_0. It will be sufficient to prove that this action is transitive and that the subgroup which leave h_0 invariant is $U(n)$. We have already seen that there are orthogonal bases $\{e_1, \ldots, e_n, J_0 e_1, \ldots, J_0 e_n\}$ and $\{e'_1, \ldots, e'_n, J_0 e'_1, \ldots, J_0 e'_n\}$ with respect to h and h' of $\mathbb{R}^{2n}$. The element T of $GL(2n;\mathbb{R})$ defined by

$$Te'_k = e_k, \quad TJ_0 e'_k = J_0 e_k \quad \text{for } k = 1, \ldots, n$$

is an element of $GL(n;\mathbb{C})$ which sends h into h'. Moreover the elements of $GL(n;\mathbb{C})$ which leave h_0 invariant is the intersection of $GL(n;\mathbb{C})$ and $O(2n)$, both being considered as subgroups of $GL(2n;\mathbb{R})$. But these elements are precisely those of $U(n)$, considered as a subgroup of $GL(2n;\mathbb{R})$. This completes the proof.

5.3 ALMOST-COMPLEX MANIFOLDS

In this section we apply the results of the previous section on real and complex vector spaces to the tangent spaces of manifolds. An *almost-complex structure* on a real differentiable manifold M is a tensor field J which at every point $x \in M$ is an endomorphism of the tangent space $T_x(M)$ such that $J^2 = -1$, where 1 denotes the identity transformation of $T_x(M)$. A manifold with such a structure is called an *almost-complex manifold*. We have seen that such a manifold must be even dimensional. Moreover it must be orientable. To prove the latter, we fix a basis $X_1, \ldots, X_n, JX_1, \ldots, JX_n$ in each $T_x(M)$. Any two such bases differ by a linear transformation with positive determinant. To fix an orientation on M we consider the family of all coordinate systems $x^1, \ldots, x^{2n}$ of M such that in each coordinate neighbourhood, the coordinate basis $(\partial/\partial x^1, \ldots, \partial/\partial x^{2n})$ of $T_x(M)$ at x differs from the chosen basis $X_1, \ldots, X_n, JX_1, \ldots, JX_n$) by a linear transformation of positive determinant. These coordinate systems determine a complete atlas for M, which is thus oriented. Thus almost complex manifolds must be even dimensional and orientable. However these two conditions are not sufficient for a manifold to have an almost complex structure. For example, Ehresmann and Hopf proved that the 4-sphere S^4 cannot have an almost complex structure (see, for example, Steenrod (1951)).

A complex manifold is a paracompact Hausdorff space which has a covering by neighbourhoods each homeomorphic to an open set in n-dimensional complex number space such that when two neighbourhoods overlap the local coordinates transform by a complex analytic transformation. If $z^1, \ldots, z^n$ are local coordinates in one such neighbourhood, and $w^1, \ldots, w^n$ are local coordinates of another neighbourhood, then when these are both defined, we have $w^i = f^i(z^1, \ldots, z^n)$, where each f^i is a holomorphic (analytic) function of the z's, and the functional determinant $\partial(w^1, \ldots, w^n)/\partial(z^1, \ldots, z^n) \neq 0$.

We shall now prove that a complex manifold admits canonically an almost complex structure. We consider first the space $\mathbb{C}^n$ of n-tuples of complex numbers $(z^1, \ldots, z^n)$ with $z^k = x^k + iy^k$, $k = 1, \ldots, n$. With respect to the coordinate system $x^1, \ldots, x^n, y^1, \ldots, y^n$, we define an almost complex structure J on $\mathbb{C}^n$ by

$$J\left(\frac{\partial}{\partial x^i}\right) = \frac{\partial}{\partial y^i}, \quad J\left(\frac{\partial}{\partial y^i}\right) = -\left(\frac{\partial}{\partial x^i}\right), \quad i = 1, \ldots, n.$$

We now prove that *a mapping f of an open set of $\mathbb{C}^n$ into $\mathbb{C}^m$ preserves the almost-complex structure J defined in the above manner, that is, $f_* \circ J = J \circ f_*$, if and only if f is holomorphic.* Let $(w^1, \ldots, w^m)$ be the natural coordinate system of $\mathbb{C}^m$ with $w^k = u^k + iv^k$. Then the map f is defined locally by

$$u^k = u^k(x^1, \ldots, x^n, y^1, \ldots, y^n)$$
$$v^k = v^k(x^1, \ldots, x^n, y^1, \ldots, y^n) \quad \text{with } k = 1, \ldots, m.$$

The mapping f is holomorphic if and only if the following Cauchy–Riemann equations are satisfied:

$$\frac{\partial u^k}{\partial x^j} = \frac{\partial v^k}{\partial y^j}, \quad \frac{\partial u^k}{\partial y^j} = -\frac{\partial v^k}{\partial x^j},$$

where $j = 1, \ldots, n$, $k = 1, \ldots, m$. However, for any f we have

5.3 Almost-complex manifolds

$$f_*\left(\frac{\partial}{\partial x^j}\right) = \sum_{k=1}^{m} \left(\frac{\partial u^k}{\partial x^j}\right) \frac{\partial}{\partial u^k} + \sum_{k=1}^{m} \left(\frac{\partial v^k}{\partial x^j}\right) \frac{\partial}{\partial v^k},$$

$$f_*\left(\frac{\partial}{\partial y^j}\right) = \sum_{k=1}^{m} \left(\frac{\partial u^k}{\partial y^j}\right) \frac{\partial}{\partial u^k} + \sum_{k=1}^{m} \left(\frac{\partial v^k}{\partial y^j}\right) \frac{\partial}{\partial v^k}.$$

We have

$$f_* \circ J\left(\frac{\partial}{\partial x^i}\right) = f_*\left(\frac{\partial}{\partial y^i}\right);$$

$$J \circ f_*\left(\frac{\partial}{\partial x^i}\right) = \sum_{k=1}^{m} \left(\frac{\partial u^k}{\partial x^j}\right) \frac{\partial}{\partial v^k} - \sum_{k=1}^{m} \left(\frac{\partial v^k}{\partial x^j}\right) \frac{\partial}{\partial u^k}.$$

Clearly

$$f_* \circ J\left(\frac{\partial}{\partial x^i}\right) = J \circ f_*\left(\frac{\partial}{\partial x^i}\right)$$

from the Cauchy–Riemann equations. Similarly we have $f_* \circ J(\partial/\partial y^i) = J \circ f_*(\partial/\partial y^i)$.

Conversely it follows that the relation $f_* \circ J = J \circ f_*$ is satisfied only when the Cauchy–Riemann equations are satisfied.

To define an almost-complex structure on a complex manifold we transfer the almost-complex structure on $\mathbb{C}^m$ to M by means of the charts. This can be done unambiguously, for, by the previous result, the process is independent of the particular charts chosen.

We have shown that a complex manifold admits an almost complex structure and therefore it must be orientable. However it is interesting to give an independent proof as follows. Let $z^1, \ldots, z^m$ be a system of local coordinates on the m-dimensional complex manifold M. Set $z^k = x^k + iy^k$ so that the x^k and y^k form a real system of local coordinates for the real $2m$-dimensional manifold. We write

$$x^k = \frac{1}{2}(z^k + \bar{z}^k) \qquad y^k = -\frac{1}{2}i(z^k - \bar{z}^k).$$

Then we see that

$$dx^1 \wedge \cdots \wedge dx^m \wedge dy^1 \wedge \cdots \wedge dy^m$$
$$= \left(-\frac{i}{2}\right)^m dz^1 \wedge \cdots \wedge dz^m \wedge d\bar{z}^1 \wedge \cdots \wedge d\bar{z}^m.$$

Thus the form

$$\Theta(z) = \left(-\frac{i}{2}\right)^m dz^1 \wedge \cdots \wedge dz^m \wedge d\bar{z}^1 \wedge \cdots \wedge d\bar{z}^m$$

is a real form of maximal order $2m$. We now prove that $\Theta(z)$ is well defined up to a positive factor. Let $w^1, \ldots, w^m$ be another system of local coordinates. Then

$$d\omega^1 \wedge \cdots \wedge d\omega^m = D\, dz^1 \wedge \cdots \wedge dz^m$$

where $D = \det\{\partial(\omega^1, \ldots, \omega^m)/\partial(z^1, \ldots, z^m)\}$. Then

$$d\bar\omega^1 \wedge \cdots \wedge d\bar\omega^m = \bar D\, d\bar z^1 \wedge \cdots \wedge d\bar z^m.$$

Hence

$$\Theta(\omega) = D\bar D\Theta(z)$$

and our claim is justified. To define Θ globally we choose a locally finite covering and a partition of unity subordinate to the covering. This gives rise to a globally defined $2m$ form over M and hence M is orientable.

We have seen that a vector space carrying a complex structure determines a splitting of the complexified space into the direct sum of vectors of type $(1, 0)$ and those of type $(0, 1)$. Conversely, a decomposition of this nature determines a complex structure on the original vector space. In our present study of almost-complex structures on manifolds, given an almost-complex structure this determines vector fields of type $(1, 0)$ and $(0, 1)$, and dually it determines 1-forms of type $(1, 0)$ and $(0, 1)$. Conversely if we are given the space of complex-valued 1-forms $T_{1,0}$ of type $(1, 0)$, and if $T_{0,1}$ denotes the space of forms which are conjugate complex to those of $T_{1,0}$ so that we have the decomposition

$$T_x^{*c} = T_{1,0} + T_{0,1},$$

then this determines the almost complex structure on the manifold.

Let x^i, $i = 1, \ldots, 2m$, be a local coordinate system. Then a basis for T_x is $\partial/\partial x^i$ and its dual basis for T^* is the set of differentials dx^i. The endomorphism J is given by

$$J_x\left(\sum_i \xi^i \frac{\partial}{\partial x^i}\right) = \sum_{i,j} h_j^i \xi^j \frac{\partial}{\partial x^i}.$$

The condition $J_x^2 = -1$ leads to the condition

$$\sum h_j^i h_k^j = -\delta_k^i, \quad 1 \leqslant i, j, k \leqslant 2n.$$

From our previous discussion it follows that the forms

$$\sum_j (h_j^i + i\delta_j^i)\, dx^j$$

are of type $(1, 0)$. They are $2m$ in number but exactly m of then are linearly independent over the ring of complex-valued functions. Thus

5.3 Almost-complex manifolds

J determines the space $T_{1,0}$, and conversely a knowledge of $T_{1,0}$ determines J.

We can now give an alternative proof that a complex manifold determines canonically an almost-complex structure. For on a complex manifold the complex-valued 1-forms expressed in terms of local coordinates z^α are linear combinations of dz^α. We define these to be the space of forms of type $(1, 0)$, and the space of forms of type $(0, 1)$ are similarly formed with $d\bar{z}^\alpha$. Since the set dz^α, $d\bar{z}^\alpha$ are linearly independent, they define an almost-complex structure.

To describe J in terms of local coordinates, let $z^\alpha = x^\alpha + iy^\alpha$. Then, using the fact that dz^α is of type $(1, 0)$ we get

$$dz^\alpha\left(\frac{\partial}{\partial x^\beta}\right) = \delta^\alpha_\beta, \quad dz^\alpha\left(\frac{\partial}{\partial y^\beta}\right) = i\delta^\alpha_\beta,$$

$$dz^\alpha\left(\frac{J\partial}{\partial x^\beta}\right) = i\delta^\alpha_\beta, \quad dz^\alpha\left(\frac{J\partial}{\partial y^\beta}\right) = -\delta^\alpha_\beta, \quad 1 \leq \alpha, \beta \leq m.$$

These relations give

$$J\left(\frac{\partial}{\partial x^\beta}\right) = \frac{\partial}{\partial y^\beta}, \quad J\left(\frac{\partial}{\partial y^\beta}\right) = -\frac{\partial}{\partial x^\beta},$$

and hence we get $J^2 = -1$ as we should.

The question arises naturally whether all almost-complex structures arise only from complex manifolds — more precisely, whether every almost-complex manifold is complex. This indeed is the case for manifolds of real dimension 2 but not in general. We now find necessary conditions that an almost-complex manifold should arise from a complex manifold.

Suppose we are given an orientable differentiable manifold of dimension $2m$, with local coordinates $x^1, \ldots, x^m$, $y^1, \ldots, y^m$. Writing $z^\alpha = x^\alpha + iy^\alpha$, $\alpha = 1, \ldots, m$, we want to find the condition that forms of type $(1, 0)$ are linear combinations of dz^α. Suppose that the almost-complex structure is determined by the forms θ^α of type $(1, 0)$, which are linearly independent over the ring of complex-valued C^∞-functions. Taking the exterior derivative we get

$$d\theta^\gamma = \sum_{\alpha,\beta} A^\gamma_{\alpha\beta}\theta^\alpha \wedge \theta^\beta + \sum_{\alpha,\beta} B^\gamma_{\alpha\beta}\theta^\alpha \wedge \bar{\theta}^\beta + \sum C^\gamma_{\alpha\beta}\bar{\theta}^\alpha \wedge \bar{\theta}^\beta$$

where $A^\gamma_{\alpha\beta}$, $B^\gamma_{\alpha\beta}$, and $C^\gamma_{\alpha\beta}$ are C^∞-complex-valued functions satisfying the relations $A^\gamma_{\alpha\beta} + A^\gamma_{\beta\alpha} = 0 = C^\gamma_{\alpha\beta} + C^\gamma_{\beta\alpha}$. The condition $d\theta^\gamma = 0$, mod θ^γ, is seen to be invariant under a linear transformation of the θ^γ's. It is certainly satisfied by $\theta^\gamma = dz^\gamma$. Thus a necessary condition for an almost-complex structure to arise from a complex structure is $C^\gamma_{\alpha\beta} = 0$. We now express this condition in terms of the endomorphism

5 Complex and almost-complex manifolds

$J = (h^i_j)$. We have the following theorem due to Eckmann-Frolicher. Let

$$h^i_{jk} = -h^i_{kj} = \frac{\partial h^i_j}{\partial x^k} - \frac{\partial h^i_k}{\partial x^j}, \qquad i, j, k = 1, \ldots, 2m.$$

$$t^i_{jk} = \sum (h^i_{jl} h^l_k - h^i_{kl} h^l_j).$$

Then the integrability condition of the almost-complex structure is $t^i_{jk} = 0$.

To prove this we note that the forms of type $(1, 0)$ are linear combinations of $\sum (h^i_j + i\delta^i_j) \, dx^j$. Thus the integrability condition can be expressed as

$$\sum dh^i_j \wedge dx^j = 0, \qquad \mathrm{mod} \sum (h^i_j + i\delta^i_j) \, dx^j,$$

that is

$$\sum h^i_{jk} \, dx^j \wedge dx^k = 0, \qquad \mathrm{mod} \sum (h^i_j + i\delta^i_j) \, dx^j.$$

If we equate to zero the forms $(h^i_j + i\delta^i_j) \, dx^j$ which are $2m$ in number, we can select a fundamental set of solutions for dx^j from $h^i_j - i\delta^i_j$, since $(h^i_j + i\delta^i_j)(h^j_k - i\delta^j_k) = 0$. The condition for integrability can therefore be written as

$$\sum h^i_{jk} (h^j_l - i\delta^j_l)(h^k_m - i\delta^k_m) = 0.$$

If we equate real and imaginary parts to zero we find that $t^i_{jk} = 0$.

For a surface we see that the skew-symmetry of $C^\gamma_{\alpha\beta}$ guarantees the condition $C^\gamma_{\alpha\beta} = 0$ is automatically satisfied. However, for a manifold of real dimension ≥ 4 the condition is non-trivial.

The condition $t^i_{jk} = 0$ is in fact equal to the vanishing of the Nijenhuis tensor N which we met earlier, where

$$N(X, Y) = [JX, JY] - [X, Y] - J[X, JY] - J[JX, Y]$$

for tangent vector fields X, Y to the manifold. In terms of local coordinates $x^1, \ldots, x^{2m}$ we have

$$N^i_{jk} = J^h_j \frac{\partial J^i_k}{\partial x^h} - J^h_k \frac{\partial J^i_j}{\partial x^h} - J^i_h \frac{\partial J^h_k}{\partial x^j} + J^i_h \frac{\partial J^h_j}{\partial x^k}.$$

It can be proved, via the theorem of Frobenius, that the condition $N = 0$ is both necessary and sufficient for an almost-complex structure to be integrable when the manifold is analytic. However the result is true under less restrictive conditions including the assumption that the manifold is C^∞, but the proof is very difficult — see, for example, Newlander and Nirenberg (1957).

5.3 Almost-complex manifolds

It is known from the results of Borel and Serre (1951), that for $n \neq 2$ or 6, S^n does not admit almost-complex structures. We shall now construct an almost-complex structure on S^6 using Cayley numbers. We remind the reader that a quaternion is a number of the form

$$q = w + xi + yj + zk$$

where w, x, y, z are real numbers. Addition follows the usual rule while multiplication is given by

$$i^2 = j^2 = k^2 = -1; \quad ij = k, \quad jk = i, \quad ki = j, \quad ij = -ji,$$
$$jk = -kj, \quad ki = -ik.$$

The conjugate of q, denoted by $\bar{q}$, is given by

$$q = w - xi - yj - zk.$$

We recall that the algebra of quaternions is associative but not commutative. A Cayley number $x = (q_1, q_2)$ is an ordered pair of quaternions. The set of Cayley numbers forms an eight-dimensional non-associative algebra over the real numbers. Addition and multiplication are defined by

$$(q_1, q_2) \pm (q_1', q_2') = (q_1 \pm q_1', q_2 \pm q_2'),$$
$$(q_1, q_2)(q_1', q_2') = (q_1 q_1' - \bar{q}_2' q_2, q_2' q_1 + q_2 \bar{q}_1').$$

The conjugate of the Cayley number $x = (q_1, q_2)$ is $\bar{x}$ where $\bar{x} = (\bar{q}_1, -q_2)$. Then $x\bar{x} = (q_1 \bar{q}_1 + \bar{q}_2 q_2, 0)$. We set $|x|^2 = q_1 \bar{q}_1 + \bar{q}_2 q_2$. Thus $|x| > 0$ unless $x = 0$. Although the associative law does not hold for the multiplication of Cayley numbers, nevertheless we have the laws

$$x(xx') = (xx)x', \quad (x'x)x = x'(xx).$$

Moreover we have

$$|xx'| = |x| |x'|$$

so that $xx' = 0$ implies either $x = 0$ or $x' = 0$.

A Cayley number $x = (q_1, q_2)$ is defined to be *real* if q_1 is real and $q_2 = 0$. It is *purely imaginary* if q_1 is a purely imaginary quaternion.

Let E^7 be the seven-dimensional real vector space formed by the purely imaginary Cayley numbers. We define an inner product (,) and a vector product $\times$ in E^7 by:

$$-(x, x') = \text{real part of } xx' \text{ for } x, x' \in E^7;$$
$$x \times x' = \text{the purely imaginary part of } xx', \text{ for } x, x' \in E^7.$$

These operators have the following properties which can be checked by direct computation:

(i) $xx = -(x, x) = -|x|^2$;
(ii) $x \times x' = -x' \times x$;
(iii) $(x \times x', x'') = (x, x' \times x'')$.

Let S^6 be the unit sphere in E^7 defined by $|x| = 1$, $x \in E^7$. The tangent space $T_x(S^6)$ to S^6 at x can be identified with the subspace $V_x = \{y \in E^7 : (x, y) = 0\}$ of E^7 parallel to it. Define a linear endomorphism J_x of V_x by

$$J_x(y) = x \times y, \quad \text{for } y \in V_x.$$

Conditions (ii) and (iii) imply that $J_x(y) \in V_x$. Also we have

$$J_x(J_x y) = J_x(x \times y) = x \times (x \times y).$$

After a lengthy but straightforward calculation we find that

$$J_x(J_x y) = -|x|^2 y = -y$$

so that $J_x^2 = -1$. Thus the family of endomorphisms J_x, $x \in S^6$, gives S^6 an almost-complex structure. Also, as a rather hairy calculation shows, the Nijenhuis tensor of this almost-complex structure does not vanish and the almost-complex structure is not derived from a complex structure. However, whether there exists a complex structure on S^6 has remained an open question for the last thirty years.

Let M be a six-dimensional orientable manifold immersed in E^7. We shall show that the almost-complex structure on S^6 induces an almost-complex structure on M. Let $g: M \to S^6$ be the Gauss map, sending $x \in M$ to the point on S^6 parallel to the unit normal to M at x. Then the tangent spaces $T_x(M)$ and $T_{g(x)}(S^6)$ are parallel and may be identified in a natural manner. Thus the almost-complex structure on S^6 induces canonically an almost-complex structure on M.

5.4 WALKER DERIVATION

A. G. Walker introduced a new derivation which he called *torsional derivation* on tensor fields on a manifold which admits an almost-complex structure. This derivation maps tensor fields of type (p, q) in the usual sense to tensorial 2-forms of type (p, q). Walker defined his derivation in terms of local coordinates. In his notation our t^i_{jk} is replaced by his H^i_{jk}. We have

$$h^i_j h^j_k = -\delta^i_k,$$

and an alternative way of writing H^i_{jk} is

$$H^i_{jk} = \tfrac{1}{4}(h^i_p \partial_{[j} h^p_{k]} - h^p_{[j} \partial_{|p|} h^i_{k]}).$$

5.4 Walker derivation

Here the square brackets denote that the skew-symmetric part must be taken with respect to the subscripts j and k, while $|p|$ merely indicates that the suffix p is unaltered by the skew-symmetrization. Then the torsional derivative $T^{i\cdots}_{j\cdots\|rs}$ of a given tensor field was defined by Walker as follows:

$$T^{i\cdots}_{j\cdots\|rs} = H^p_{rs}\partial_p T^{i\cdots}_{j\cdots} + T^{p\cdots}_{j\cdots}h^i_{prs} + \cdots - T^{i\cdots}_{p\cdots}h^p_{jrs} - \cdots,$$

where the right-hand member contains a term like $T^{p\cdots}_{j\cdots}h^i_{prs}$ for each contravariant suffix of $T^{i\cdots}_{j\cdots}$ and a term like $-T^{i\cdots}_{p\cdots}h^p_{jrs}$ for each covariant suffix of $T^{i\cdots}_{j\cdots}$ and where

$$h^i_{jrs} = -\tfrac{1}{2}\partial_j H^i_{rs} + \tfrac{1}{2}h^i_p(h^q_j\partial_q H^p_{rs} - H^q_{rs}\partial_q h^p_j + H^p_{qs}\partial_r h^q_j - H^p_{qr}\partial_s h^q_j).$$

Walker arrived at this rather strange formula as a result of his work on parallel r-planes. However the following alternative definition was first discovered by A. Nijenhuis. We define the action of the derivation on a scalar field f by

$$D: f \to \mathrm{d}f \barwedge H$$

where $(Df \barwedge H)(u_1, u_2) = H(u_1, u_2)f$. In terms of local coordinates

$$D: f \to H^i_{rs}\partial_i f.$$

For the action on contravariant vector fields we have

$$D: u \to \tfrac{1}{2}h[hu, H] - \tfrac{1}{2}[u, H]$$

where $[u, H]$ denotes the Lie derivative of H with respect to the vector field u, that is, in terms of local coordinates

$$([u, H])^i_{jk} = u^p\partial_p H^i_{jk} - \partial_p u^i \cdot H^p_{jk} + \partial_j u^p \cdot H^i_{pk} + \partial_k u^p \cdot H^i_{jp}.$$

Making use of the relation $h^2 = -1$, it is easily checked that

$$D(fu) = Df \otimes u + fDu,$$

and this enables the domain of D to be extended to include tensor fields of arbitrary type (p, q) in such a way that

$$D(T_1 \otimes T_2) = DT_1 \otimes T_2 + T_1 \otimes DT_2.$$

We have assumed that the derivation commutes with contraction, as was the case in Walker's definition.

It is easily verified that $h^i_{j\|rs} = 0$ and $\delta^i_{j\|rs} = 0$.

With the non-commutability of covariant differentiation as a guide, Walker defined the *torsional curvature tensor* as follows. For an arbitrary vector field u we have

$$u^i_{\|rs\|pq} - u^i_{\|pq\|rs} = H^i_{trspq}u^t$$

where the tensor H^i_{trspq} depends only on h and its derivatives. This can be considered as the intrinsic curvature tensor of the almost-complex structure. This suggests a possible classification of almost-complex structures. For example, what are the special properties of almost-complex structures for which $H^i_{trspq\|uv} = 0$? This and similar questions remain research problems – almost nothing is known.

5.5 HERMITIAN METRICS AND KÄHLER METRICS

A Hermitian metric on an almost-complex manifold M is a Riemannian metric g which is invariant with respect to the almost-complex structure J, that is, $g(JX, JY) = g(X, Y)$ for any vector fields X, Y on M. The manifold is then called *almost Hermitian*.

The existence of a Hermitian metric imposes no extra condition on an almost-complex manifold. For, if we take any Riemannian metric g on M we get a Hermitian metric h by setting

$$h(X, Y) = g(X, Y) + g(JX, JY).$$

We know from our previous examination of complex vector spaces that every Hermitian metric g on an almost-complex manifold can be uniquely extended to a complex symmetric tensor field covariant of degree 2, denoted also by g, with the properties:

(i) $g(\bar{Z}, \bar{W}) = \overline{g(Z, W)}$ for complex vector fields Z and W.
(ii) $g(Z, \bar{Z}) > 0$ for any non-zero complex vector Z.
(iii) $g(Z, \bar{W}) = 0$ for any vector field Z of type $(1, 0)$ and any vector field W of type $(0, 1)$.

Conversely every complex symmetric tensor field, covariant of degree 2 with these properties is the complex extension of a Hermitian metric on M.

We have already associated a skew-symmetric bilinear form on a vector space with a Hermitian inner product. The same process leads to the fundamental 2-form Φ of an almost Hermitian manifold, defined by $\Phi(X, Y) = g(X, JY)$ for all vector fields X and Y. Clearly $\Phi(JX, JY) = \Phi(X, Y)$.

Due to the commonness of Hermitian metrics, it seems natural to impose further conditions on such metrics. One way of doing this is to impose the condition that the form Φ be closed. Another way is to impose the condition $\nabla_X J = 0$, where ∇ is the Levi-Civita connexion obtained from g. We shall see that these conditions are related, and that for complex manifolds they are equivalent. A Hermitian metric which satisfies $d\Phi = 0$ is called a *Kähler metric*. An almost complex manifold

5.5 Hermitian metrics and Kähler metrics

with a Kähler metric is called an *almost-Kähler manifold*. If in addition the Nijenhuis tensor vanishes, the manifold is called a *Kähler manifold*.

Many years ago I was introduced to Professor Kähler and I said that, for me, it was a great honour to meet the author of Kähler manifolds. In broken English he denied all knowledge of Kähler manifolds and remarked that he was an expert only on partial differential equations. I have never forgotten this experience!

We indicate how to prove that the conditions $d\Phi = 0$ and $\nabla_X J = 0$ are equivalent for Kähler manifolds. First we have

$$g((\nabla_X J)Y, Z) = g(\nabla_X (JY), Z) - g(J(\nabla_X Y), Z)$$
$$= g(\nabla_X (JY), Z) + g(\nabla_X Y, JZ).$$

Moreover, from the definition of the Christoffel symbols of a Riemannian metric we have

$$2g(\nabla_X Y, Z) = X(g(Y, Z)) + Y(g(X, Z)) - Z(g(X, Y))$$
$$+ g([X, Y], Z) + g([Z, X], Y) + g(X, [Z, Y]).$$

Using this formula for the two terms on the right we get a total of 12 terms. We next apply the formula for the exterior derivative of a 2-form, namely,

$$3 d\Phi(X, Y, Z) = X(\Phi(Y, Z)) + Y(\Phi(Z, X)) + Z(\Phi(X, Y))$$
$$- \Phi([X, Y], Z) - \Phi([Z, X], Y) - \Phi([Y, Z], X)$$

to the expression

$$6 d\Phi(X, JY, JZ) - 6 d\Phi(X, Y, Z)$$

to get a further 12 terms. When we substitute in the expression

$$4g((\nabla_X J)Y, Z) - 6 d\Phi(X, JY, JZ) + 6 d\Phi(X, Y, Z)$$

we find that 20 of the 24 terms cancel, leaving four terms which make up $g(N(Y, Z), JX)$. In this way we establish

$$4g((\nabla_X J)Y, Z) = 6 d\Phi(X, JY, JZ) - 6 d\Phi(X, Y, Z)$$
$$+ g(N(Y, Z), JX).$$

From this we clearly see that $N = 0$ and $d\Phi = 0$ imply $\nabla_X J = 0$.

To prove the converse, we choose normal coordinates centred at an arbitrary point $P \in M$ so that the Christoffel symbols evaluated at P are zero. Then $\nabla_X J = 0$ implies that $\partial_k h_j^i = 0$ at P, and hence the Nijenhuis tensor vanishes at P. Since P was arbitrary it follows that $N = 0$. Also since $\Phi(X, Y) = g(X, JY)$ and both g and J have zero covariant derivative, it follows that $\nabla \Phi = 0$. Again using normal coordinates, this implies that at P, $d\Phi = 0$. Since P was arbitrary it follows that $d\Phi = 0$.

5.6 COMPLEX MANIFOLDS

From now on we concentrate on complex manifolds and, in particular, on Kähler manifolds. First however we consider examples of complex manifolds.

The simplest example is the complex number space $\mathbb{C}_m$ whose points are ordered m-tuples of complex numbers $(z^1, \ldots, z^m)$.

This is covered by a single coordinate system. In particular $\mathbb{C}_1$ is called the Gauss plane.

Our second example is the complex projective space $\mathbb{C}P^m$. To define this we consider the space $\mathbb{C}_{m+1} \setminus \{0\}$, and impose an equivalence relation by identifying those point $(z^0, \ldots, z^n)$ which differ from each other by a non-zero complex factor. The resulting quotient space is $\mathbb{C}P^m$. We claim that the open sets U_i, defined by $z^i \neq 0$, $i = 0, \ldots, m$, form a covering of $\mathbb{C}P^m$. In U_i we have local coordinates

$$_i\zeta^k = \frac{z^k}{z^i}, \qquad 0 \leqslant k \leqslant m, \, k \neq i.$$

The transformation of local coordinates in $U_i \cap U_j$ is given by

$$_j\zeta^h = \frac{_i\zeta^h}{_i\zeta^j}, \qquad 0 \leqslant h \leqslant m, \, h \neq j,$$

which are holomorphic functions. In particular $\mathbb{C}P^1$ is the Riemann sphere and is homeomorphic to S^2.

By assigning to a point of $\mathbb{C}_{m+1} \setminus \{0\}$, the point determined in the quotient space $\mathbb{C}P^m$, we get a projection map $\psi: \mathbb{C}_{m+1} \setminus \{0\} \to \mathbb{C}P^m$, for which the inverse image of each point is $C^* = C_1 \setminus \{0\}$. This is a simple example of a holomorphic line bundle, a concept which plays an important role in more advanced complex manifold theory and also in algebraic geometry.

In $\psi^{-1}(U_i)$ we can use coordinates

$$_i\zeta^h = \frac{z^h}{z^i}, \qquad 0 \leqslant h \leqslant m, \, h \neq i,$$

together with the complex number z^i which specifies the fibre coordinate relative to U_i. This shows that $\psi^{-1}(U_i)$ is homeomorphic to $U_i \times C^*$. Moreover in $U_i \cap U_j$ the fibre coordinates z^i, z^j are related by

5.6 Complex manifolds

$$z^i = z^i_j \zeta^i = \frac{z^j}{i \zeta^j}.$$

Thus the change of fibre coordinate is effected by multiplication by a non-zero holomorphic function.

To each point $p \in \mathbb{C}P^m$ we may associate the coordinates of $\psi^{-1}(p)$, called its homeogeneous coordinates. These can be normalized so that

$$\sum_{k=0}^{m} z^k \bar{z}^k = 1.$$

If we write $z^k = x^k + iy^k$, then this equation represents a sphere S^{2m+1} as a submanifold of $\mathbb{R}^{2m+2}$. The restriction of ψ to S^{2m+1} gives a mapping $\psi: S^{2m+1} \to \mathbb{C}P^m$ under which the inverse image of each point is a circle. This map was first studied by Hopf, and gives rise to what is now called the *Hopf fibration of S^{2m+1}*. Perhaps this is the simplest non-trivial example of a circle bundle.

Further examples of complex manifolds can be obtained as submanifolds of $\mathbb{C}P^m$ and as quotient manifolds of $\mathbb{C}^m$.

For example we have the non-singular hyperquadric of $\mathbb{C}P^m$ given by

$$(z^0)^2 + \cdots + (z^m)^2 = 0.$$

There is an important theorem of Chow—see for example, Gunning and Rossi (1965) page 70.

Theorem (5.6.1) (Chow's theorem) *Every compact submanifold imbedded in $\mathbb{C}P^m$ is an algebraic variety; that is, it is the space defined by a finite number of homogeneous polynomial equations.*

On the other hand there is no point in considering compact submanifolds of $\mathbb{C}^m$ because of the following theorem.

Theorem (5.6.2) *A connected compact submanifold of $\mathbb{C}^m$ must be a single point.*

The proof of the theorem depends on the following lemma.

Lemma (5.6.3) *Let f be a holomorphic function on a complex manifold M. Suppose that $p_0 \in M$ has the property that $|f(p)| \leq |f(p_0)|$ for all p in a neighbourhood of p_0. Then $f(p) = f(p_0)$ in some neighbourhood of p_0.*

For $m = 1$ the result follows from classical complex variable theory in the form of the maximum modulus principle. In the more general case of m dimensions we apply the previous argument to lines through p_0.

Now let M be a connected compact submanifold of $\mathbb{C}^m$. Each coordinate of $\mathbb{C}^m$ is a holomorphic function on M and from the lemma it follows that it must be constant on each connected component of M. Hence M must be a point and the theorem is proved.

However examples of complex manifolds occur as quotient manifolds of $\mathbb{C}^m$.

5.7 KÄHLER METRICS IN LOCAL COORDINATE SYSTEMS

In this section we follow the treatment of Kobayashi and Nomizu (1963), Chapter 9, which itself follows closely Chapter 8 of Yano and Bochner (1953).

Let M be an m-dimensional complex manifold and let $z^1, \ldots, z^m$ be a local complex coordinate system in M. We let Greek suffixes take values $1, \ldots, m$ and Roman capital suffixes take values $1, \ldots, m, \bar{1}, \ldots, \bar{m}$. Set

$$Z_\alpha = \frac{\partial}{\partial z^\alpha}, \qquad Z_{\bar\alpha} = \frac{\partial}{\partial \bar z^\alpha}. \tag{5.1}$$

As previously described, given a Hermitian metric g on M, we extend the inner product in each tangent space $T_x(M)$ defined by g to a complex symmetric bilinear form in the complex tangent space $T_x^c(M)$. We set

$$g_{AB} = g(Z_A, Z_B). \tag{5.2}$$

We now show that the Hermitian property gives rise to the relations

$$g_{\alpha\beta} = g_{\bar\alpha\bar\beta} = 0, \tag{5.3}$$

and that $(g_{\alpha\bar\beta})$ is an $n \times n$ Hermitian matrix. Since g is J-invariant we have

$$g\left(\frac{\partial}{\partial z^\alpha}, \frac{\partial}{\partial z^\beta}\right) = g\left(J\frac{\partial}{\partial z^\alpha}, J\frac{\partial}{\partial z^\beta}\right) = g\left(i\frac{\partial}{\partial z^\alpha}, i\frac{\partial}{\partial z^\beta}\right) = -g\left(\frac{\partial}{\partial z^\alpha}, \frac{\partial}{\partial z^\beta}\right)$$

and so $g_{\alpha\beta} = 0$. Similarly

$$g\left(\frac{\partial}{\partial \bar z^\alpha}, \frac{\partial}{\partial \bar z^\beta}\right) = g\left(J\frac{\partial}{\partial \bar z^\alpha}, J\frac{\partial}{\partial \bar z^\beta}\right) = g\left(-i\frac{\partial}{\partial \bar z^\alpha}, -i\frac{\partial}{\partial \bar z^\beta}\right)$$

$$= -g\left(\frac{\partial}{\partial \bar z^\alpha}, \frac{\partial}{\partial \bar z^\beta}\right)$$

giving $g_{\bar\alpha\bar\beta} = 0$. We write the metric as

$$ds^2 = 2\sum_{\alpha,\beta} g_{\alpha\bar\beta}\, dz^\alpha\, d\bar z^\beta. \tag{5.4}$$

5.7 Kähler metrics in local coordinate systems

The skew-symmetric bilinear form previously obtained for a Hermitian manifold, namely,

$$\phi(Z, W) = -i \sum_{j,k=1}^{m} h_{j\bar{k}} (\xi^j(Z)\bar{\xi}^k(W) - \xi^j(W)\bar{\xi}^k(Z))$$

now gives rise to the fundamental 2-form Φ where

$$\Phi = -2i \sum_{\alpha,\beta} g_{\alpha\bar{\beta}} \, dz^\alpha \wedge d\bar{z}^\beta. \tag{5.5}$$

We have on taking the exterior derivative.

$$d\Phi = -2i \sum_{\alpha,\beta,\gamma} \left[\frac{\partial g_{\alpha\bar{\beta}}}{\partial z^\gamma} dz^\gamma \wedge dz^\alpha \wedge d\bar{z}^\beta + \frac{\partial g_{\alpha\bar{\beta}}}{\partial \bar{z}^\gamma} d\bar{z}^\gamma \wedge dz^\alpha \wedge d\bar{z}^\beta \right] = 0,$$

giving rise to the conditions

$$\frac{\partial g_{\alpha\bar{\beta}}}{\partial z^\gamma} = \frac{\partial g_{\gamma\bar{\beta}}}{\partial z^\alpha}; \qquad \frac{\partial g_{\alpha\bar{\beta}}}{\partial \bar{z}^\gamma} = \frac{\partial g_{\alpha\bar{\gamma}}}{\partial \bar{z}^\beta}, \tag{5.6}$$

which are clearly necessary and sufficient for the metric to be Kähler.

For any affine connexion with covariant derivative operator ∇ on M, we set

$$\nabla_{Z_B} Z_C = \sum_A \Gamma^A_{BC} Z_A \tag{5.7}$$

to obtain by complex linearity an operator on complex vector fields. If we make the convention that $\bar{\bar{\alpha}} = \alpha$, we have

$$\bar{\Gamma}^A_{BC} = \Gamma^{\bar{A}}_{\bar{B}\bar{C}}. \tag{5.8}$$

We now prove that the Kähler condition implies the relations

$$\Gamma^\alpha_{B\bar{\gamma}} = \Gamma^{\bar{\alpha}}_{B\gamma} = 0. \tag{5.9}$$

We have

$$\nabla_{\frac{\partial}{\partial z^\alpha}} \left(J \frac{\partial}{\partial z^\beta} \right) = i \nabla_{\frac{\partial}{\partial z^\alpha}} \left(\frac{\partial}{\partial z^\beta} \right) = i \Gamma^\gamma_{\alpha\beta} \frac{\partial}{\partial z^\gamma} + i \Gamma^{\bar{\gamma}}_{\alpha\beta} \frac{\partial}{\partial \bar{z}^\gamma},$$

$$J \nabla_{\frac{\partial}{\partial z^\alpha}} \left(\frac{\partial}{\partial z^\beta} \right) = J \left(\Gamma^\gamma_{\alpha\beta} \frac{\partial}{\partial z^\gamma} + \Gamma^{\bar{\gamma}}_{\alpha\beta} \frac{\partial}{\partial \bar{z}^\gamma} \right) = i \Gamma^\gamma_{\alpha\beta} \frac{\partial}{\partial z^\gamma} - i \Gamma^{\bar{\gamma}}_{\alpha\beta} \frac{\partial}{\partial \bar{z}^\gamma}$$

giving $\Gamma^{\bar{\gamma}}_{\alpha\beta} = 0$. Similarly

$$\nabla_{\frac{\partial}{\partial z^\alpha}} \left(J \frac{\partial}{\partial \bar{z}^\beta} \right) = -i \nabla_{\frac{\partial}{\partial z^\alpha}} \frac{\partial}{\partial \bar{z}^\beta} = -i \Gamma^\gamma_{\alpha\bar{\beta}} \frac{\partial}{\partial z^\gamma} - i \Gamma^{\bar{\gamma}}_{\alpha\bar{\beta}} \frac{\partial}{\partial \bar{z}^\gamma},$$

$$J \nabla_{\frac{\partial}{\partial z^\alpha}} \left(\frac{\partial}{\partial \bar{z}^\beta} \right) = J \left(\Gamma^\gamma_{\alpha\bar{\beta}} \frac{\partial}{\partial z^\gamma} + \Gamma^{\bar{\gamma}}_{\alpha\bar{\beta}} \frac{\partial}{\partial \bar{z}^\gamma} \right) = i \Gamma^\gamma_{\alpha\bar{\beta}} \frac{\partial}{\partial z^\gamma} - i \Gamma^{\bar{\gamma}}_{\alpha\bar{\beta}} \frac{\partial}{\partial \bar{z}^\gamma}$$

so that $\Gamma^\gamma_{\alpha\bar\beta} = 0$.

The condition that the connexion has zero torsion, that is,

$$\nabla_X Y - \nabla_Y X = [X, Y],$$

leads to the relations

$$\Gamma^\alpha_{\beta\gamma} = \Gamma^\alpha_{\gamma\beta}, \qquad \Gamma^{\bar\alpha}_{\bar\beta\bar\gamma} = \Gamma^{\bar\alpha}_{\bar\gamma\bar\beta} \qquad (5.10)$$

while all other components

$$\Gamma^A_{BC} = 0. \qquad (5.11)$$

The condition $\nabla_C g_{AB} = 0$, together with the above relations for a Kähler metric, lead to

$$\sum_\alpha g_{\alpha\bar\epsilon}\Gamma^\alpha_{\beta\gamma} = \frac{\partial g_{\bar\epsilon\beta}}{\partial z^\gamma}, \qquad \sum_\alpha g_{\bar\alpha\epsilon}\Gamma^{\bar\alpha}_{\bar\beta\bar\gamma} = \frac{\partial g_{\epsilon\bar\beta}}{\partial \bar z^\gamma}, \qquad (5.12)$$

which, in fact, determine all the Γ^A_{BC}.

We set

$$R(Z_C, Z_D) = \sum_A K^A_{BCD} Z_A. \qquad (5.13)$$

We write

$$K_{ABCD} = g(R(Z_C, Z_D) Z_B, Z_A) \qquad (5.14)$$

so that

$$K_{ABCD} = \sum_E g_{AE} K^E_{BCD}. \qquad (5.15)$$

Since $\nabla J = 0$, it follows that $R(Z_C, Z_D)$ commutes with J and hence

$$K^\alpha_{\bar\beta CD} = K^{\bar\alpha}_{\beta CD} = 0. \qquad (5.16)$$

Since $g_{\alpha\beta} = g_{\bar\alpha\bar\beta} = 0$ it follows that

$$K_{\alpha\beta CD} = K_{\bar\alpha\bar\beta CD}. \qquad (5.17)$$

The usual symmetry rules give

$$K_{ABCD} = -K_{ABDC} = -K_{BACD} = K_{CDAB}.$$

Hence we have

$$K^A_{B\gamma\delta} = K^A_{B\bar\gamma\bar\delta} = 0 \qquad (5.18)$$

and

$$K_{AB\gamma\delta} = K_{AB\bar\gamma\bar\delta} = 0. \qquad (5.19)$$

Consequently for any Kähler manifold, the only non-zero components of the curvature tensor are necessarily of the form

5.7 Kähler metrics in local coordinate systems

$$K^\alpha_{\beta\gamma\bar\delta}, \quad K^\alpha_{\bar\beta\gamma\bar\delta}, \quad K^{\bar\alpha}_{\beta\gamma\bar\delta}, \quad K^{\bar\alpha}_{\bar\beta\gamma\bar\delta},$$

$$K_{\alpha\bar\beta\gamma\bar\delta}, \quad K_{\alpha\bar\beta\gamma\bar\delta}, \quad K_{\bar\alpha\bar\beta\gamma\bar\delta}, \quad K_{\bar\alpha\bar\beta\gamma\bar\delta}.$$

In the formula

$$R(X, Y)Z = \nabla_X \nabla_Y Z - \nabla_Y \nabla_X Z - \nabla_{[X, Y]} Z$$

we put $X = Z_\gamma$, $Y = Z_{\bar\delta}$, $Z = Z_\beta$ and after making use of (5.7), (5.9), (5.10) and (5.11) we get

$$K^\alpha_{\beta\gamma\bar\delta} = -\frac{\partial \Gamma^\alpha_{\beta\gamma}}{\partial \bar z^\delta}. \tag{5.20}$$

Using (5.3), (5.12), and (5.15) in (5.20) we get

$$K_{\alpha\bar\beta\gamma\bar\delta} = \frac{\partial^2 g_{\alpha\bar\beta}}{\partial z^\gamma \partial \bar z^\delta} - \sum_{\lambda,\mu} g^{\mu\bar\lambda} \frac{\partial g_{\alpha\bar\mu}}{\partial z^\gamma} \frac{\partial g_{\bar\beta\lambda}}{\partial \bar z^\delta} \tag{5.21}$$

where $(g^{\alpha\bar\beta})$ is the inverse matrix of $(g_{\alpha\bar\beta})$.

The components K_{AB} of the Ricci tensor are given by

$$K_{AB} = \sum_C K^C_{ACB}.$$

If we use (5.16), (5.18), and (5.20) we find

$$K_{\alpha\bar\beta} = -\frac{\partial \Gamma^\gamma_{\alpha\gamma}}{\partial \bar z^\beta}; \quad K_{\alpha\bar\beta} = \bar K_{\alpha\bar\beta}; \quad K_{\alpha\beta} = K_{\bar\alpha\bar\beta} = 0. \tag{5.22}$$

We denote by G the determinant of the matrix $(g_{\alpha\bar\beta})$, and on differentiation we get

$$\frac{\partial G}{\partial z^\alpha} = \sum_{\beta,\gamma} G g^{\beta\bar\gamma} \frac{\partial g_{\beta\bar\gamma}}{\partial z^\alpha},$$

that is

$$\frac{\partial (\log G)}{\partial z^\alpha} = \sum_\gamma \Gamma^\gamma_{\alpha\gamma}, \tag{5.23}$$

where we have used (5.12). Using this equation together with (5.22) gives

$$K_{\alpha\bar\beta} = -\frac{\partial^2 \log G}{\partial z^\alpha \partial \bar z^\beta}. \tag{5.24}$$

The local expression for the *Ricci form* ρ is

$$\rho = -2i \sum K_{\alpha\bar\beta} \, dz^\alpha \wedge d\bar z^\beta. \tag{5.25}$$

This Ricci form is real and closed.

174   5   *Complex and almost-complex manifolds*

Let F be a real-valued function in a coordinate neighbourhood of M and set

$$g_{\alpha\bar{\beta}} = \frac{\partial^2 F}{\partial z^\alpha \partial \bar{z}^\beta}. \tag{5.26}$$

Then $(g_{\alpha\bar{\beta}})$ is Hermitian, and also satisfies equation (5.6), which is a necessary and sufficient condition for the metric to be Kähler. It follows that any such F gives rise to a Kähler metric. In fact the converse is also true, namely, every Kähler metric can be written $\mathrm{d}s^2 = 2g_{\alpha\bar{\beta}}\,\mathrm{d}z^\alpha\,\mathrm{d}\bar{z}^\beta$ where $(g_{\alpha\bar{\beta}})$ is given by (5.26). However we do not prove this result.

The curvature form is a 2-form with values in the endomorphisms of the set $T^{1,0}(M)$ and is defined by

$$\Theta = K_{\alpha\bar{\beta}\delta}^{\phantom{\alpha\bar{\beta}}\gamma}\,\mathrm{d}z^\alpha \wedge \mathrm{d}\bar{z}^\beta.$$

If we define $c_1,\ldots,c_m$ by

$$\det(1 + it\Theta) = 1 + tc_1 + \cdots + t^m c_m,$$

then it turns out that c_i is a real closed $2i$-form and is independent of the choice of the Kähler metric. Thus c_i defines a cohomology class $c_i(M)$ in the de Rham cohomology group $H^{2i}(M)$. This is called the *i*th *Chern class* and the c_i is called the *i*th *Chern form with respect to g*. Clearly the first Chern form is the Ricci form. We do not make use of these important forms since this would carry us outside the range of the present text.

5.8 HOLOMORPHIC SECTIONAL CURVATURE

We have proved that in an almost-Hermitian manifold the vector fields X and JX are orthogonal. It follows that, at a point p, the vectors X and JX determine a two-dimensional subspace and hence a sectional curvature defined by this subspace. This is called the *holomorphic sectional curvature* at p in the direction X. If X is a unit vector, the holomorphic sectional curvature is given by $R(X, JX, X, JX)$. It may happen that at the point p the holomorphic sectional curvature is independent of the vector X. In that case we say that the manifold has constant holomorphic sectional curvature at p. Indeed, we have a result which is analogous to Schur's theorem in the case of real Riemannian manifolds, namely, if the holomorphic sectional curvature is constant at each point p of a Kähler manifold, then it has the same constant value over the whole manifold. Such a manifold is said to have *constant holomorphic sectional curvature*. As examples of such manifolds we have the following two cases.

5.8 Holomorphic sectional curvature

(1) It can be shown that the complex projective space $\mathbb{C}P^n$ carries globally the metric described in terms of inhomogeneous coordinates $z^1, \ldots, z^n$ by

$$ds^2 = \frac{4}{c} \cdot \frac{\left(1 + \sum z^\alpha \bar{z}^\alpha\right)\left(\sum dz^\alpha \, d\bar{z}^\alpha\right) - \left(\sum \bar{z}^\alpha \, dz^\alpha\right)\left(\sum z^\alpha \, d\bar{z}^\alpha\right)}{\left(1 - \sum z^\alpha \bar{z}^\alpha\right)^2}$$

where c is a positive constant. It is not difficult to check that the holomorphic sectional curvature is constant; in fact it is equal to c.

(2) For any negative number c, the open unit ball $D_n = \{(z^1, \ldots, z^n) \mid \Sigma z^u \bar{z}^u < 1\}$ in C^n carries globally a metric of constant holomorphic sectional curvature c. With respect to the coordinate system $z^1, \ldots, z^n$ of $\mathbb{C}^n$ it is given by

$$ds^2 = -\frac{4}{c} \cdot \frac{\left(1 - \sum z^\alpha \bar{z}^\alpha\right)\left(\sum dz^\alpha \, d\bar{z}^\alpha\right) - \left(\sum \bar{z}^\alpha \, dz^\alpha\right)\left(\sum z^\alpha \, d\bar{z}^\alpha\right)}{\left(1 - \sum z^\alpha \bar{z}^\alpha\right)^2}$$

We note that this metric is invariant under the identification of $(z^1, \ldots, z^n)$ with $(-z^1, \ldots, -z^n)$. So effectively we have defined a metric of constant homomorphic sectional curvature over the complex hyperbolic space.

It may be checked that the metrics in both (1) and (2) are Kähler.

We have already said that the analogue of Schur's theorem holds for Kähler manifolds. In fact it holds for a wider class of Hermitian manifolds but not for all Hermitian manifolds. For detailed information on this matter the reader is referred to the paper by Gray and Vanhecke (1979).

Kähler manifolds of constant holomorphic sectional curvature are natural analogues of Riemannian manifolds of constant sectional curvature. It is known that a Kähler manifold of constant holomorphic sectional curvature c is locally isometric to either the complex projective space $\mathbb{C}P^n(c)$, the complex hyperbolic space $\mathbb{C}D^n(c)$, or to complex Euclidean space $\mathbb{C}^n$. This result is due to Hawley (1953), and independently to Igusa (1954).

In the paper by Gray and Vanhecke mentioned above the authors show that the sphere $S^6(c)$ with its usual almost-complex structure has constant holomorphic sectional curvature, but it is not a Kähler manifold. Perhaps one of the most striking results of their paper is the following.

Theorem (5.8.1) Let ds^2 be the usual metric on $\mathbb{C}^n$ and let $f: \mathbb{C}^n \to \mathbb{C}$ be any non-linear holomorphic function. Then $\mathbb{C}^n$ with the metric $(1 + \operatorname{Re} f(z))^{-2} ds^2$ is a Hermitian manifold with pointwise constant holomorphic sectional curvature which is not the same for the whole manifold.

In earlier papers Gray introduced a useful generalization of the Kähler condition — a *nearly Kähler* manifold is one for which $(\nabla_X J)X = 0$, as distinct from the Kähler condition $(\nabla_X J)Y = 0$. In fact there are five natural classes of almost-Hermitian manifolds which we denote by K, NK, AK, QK, and H. The defining conditions for these classes are as follows:

$$K: (\nabla_X J)Y = 0,$$

$$NK: (\nabla_X J)X = 0,$$

$$AK: dF = 0,$$

$$QK: (\nabla_X J)Y + (\nabla_{JX} J)JY = 0,$$

$$H: (\nabla_X J)Y - (\nabla_{JX} J)JY = 0, \qquad X, Y \in \mathfrak{X}(M),$$

of which the condition defining H is equivalent to the vanishing of the Nijenhuis tensor. In each of the above classes there exists a curvature identity. In particular we consider the following curvature identities:

(1) $R_{WXYZ} = R_{WXJYJZ}$,

(2) $R_{WXYZ} = R_{JWJXYZ} + R_{JWXJYZ} + R_{JWXYJZ}$,

(3) $R_{WXYZ} = R_{JWJXJYJZ}$.

Following Gray and Vanhecke, for a given class L of almost-Hermitian manifolds, let L_i be the subclass of manifolds whose curvature operator satisfies identity (i), $i = 1, 2, 3$. Certain equalities occur among the various classes. In particular we have

Theorem (5.8.2)

(i) $K_1 = K_2 = K_3 = K$,

(ii) $K = NK_1$,

(iii) $NK_2 = NK_3 = NK$,

(iv) $K = AK_1$,

(v) $H_2 = H_3$.

Result (i) is easy to prove. Results (ii), (iii), and (iv) are proved in Gray (1969). Result (v) is proved in Gray (1976).

5.9 CURVATURE AND CHERN FORMS OF KÄHLER MANIFOLDS

Let M be a Riemannian manifold and let R be the curvature tensor of M with components R_{ijkh} where the indices i, j, k, h represent vectors in a chosen orthonormal basis at $m \in M$. If w, x, y, z are general vectors in M_m, then $R(w, x, y, z)$ denotes the value of the curvature tensor on these vectors. The curvature forms are given by

$$\Omega_{ij}(x, y) = R(i, j, x, y)$$

Now assume that M is a Kähler manifold of real dimension $2n$ with almost-complex structure J. For $m \in M$ we can choose an orthonormal basis $\{1, \ldots, n, 1*, \ldots, n*\}$ where $Ji = i^*, Ji^* = -i$.
For w, x belonging to $T^c_m(M)$ we put

$$\Xi_{ij}(w, x) = \Omega_{ij}(w, x) - \sqrt{(-1)} \Omega_{ij^*}(w, x).$$

Then, using the fact that in a Kähler manifold $R(X, Y) = R(JX, JY)$ we see that Ξ is a skew-symmetric Hermitian matrix of complex 2-forms. It follows that

$$\det_{1 \leq i,j \leq n} \left(\delta_{ij} - \frac{1}{2\pi\sqrt{-1}} \Xi_{ij} \right) = 1 + \gamma_1 + \cdots + \gamma_n$$

is a globally closed form which represents the *total Chern class* of M via de Rham's theorem (for complete details see Kobayashi and Nomizu (1963) Vol. 2, p. 307). In particular the first and second Chern forms are given by

$$2\pi\gamma_1(x, y) = \sqrt{(-1)} \sum_{i=1}^{n} \Xi_{ii}(x, y) = \sum_{i=1}^{n} \Omega_{ii^*}(x, y),$$

$$4\pi^2\gamma_2(w, x, y, z) = \sum_{1 \leq i < j \leq n} \{\Omega_{ii^*} \wedge \Omega_{jj^*} - \Omega_{ij} \wedge \Omega_{ij} - \Omega_{ij^*} \wedge \Omega_{ij^*}\}(w, x, y, z).$$

Thus the first Chern class is closely related to the Ricci tensor — in fact we have

$$R(x, y) = 2\pi\gamma_1(x, Jy).$$

It follows that a Kähler manifold for which the Ricci tensor is positive on holomorphic plane sections (generated by X and JX) satisfies the condition that γ^1 is positive on such sections. Details of relations between curvature and higher Chern forms will be found in Gray (1978a).

The nth Chern class $c_n(M)$ is related to the Euler-Poincaré characteristic χ of a compact Kähler manifold M by

$$c_n(T(M)) \cdot M = \chi(M)$$

where the left-hand side stands for the value of $c_n(M)$ taken over the fundamental cycle of M.

If the Ricci form is a multiple of the Kähler form, the manifold is said to be *Kähler–Einstein*. One of the most important results obtained in the last 15 years is due to S. T. Yau, the simplest form of which states that in a Kähler manifold whose first Chern class is zero another Kähler metric can be found with respect to which the manifold becomes Kähler–Einstein.

In the remainder of this chapter we shall consider the elements of twistor theory of a Riemannian manifold involving the bundle of almost-Hermitian structures. Much of this material was covered in a course of lectures at Durham given by Dr. David Wilkins, and I am grateful to him for permission to make use of this here. We do not probe deeply into the subject but the reader may find it a useful preliminary to the more detailed book by Burstall and Rawnsley (1990).

5.10 REVIEW OF COMPLEX STRUCTURES ON VECTOR SPACES

Let V be an even-dimensional inner product vector space over the reals. A complex structure on V is an endomorphism $J: V \to V$ satisfying $J^2 = -I$. The map J is compatible with the inner product if and only if $J^T = J^{-1}$, where J^T is the transpose of J. We see immediately that $SO(V) \cap so(V)$ is precisely the set of all complex structures on V compatible with the inner product. Here $SO(V)$ denotes the special orthogonal group acting on V and $so(V)$ is its Lie algebra (that is, the set of skew-symmetric $2n \times 2n$ matrices).

Every complex structure J on V determines an orientation of V. This orientation has the property that if the vectors $e_1, \ldots, e_n \in V$ are such that $\{e_1, Je_1, \ldots, e_n, Je_n\}$ is a basis of V, then this basis is positively oriented. If we specify an orientation of V then a complex structure J is said to be compatible with the orientation if the orientation of V determined by J agrees with the specified orientation.

Let W denote the subset of $so(V)$ consisting of all complex structures on V compatible with the inner product and orientation of V. Then W is a homogeneous space. Indeed, if we choose $J_0 \in W$ and define a map

$$f: SO(V) \to W: A \mapsto AJ_0A^{-1}$$

then f is surjective and thus defines a diffeomorphism from $SO(V)/U(V, J_0)$ onto W where

5.10 Review of complex structures on vector spaces

$$U(V, J_0) \equiv \{A \in SO(V) : AJ_0 = J_0 A\}.$$

Given $J \in W$ we have an involution of $so(V)$ mapping $a \in so(V)$ to JaJ^{-1}. Let $u(V, J)$ denote the positive eigenspace of this involution and let $w(V, J)$ denote the negative eigenspace. We claim that $w(V, J)$ is the tangent space to W at J. To see this, let us suppose that $t \to J(t)$ is a curve in $so(V)$ which is contained in W. Since $J(t)^2 = -I$, it follows that

$$0 = \frac{d}{dt}(J(t)^2) = \frac{dJ}{dt}(t)J(t) + J(t)\frac{dJ}{dt}(t)$$

and hence

$$\frac{d}{dt}J(t) \in w(V, J(t)).$$

Conversely let $t \to J(t)$ be a curve in $so(V)$ with the property that $J(0)^2 = -I$, and

$$\frac{dJ}{dt}(t) \in w(V, J(t)).$$

Then

$$\frac{d}{dt}(J(t)^2) = 0$$

and hence $J(t)^2 = -I$, showing that the curve $t \to J(t)$ does indeed lie in W. Thus $w(V, J)$ is the tangent space to W at J.

Let us choose an $SO(V)$-invariant inner product $\langle \ , \ \rangle$ on $so(V)$, such as the inner product

$$\langle a, b \rangle \mapsto \text{trace}(ab^T).$$

Such an inner product restricts to a Riemannian metric on W.

We now define an almost-complex structure on W. Given an element a of the tangent space $w(V, J)$ to W at J, define

$$\Im a = \tfrac{1}{2}[J, a].$$

Note that

$$\Im a = \tfrac{1}{2}(Ja - aJ) = Ja = -aJ,$$

since $a \in w(V, J)$, and hence

$$J(\Im a)J^{-1} = -J(aJ)J^{-1} = -Ja = -\Im a,$$

and also

$$\Im^2 a = J(Ja) = -a.$$

Thus $\mathfrak{J}$ does indeed define a complex structure on the tangent space to W at J. By defining $\mathfrak{J}$ in this fashion on every tangent space to W, we thus define an almost-complex structure $\mathfrak{J}$ on the Riemannian manifold W. We have the following theorem.

Theorem (5.10.1) *The space $(W, \mathfrak{J})$ is a symmetric space and a Kähler manifold.*

To prove the theorem we must first check that $\mathfrak{J}$ is compatible with the Riemannian metric on W. Since the inner product on $\mathrm{so}(V)$ is chosen to be $\mathrm{SO}(V)$-invariant, we see that

$$\langle [c,a], b \rangle + \langle a, [c,b] \rangle = 0$$

for all $a, b, c \in \mathrm{so}(V)$. In particular if $J \in W$ and $a, b \in w(V, J)$ then

$$\langle [\tfrac{1}{2}J, a], b \rangle + \langle a, [\tfrac{1}{2}J, b] \rangle = 0$$

showing that

$$\langle \mathfrak{J}a, b \rangle + \langle a, \mathfrak{J}b \rangle = 0,$$

whenever a and b belong to the tangent space to W at J. Replacing b by $\mathfrak{J}b$, we see that

$$\langle \mathfrak{J}a, \mathfrak{J}b \rangle = \langle a, b \rangle$$

as required.

If $J \in W$, $a \in w(V, J)$, and $A \in \mathrm{SO}(V)$, then

$$A(\mathfrak{J}a)A^{-1} = \tfrac{1}{2} A[J, a]A^{-1}$$
$$= \tfrac{1}{2} [AJA^{-1}, AaA^{-1}]$$
$$= \mathfrak{J}(AaA^{-1})$$

on considering AaA^{-1} as an element of the tangent space to W at the complex structure $AJA^{-1} \in W$. Thus the almost-complex structure $\mathfrak{J}$ on W is $\mathrm{SO}(V)$-invariant.

Given $J \in W$, the map sending $J_1 \in W$ to JJ_1J^{-1} is an isometry of W which reverses all geodesics through J. Thus W is a symmetric space.

If $\nabla \mathfrak{J}$ is the covariant derivative of the almost-complex structure $\mathfrak{J}$, then $\nabla \mathfrak{J}$ is an $\mathrm{SO}(V)$-invariant tensor of odd order. In particular $\nabla \mathfrak{J}$ is invariant under the isometry $J_1 \to JJ_1J^{-1}$, for all $J \in W$. Since $\nabla \mathfrak{J}$ is of odd order it follows immediately that $\nabla \mathfrak{J} = 0$. Hence $(W, \mathfrak{J})$ is a Kähler manifold, and the theorem is proved.

5.11 THE TWISTOR SPACE OF A RIEMANNIAN MANIFOLD

Let M be an oriented even-dimensional Riemannian manifold, and let $\mathrm{so}(M) \to M$ be the vector bundle whose fibre over $m \in M$ is the vector

5.12 The twistor space of S^4

space of skew-symmetric endomorphisms of M_m. Let $\pi: \mathfrak{T}_+(M) \to M$ be the fibre bundle over M whose fibre over $m \in M$ consists of all complex structures on M_m compatible with the metric and orientation. Note that we have an inclusion $\mathfrak{T}_+(M) \hookrightarrow \operatorname{so}(M)$ of fibre bundles over M.

The Levi-Civita connexion on M determines a connexion on the vector bundle $\operatorname{so}(M) \to M$. For all $a \in \operatorname{so}(M)$, denote the vertical and horizontal spaces of $\operatorname{so}(M)_a$ by $V_a \operatorname{so}(M)$ and $H_a \operatorname{so}(M)$, respectively. The splitting of the tangent bundle of $\operatorname{so}(M)$ into vertical and horizontal subbundles induces a corresponding splitting of the tangent bundle of $\mathfrak{T}_+(M)$. For all $J \in \mathfrak{T}_+(M)$ let $V_J \mathfrak{T}_+(M)$ and $H_J \mathfrak{T}_+(M)$ denote the vertical and horizontal subspaces of $T_J \mathfrak{T}_+(M)$, respectively.

We now define two almost-complex structures $\mathfrak{J}_1$ and $\mathfrak{J}_2$ on the twistor space $\mathfrak{T}_+(M)$. First we define $\mathfrak{J}_1$ and $\mathfrak{J}_2$ on horizontal vectors. Let $J \in \mathfrak{T}_+(M)$ be a complex structure on the tangent space M_m to M at m. Let $X \in H_J \mathfrak{T}_+(M)$ be the horizontal lift at J of some tangent vector $Y \in M_m$. We define both $\mathfrak{J}_1 X$ and $\mathfrak{J}_2 X$ to be the horizontal lift of JY at J. Next we define $\mathfrak{J}_1$ and $\mathfrak{J}_2$ on vertical vectors. Note that the fibre of $\mathfrak{T}_+(M) \to M$ over m is the manifold of complex structures on M_m, on which we have already defined a complex structure. We define $\mathfrak{J}_1$ on vertical vectors so as to agree with this structure. Thus if J is an element of the fibre of $\mathfrak{T}_+(M) \to M$ and if $X \in V_J \mathfrak{T}_+(M)$ corresponds to some element a of $\operatorname{so}(M_m)$ (so that X is tangent to the line $t \to J + ta$ at $t = 0$ where a satisfies $aJ + Ja = 0$) then we define both $\mathfrak{J}_1 X$ and $-\mathfrak{J}_2 X$ to be the vertical vector at J corresponding to the element $\frac{1}{2}[J, a]$ of $\operatorname{so}(M_m)$. This defines $\mathfrak{J}_1$ and $\mathfrak{J}_2$ on both horizontal and vertical vectors, and hence defines $\mathfrak{J}_1$ and $\mathfrak{J}_2$ on all tangent vectors to $\mathfrak{T}_+(M)$ by linearity.

Let $\phi: M \to M$ be an isometry of M. Then ϕ lifts to a fibre preserving map $\tilde\phi: \mathfrak{T}_+(M) \to \mathfrak{T}_+(M)$ covering ϕ. If $J \in \mathfrak{T}_+(M)$ is a complex structure on the tangent space to M at m, then $\tilde\phi(J)$ is defined to be the complex structure $\phi_* \circ J \circ \phi_*^{-1}$ on the tangent space to M at $\phi(m)$. Here ϕ_* denotes the derivative of ϕ at m.

In particular, suppose that M is a symmetric space, and for all $m \in M$ let us denote by τ_m the inversion about m. Let $J \in \mathfrak{T}_+(M)$ be a complex structure on the tangent space to M at m. Then the derivative at J of the lift $\tilde\tau_m$ of τ_m is an involution of the tangent space to $\mathfrak{T}_+(M)$ at J. The positive and negative eigenspaces of this involution are the vertical and the horizontal spaces at J.

5.12 THE TWISTOR SPACE OF S^4

We shall show that the twistor space of S^4 is $\mathbb{C}P^3$ and that the twistor map is the quotient map from $\mathbb{C}P^3$ onto $\mathbb{H}P^1$, where we identify quaternionic projective space $\mathbb{H}P^1$ with S^4.

We identify $\mathbb{H}$ with R^4 in the standard fashion, and we regard $\mathbb{R}$ and $\mathbb{C}$ as being embedded in $\mathbb{H}$. We identify $\mathbb{H}$ with $\mathbb{C}^2$ via the isomorphism of vector spaces over $\mathbb{C}$ which maps to $(z, w) \in \mathbb{C}^2$ the quaternion $z + w\mathrm{j}$. The standard orientation of $\mathbb{H}$ is that in which $(1, \mathrm{i}, \mathrm{j}, \mathrm{k})$ is a positively oriented basis of $\mathbb{H}$ as a vector space over $\mathbb{R}$.

Every complex structure on $\mathbb{H}$ compatible with the inner product and the standard orientation is of the form $q \to rq$ for some uniquely determined purely imaginary quaternion r of unit norm. To see this, let us denote by J_0 the standard complex structure. J on $\mathbb{H}$ compatible with the inner product and the orientation is then of the form $A^{-1}J_0 A$ for some $A \in SO(4)$. But the double covering of $SO(4)$ is $Sp(1) \times Sp(1)$, where the covering homomorphism sends $(u, v) \in Sp(1) \times Sp(1)$ to the rotation $q \to uq\bar{v}$. In particular there exist $u, v \in Sp(1)$ such that $Aq = uq\bar{v}$. Then

$$J_q = A^{-1}J_0 Aq = \bar{u}\mathrm{i}(uq\bar{v})v = (\bar{u}\mathrm{i}u)q.$$

Now the map $u \to \bar{u}\mathrm{i}u$ sends $Sp(1)$ onto the space of purely imaginary quaternions of unit norm. This shows that every complex structure on $\mathbb{H}$ compatible with the inner product and the positive orientation is of the form $q \to rq$ for some purely imaginary quaternion r of unit norm, as required. In particular we note that the manifold of all such complex structures is isomorphic to the 2-sphere, namely the unit sphere in the space of imaginary quaternions.

We now construct the twistor space of S^4. We let $\mathbb{H}P^1$ be the quotient space of $\mathbb{H}^2 \setminus \{0\}$ under the equivalence relation in which (q_1, q_2) and (r_1, r_2) are defined to be in the same equivalence class if and only if there exists $q \in \mathbb{H} \setminus \{0\}$ with the property that $r_1 = qq_1$ and $r_2 = qq_2$. We denote the equivalence class of (q_1, q_2) by $(q_1 : q_2)$.

We denote by S^7 the space of all elements of $\mathbb{H}^2$ with unit norm. Then $\mathbb{H}P^1$ is the quotient space of S^7 on the left by $Sp(1)$, where $Sp(1)$ acts by left multiplication. We denote the quotient map by $\rho: S^7 \to \mathbb{H}P^1$.

$Sp(1)$ acts as a group of isometries of S^7, and thus there is a unique metric on $\mathbb{H}P^1$ with the property that $\rho: S^7 \to \mathbb{H}P^1$ is a Riemannian submersion. The space $\mathbb{H}P^1$ with this metric may be identified with S^4.

For each $x \in S^7$ let $\lambda_x: S^4_{\rho(x)} \to S^7_x$ be the monomorphism which maps $S^4_{\rho(x)}$ isometrically onto the orthogonal complement in S^7_x of the subspace tangent to the fibre through x of $\rho: S^7 \to S^4$. The image of λ_x is an $\mathbb{H}$-invariant subspace of S^7_x, so that each $x \in S^7$ determines a unique complex structure J_x on $S^4_{\rho(x)}$ with the property that

$$\lambda_x(J_x v) = \mathrm{i}\lambda_x(v)$$

for all $v \in S^4_{\rho(x)}$. The map J_x is compatible with metric and orientation.

5.12 The twistor space of S^4

Thus we see that ρ lifts to a map $\tilde{\rho}: S^7 \to \mathfrak{T}_+(S^4)$ which sends $x \in S^7$ to $J_x: S^4_{\rho(x)} \to S^4_{\rho(x)}$.

Let $x, y \in S^7$ belong to the same fibre of $\rho: S^7 \to S^4$. Then there exists $q \in \mathrm{Sp}(1)$ such that $y = qx$. Then

$$\lambda_x(J_y v) = \bar{q}\lambda_y(J_y v) = \bar{q} i \lambda_y(v) = \bar{q} i q \lambda_x(v).$$

Using the fact that every complex structure on $\mathbb{H}$ is of the form $r \to sr$ for some purely imaginary quaternion s of unit norm, we see that $\tilde{\rho}: S^7 \to \mathfrak{T}_+(S^4)$ is surjective. Also $\tilde{\rho}(x) = \tilde{\rho}(y)$ if and only if $y = qx$ for some $q \in U(1)$. Hence

$$\mathfrak{T}_+(S^4) \cong \frac{S^7}{U(1)} = \mathbb{C}P^3,$$

and the twistor map from $\mathbb{C}P^3$ to S^4 is the map induced by $\rho: S^7 \to S^4$.

We now show that the standard complex structure on $\mathbb{C}P^3$ preserves the vertical and horizontal bundles with respect to the twistor map $\pi: \mathbb{C}P^3 \to S^4$ just constructed. We use the fact that S^4 is a symmetric space.

Let $x \in S^7$. On regarding x as an element of $\mathbb{H}^2$, we see that x generates a quaternionic subspace V of $\mathbb{H}^2$, with orthogonal complement $V^\perp$. Let $\sigma: \mathbb{H}^2 \to \mathbb{H}^2$ be the $\mathbb{H}$-linear involution of $\mathbb{H}^2$ whose positive and negative eigenspaces are V and $V^\perp$, respectively. The map σ induces an isometric involution τ of $\mathbb{H}P^1$, and this involution is the inversion of $\mathbb{H}P^1$ about the point $\rho(x)$. On identifying $\mathbb{H}^2$ with $\mathbb{C}^4$, we see that V and $V^\perp$ are two-dimensional complex subspaces of $\mathbb{C}^4$. We deduce that σ induces a holomorphic involution $\hat{\tau}$ of $\mathbb{C}P^3$. It is easy to see that $\hat{\tau}: \mathbb{C}P^3 \to \mathbb{C}P^3$ is the lift of $\tau: \mathbb{H}P^1 \to \mathbb{H}P^1$ to the twistor space. Since $\hat{\tau}$ is holomorphic, the eigenspaces of the derivative of $\hat{\tau}$ at $[x] \in \mathbb{C}P^3$ are complex subspaces of $\mathbb{C}P^3_{[x]}$, where $[x]$ denotes the image of x under the natural projection $S^7 \to \mathbb{C}P^3$. We deduce that the vertical and the horizontal subspace of $\mathbb{C}P^3_{[x]}$ are preserved by the standard complex structure on $\mathbb{C}P^3$. Moreover these subspaces are mutually orthogonal with respect to the standard metric on $\mathbb{C}P^3$, and the image of $\lambda_x: S^4_{\rho(x)} \to S^7_x$ is mapped onto the horizontal subspace of $\mathbb{C}P^3_{[x]}$ by the derivative of the natural projection from S^7 to $\mathbb{C}P^3$. We conclude from this that if $v \in \mathbb{C}P^3_{[x]}$ is horizontal, then

$$\pi_*(iv) = J_x \pi_*(v)$$

where $\pi_*: \mathbb{C}P^3_{[x]} \to S^4_{\rho(x)}$ is the derivative of the twistor map $\pi: \mathbb{C}P^3 \to S^4$ at $[x]$.

We claim that the standard complex structure on $\mathbb{C}P^3$ corresponds to the almost-complex structure $\mathfrak{J}_1$ defined on the twistor space $\mathfrak{T}_+(S^4)$ of S^4. We have seen that the standard complex structure preserves the vertical and horizontal spaces and agrees with $\mathfrak{J}_1$ on the horizontal space. Thus it only remains to show that the standard complex structure on $\mathbb{C}P^3$ agrees with $\mathfrak{J}_1$ on the vertical space. Let $[x] \in \mathbb{C}P^3$ be the equivalence class of $x \in S^7$. We may identify $\mathbb{C}P^3_{[x]}$ with the orthogonal complement of $\mathbb{C}x$ in $\mathbb{C}^4$. Under this identification the vertical subspace of $\mathbb{C}P^3_{[x]}$ corresponds to the complex subspace of $\mathbb{C}^4$, spanned as a real vector space by jx and kx. A straightforward calculation shows that the derivative of $\tilde{\rho}$ at x maps jx and kx to the endomorphisms a and b of $S^4_{\rho(x)}$ defined by the property that, for all $v \in S^4_{\rho(x)}$,

$$\lambda_x(av) = (ij - ji)\lambda_x(v) = 2k\lambda_x(v),$$
$$\lambda_x(bv) = (ik - ki)\lambda_x(v) = -2j\lambda_x(v).$$

The elements a and b are skew-symmetric and anti-commute with J_x. Then

$$\lambda_x(\tfrac{1}{2}[J_x, a]v) = \lambda_x(J_x av),$$
$$= i\lambda_x(av),$$
$$= 2ik\lambda_x(v),$$
$$= \lambda_x(bv).$$

Thus $b = \tfrac{1}{2}[J_x, a]$. We deduce from this that the standard complex structure on $\mathbb{C}P^3$ (which maps jx to kx) corresponds to the almost-complex structure $\mathfrak{J}_1$, as required.

5.13 TWISTOR SPACES OF SPHERES

Let S^{2n} be a sphere of dimension $2n$. We shall show that the twistor space $\mathfrak{T}_+(S^{2n})$ of S^{2n} is isomorphic to the space W_{2n+2} of complex structures on the vector space $\mathbb{R}^{2n+2}$ compatible with metric and orientation.

To see this, choose a unit vector $e \in \mathbb{R}^{2n+2}$. Regard S^{2n} as the unit sphere in the orthogonal complement E of the line generated by e. Let $J \in W_{2n+2}$ be any complex structure on $\mathbb{R}^{2n+2}$ compatible with metric and orientation. Thus Je is a unit vector orthogonal to e. Thus we have a map $f: W_{2n+2} \to S^{2n}$ which sends $J \in W_{2n+2}$ to Je. The orthogonal complement F_J of the subspace spanned by e and Je may be identified with the tangent space to S^n at Je in the usual manner. Now F_J is invariant under J, hence J restricts to a complex structure $\hat{J}: F_J \to F_J$ on the tangent space to S^{2n} at Je. We may thus define $\tilde{f}: W_{2n+2} \to \mathfrak{T}_+(S^{2n})$ to be the map sending $J \in W_{2n+2}$ to $\hat{J}: F_J \to F_J$. It is easy to see that $\tilde{f}$ is an isomorphism of W^{2n+2} with $\mathfrak{T}_+(S^{2n})$, and that $\pi \circ \tilde{f} = f$, where $\pi: \mathfrak{T}_+(S^{2n}) \to S^{2n}$ is the natural projection.

5.14 Riemannian immersions and Codazzi's equation

Recall that the tangent bundle of the twistor space splits into the Whitney sum of its vertical and horizontal subbundles. We shall now identify the subbundles of the tangent bundle of W_{2n+2} corresponding to these vertical and horizontal bundles. We use the fact that S^{2n} is a symmetric space.

For all $p \in S^{2n}$, let $\sigma_p : \mathbb{R}^{2n+2} \to \mathbb{R}^{2n+2}$ be the involution of $\mathbb{R}^{2n+2}$ whose positive eigenspace is the space $\langle e, p \rangle$ spanned by e and p, and whose negative eigenspace is the orthogonal complement of the positive eigenspace. Then the restriction of σ_p to S^{2n} is the inversion of S^{2n} about p. On regarding W_{2n+2} as the twistor space of S^{2n}, it is easy to see that the map sending $J \in W_{2n+2}$ to $\sigma_p \circ J \circ \sigma_p^{-1}$ is the lift of $\sigma_p | S^{2n}$ to the twistor space of S^{2n}. The tangent space of W_{2n+2} at J is identified with the vector space $w(2n+2, J)$ of all skew-symmetric endomorphisms of $\mathbb{R}^{2n+2}$ that anti-commute with J. If $p = f(J)$, then the vertical space $w_V(2n+2, J)$ at J consists of all $a \in w(2n+2, J)$ such that $\sigma_p a \sigma_p^{-1} = a$, and the horizontal space $w_H(2n+2, J)$ at J consists of all $a \in w(2n+2, J)$ such that $\sigma_p a \sigma_p^{-1} = -a$.

If $a \in w(2n+2, J)$ is vertical, then the subspaces $\langle e, Je \rangle$ and F_J are invariant under J. But a is skew-symmetric and anti-commutes with J, hence $a | \langle e, Je \rangle = 0$. We deduce that if a is vertical then a maps $\mathbb{R}^{2n+2}$ into F_J. If a is horizontal then a maps $\langle e, Je \rangle$ into F_J and maps F_J into $\langle e, Je \rangle$.

We recall that we have defined a complex structure $\mathfrak{J}$ on W_{2n+2} such that if $J \in W_{2n+2}$ then the restriction of $\mathfrak{J}$ to the tangent space to W_{2n+2} at J maps $a \in w(2n+2, J)$ to $\frac{1}{2}[J, a]$. The map $\mathfrak{J}$ preserves the vertical and horizontal subspaces of the tangent space to W_{2n+2} at J, since the subspaces $\langle e, Je \rangle$ and F_J of $\mathbb{R}^{2n+2}$ are invariant under J. Using the fact that $\frac{1}{2}[J, a] = Ja$, we see that if $a \in w(2n+2, J)$ is vertical, then $\mathfrak{J}$ maps a to $\hat{J}a$ (recall that $\hat{J}$ is the restriction of J to F_J). If a is horizontal, then the derivative of the twistor map $f: W_{2n+2} \to S^{2n}$ at J maps a to ae and maps $\mathfrak{J}a$ to $\hat{J}(ae)$. This follows because every element a of $w(2n+2, J)$ maps $\langle e, Je \rangle$ into F_J so that $ae \in F_J$. We deduce that $\mathfrak{J}$ corresponds to the almost-complex structure $\mathfrak{J}_1$ defined on the twistor space $\mathfrak{T}_+(S^{2n})$ of S^{2n}.

5.14 RIEMANNIAN IMMERSIONS AND CODAZZI'S EQUATION

We recall previous results about Riemannian immersions. Let $i: M \hookrightarrow \tilde{M}$ be an isometric immersion of Riemannian manifolds. We denote by NM the normal bundle of M in $\tilde{M}$, consisting of all vectors based at some point of M which are orthogonal to the submanifold M. We denote the Levi-Civita connexions of M and $\tilde{M}$ by ∇ and $\tilde{\nabla}$ respectively. Given vector fields X and Y tangential to M, we may split $\tilde{\nabla}_X Y$ into its tangential and normal components. Then

$$\tilde{\nabla}_X Y = \nabla_X Y + S(X, Y)$$

where $S = TM \otimes TM \to NM$ is the second fundamental tensor of the immersion. This tensor has the property that

$$S(X, Y) = S(Y, X)$$

for all vectors X and Y tangential to M. Given a vector field X tangential to M and another vector field ν normal to M, we may split $\tilde{\nabla}_X \nu$ into its tangential and normal components. Then

$$\tilde{\nabla}_X \nu = -A_\nu X + \mathrm{D}_X \nu$$

where D denotes the covariant derivative defined on sections of the normal bundle $NM \to M$ of M, and where $A_\nu : TM \to TM$ is the tensor field defined by the property that

$$\langle A_\nu X, Y \rangle = \langle S(X, Y), \nu \rangle$$

for all ν, X, and Y. We define the covariant derivative D on tensors $T: TM \otimes TM \to NM$ such that

$$(\mathrm{D}_X T)(Y, Z) = \mathrm{D}_X(T(Y, Z)) - T(\nabla_X Y, Z) - T(Y, \nabla_X Z)$$

for all vector fields X, Y, Z tangential to M.

Let $\tilde{R}$ denote the Riemann curvature tensor of $\tilde{M}$. Codazzi's equation states that, for all vector fields X, Y, Z tangential to M, the normal component $(\tilde{R}(X, Y)Z)^\perp$ of $\tilde{R}(X, Y)Z$ is given by

$$(\tilde{R}(X, Y)Z)^\perp = (\mathrm{D}_X S)(Y, Z) - (\mathrm{D}_Y S)(X, Z).$$

In particular, if $\tilde{M}$ has constant sectional curvature then

$$(\mathrm{D}_X S)(Y, Z) = (\mathrm{D}_Y S)(X, Z).$$

The mean curvature vector field of the immersion $i: M \to \tilde{M}$ is the normal vector field $H: M \to NM$ defined by

$$H = \frac{1}{(\dim M)} \operatorname{trace} S.$$

Thus if $(E_1, \ldots, E_n)$ is an orthonormal moving frame on M, then

$$H = \frac{1}{n} \sum_{j=1}^{n} S(E_j, E_j).$$

We say that the mean curvature vector *is parallel in the normal bundle* if $\mathrm{D}_X H = 0$ for all tangential vector fields X on M.

5.15 ISOTROPIC IMMERSIONS

Let $i: M \hookrightarrow \tilde{M}$ be a conformal immersion of a Riemann surface M into a Riemannian manifold $\tilde{M}$. The canonical bundle of M is, by definition, the cotangent bundle $T^*M \to M$ of M, regarded as a holomorphic line bundle over M. Let $E_4 \to M$ be the holomorphic line bundle which is the fourth tensor power of the canonical bundle of M. We now construct a section $\Lambda: M \to E_4$ of this bundle.

Let $z = x + iy$ be a local complex coordinate on M. We define Λ on the domain of definition of z by

$$\Lambda = \left\langle S\left(\frac{\partial}{\partial z}, \frac{\partial}{\partial z}\right), S\left(\frac{\partial}{\partial z}, \frac{\partial}{\partial z}\right) \right\rangle (dz)^4$$

where $S: TM \times TM \to NM$ is the second fundamental form of the immersion. We define an *isotropic immersion* to be one for which $\Lambda \equiv 0$. Now

$$S\left(\frac{\partial}{\partial z}, \frac{\partial}{\partial z}\right) = S\left(\tfrac{1}{2}\left(\frac{\partial}{\partial x} - i\frac{\partial}{\partial y}\right), \tfrac{1}{2}\left(\frac{\partial}{\partial x} - i\frac{\partial}{\partial y}\right)\right)$$

$$= \tfrac{1}{4}\left(S\left(\frac{\partial}{\partial x}, \frac{\partial}{\partial x}\right) - S\left(\frac{\partial}{\partial y}, \frac{\partial}{\partial y}\right) - 2iS\left(\frac{\partial}{\partial x}, \frac{\partial}{\partial y}\right)\right).$$

We see therefore that the immersion is isotropic if and only if the vectors

$$S\left(\frac{\partial}{\partial x}, \frac{\partial}{\partial x}\right) - S\left(\frac{\partial}{\partial y}, \frac{\partial}{\partial y}\right) \quad \text{and} \quad 2S\left(\frac{\partial}{\partial x}, \frac{\partial}{\partial y}\right)$$

are orthogonal and have the same magnitude. We conclude therefore that a conformal immersion $i: M \hookrightarrow \tilde{M}$ is isotropic if and only if for every moving frame (E_1, E_2) defined on some open set in M, the vectors $S(E_1, E_1) - S(E_2, E_2)$ and $2S(E_1, E_2)$ are orthogonal and have the same length.

Theorem (5.15.1) *Let $i: M \hookrightarrow \tilde{M}$ be a conformal immersion of a Riemann surface M into a space $\tilde{M}$ of constant sectional curvature. If the mean curvature vector of the immersion is parallel in the normal bundle of M, then the section Λ of the holomorphic line bundle $E_4 \to M$ is a holomorphic section.*

Proof. Let us define

$$Z = \frac{\partial}{\partial z} = \tfrac{1}{2}\left(\frac{\partial}{\partial x} - i\frac{\partial}{\partial y}\right),$$

$$\bar{Z} = \frac{\partial}{\partial \bar{z}} = \tfrac{1}{2}\left(\frac{\partial}{\partial x} + i\frac{\partial}{\partial y}\right).$$

Then
$$\nabla_{\bar{Z}}Z = \nabla_Z \bar{Z} = 0,$$
$$\nabla_Z Z = \langle \bar{Z}, Z \rangle^{-1} \langle \bar{Z}, \nabla_Z Z \rangle Z$$

since $\nabla_Z Z$ is a holomorphic vector field and is of the form fZ for some holomorphic function f. Now

$$S(\bar{Z}, Z) = \tfrac{1}{4} S\left(\frac{\partial}{\partial x} + i\frac{\partial}{\partial y}, \frac{\partial}{\partial x} - i\frac{\partial}{\partial y}\right)$$
$$= \tfrac{1}{4}\left(S\left(\frac{\partial}{\partial x}, \frac{\partial}{\partial x}\right) + S\left(\frac{\partial}{\partial y}, \frac{\partial}{\partial y}\right)\right)$$
$$= H\langle \bar{Z}, Z \rangle.$$

Thus
$$(D_Z S)(\bar{Z}, Z) = D_Z(S(\bar{Z}, Z)) - S(\nabla_Z \bar{Z}, Z) - S(\bar{Z}, \nabla_Z Z)$$
$$= D_Z(H\langle \bar{Z}, Z \rangle) - \langle \bar{Z}, Z \rangle^{-1}\langle Z, \nabla_Z Z \rangle S(\bar{Z}, Z)$$
$$= D_Z H \langle \bar{Z}, Z \rangle + H\langle \bar{Z}, \nabla_Z Z \rangle - H\langle \bar{Z}, \nabla_Z Z \rangle$$
$$= D_Z H \langle \bar{Z}, Z \rangle.$$

But $\tilde{M}$ is a space of constant curvature, so we have
$$(D_Z S)(\bar{Z}, Z) = (D_{\bar{Z}} S)(Z, Z),$$
by Codazzi's equation. Thus if H is parallel in the normal bundle, then
$$\frac{\partial}{\partial \bar{z}}\langle S(Z, Z), S(Z, Z) \rangle = 2\langle S(Z, Z), (D_{\bar{Z}} S)(Z, Z) \rangle$$
$$= 2\langle S(Z, Z), (D_Z S)(\bar{Z}, Z) \rangle$$
$$= 2\langle \bar{Z}, Z \rangle \langle S(Z, Z), D_Z H \rangle$$
$$= 0$$

and hence Λ is holomorphic.

Corollary (5.15.2) Let $i: S^2 \hookrightarrow \tilde{M}$ be a conformal immersion of S^2 in a space $\tilde{M}$ of constant sectional curvature. If the mean curvature vector of the immersion is parallel, then the immersion is isotropic.

Proof. The canonical bundle of S^2 has degree -2. Hence the holomorphic bundle $E_4 \to S^2$ has degree -8. Therefore the only holomorphic section of this bundle is the zero section. But Λ is holomorphic by the above theorem, and so is identically zero. Hence the immersion is isotropic and the corollary is proved.

5.16 THE BUNDLE OF ORIENTED ORTHONORMAL FRAMES

Let M^{2n} be an oriented even-dimensional Riemannian manifold. Let $\tilde{\pi}: F(M) \to M$ be the principal $SO(2n)$ bundle over M whose fibre over $m \in M$ consists of all oriented orthonormal frames $f: \mathbb{R}^{2n} \to M_m$. The group $SO(2n)$ acts on $F(M)$ on the right, sending $(f, A) \in F(M) \times SO(2n)$ to the composition $f \circ A$. For all $a \in so(2n)$, let $\sigma(a)$ denote the vector field on $F(M)$ generating the one-parameter group of diffeomorphisms

$$(f, t) \to f \circ \exp(ta).$$

We say that $\sigma(a)$ is the *fundamental vertical vector field* on $F(M)$ determined by a. We denote the value of $\sigma(a)$ at $f \in F(M)$ by $\sigma_f(a)$.

The Levi-Civita connexion on M determines a splitting of the tangent space $TF(M)$ as a Whitney sum $VF(M) \oplus HF(M)$ of vector bundles over $F(M)$. The *vertical bundle* $VF(M)$ consists of all vectors tangent to the fibres of $\tilde{\pi}: F(M) \to M$. The bundle $HF(M)$ is the *horizontal bundle*.

Let $f: \mathbb{R}^{2n} \to M_m$ be an element of $F(M)$ and let $\xi \in \mathbb{R}^{2n}$. Then there is a unique vector $B_f(\xi) \in H_f F(M)$ with the property that

$$\tilde{\pi}_*(B_f(\xi)) = f(\xi),$$

(where $\tilde{\pi}_*$ is the derivative of $\tilde{\pi}$ at f). The map $f \to B_f(\xi)$ is a smooth vector field on $F(M)$ which we denote by $B(\xi)$. We call $B(\xi)$ the *basic horizontal vector field* on $F(M)$ determined by ξ.

Let $a, b \in so(2n)$ and $\xi, \eta \in \mathbb{R}^{2n}$. The structural equations for the Levi-Civita connexion, expressed in terms of Lie brackets of fundamental and basic vector fields, state that

$$[\sigma(a), \sigma(b)] = \sigma([a, b]),$$
$$[\sigma(a), B(\xi)] = B(a \cdot \xi)$$
$$[B(\xi), B(\eta)] = -\tilde{\Omega}(B(\xi), B(\eta))$$

where $\tilde{\Omega}: HF(M) \otimes HF(M) \to VF(M)$ is the tensor on $F(M)$ given by

$$\tilde{\Omega}(X, Y) = \sigma_f(f^{-1} \circ \mathcal{R}(\tilde{\pi}_* X, \tilde{\pi}_* Y) \circ f)$$

for all $X, Y \in H_f F(M)$, where $\mathcal{R}: TM \otimes TM \to so(TM)$ is the Riemann curvature operator on M.

5.17 THE RELATIONSHIP BETWEEN THE FRAME BUNDLE AND THE TWISTOR SPACE

Let M be an oriented even-dimensional Riemannian manifold, with dimension $2n$. The manifold $F(M)$ of oriented orthonormal frames is the total space of a principal $SO(2n)$ bundle over M. But we can also regard $F(M)$ as the total space of a principal $U(n)$ bundle over the twistor space $\mathfrak{T}_+(M)$.

Choose a standard complex structure J_0 on $\mathbb{R}^{2n}$ compatible with inner product and orientation. Then, for any element $f: \mathbb{R}^{2n} \to M_m$ of $F(M)$, the endomorphism $f \circ J_0 \circ f^{-1}: M_m \to M_m$ is a complex structure on M_m compatible with metric and orientation, and thus represents an element of the fibre of $\pi: \mathfrak{T}_+(M) \to M$ over m. We denote $f \circ J_0 \circ f^{-1}$ by J_f, and we define $\phi: F(M) \to \mathfrak{T}_+(M)$ to be the map sending f to J_f.

Observe that if $f_1, f_2 \in F(M)$, then $\phi(f_1) = \phi(f_2)$ if and only if $f_2 = f_1 \circ A$ for some element A of $U(n)$, where $U(n)$ denotes the subgroup of $SO(2n)$ consisting of all matrices which commute with J_0. The group $U(n)$ acts freely on $F(M)$. It follows directly from the local triviality of $\tilde{\pi}: F(M) \to M$ and $SO(2n) \to SO(2n)/U(n)$ that the projection $F(M) \to F(M)/U(n)$ is locally trivial. We may identify $F(M)/U(n)$ with $\mathfrak{T}_+(M)$ via the map $\phi: F(M) \to \mathfrak{T}_+(M)$. We deduce that $\phi: F(M) \to \mathfrak{T}_+(M)$ is a principal $U(n)$ bundle.

The derivative $\phi_*: TF(M) \to T\mathfrak{T}(M)$ of ϕ maps the vertical and horizontal subbundles of $TF(M)$ onto the vertical and horizontal subbundles of $T\mathfrak{T}_+(M)$. The embedding $\mathfrak{T}_+(M) \hookrightarrow so(\mathfrak{T}M)$ enables us to identify the vertical space at $J_f \in \mathfrak{T}_+(M)$ with the subspace $w(M_m, J_f)$ of $so(M_m)$ consisting of all skew-symmetric endomorphisms of M_m which anti-commute with J_f. We claim that if $a \in so(2n)$ then the derivative φ_* of ϕ at f maps the vertical vector $\sigma_f(a)$ corresponding to a at the frame f to the element of the vertical space to $\mathfrak{T}_+(M)$ at J_f corresponding to $[faf^{-1}, J_f] \in w(M_m, J_f)$. To see this, let $\gamma: \mathbb{R} \to F(M)$ be the curve given by $\gamma(t) = f \circ \exp(ta)$. Then $\sigma_f(a)$ is the tangent vector to γ at $t = 0$. Then $\phi_*(\sigma_f(a))$ is the tangent vector to the curve $t \to \gamma(t) J_0 \gamma(t)^{-1}$ at $t = 0$.

Hence
$$\phi_*(\phi_f(a)) = f(aJ_0 - J_0 a)f^{-1}$$
$$= f[a, J_0]f^{-1}$$
$$= [faf^{-1}, J_f]$$

as required.

Note that if $a \in u(n)$, where $u(n)$ denotes the Lie algebra of $U(n)$, then $\phi_*(\sigma_f(a)) = 0$. Also note that, for any $a \in so(2n)$,

5.17 The frame bundle and the twistor space

$$\phi_*(\sigma_f(\tfrac{1}{2}[J_0, a])) = \tfrac{1}{2}[J_f, \phi_*(\sigma_f(a))].$$

Now let us define tensor fields $\Sigma_1: TF(M) \to TF(M)$ and $\Sigma_2: TF(M) \to TF(M)$ on $F(M)$ such that

$$\sum_1 (B(\xi)) = \sum_2 (B(\xi)) = B(J_0\xi),$$

$$\sum_1 (\sigma(a)) = -\sum_2 (\sigma(a)) = \tfrac{1}{2}\sigma([J_0, a]),$$

for all $\xi \in \mathbb{R}^{2n}$ and $a \in so(2n)$. Then we observe that

$$\phi_* \circ \sum_1 = \mathfrak{J}_1 \circ \phi_*,$$

$$\phi_* \circ \sum_2 = \mathfrak{J}_2 \circ \phi_*,$$

where $\mathfrak{J}_1$ and $\mathfrak{J}_2$ are the two standard almost-complex structures defined on the twistor space. Indeed $\phi_*(\Sigma_1(B(\xi)))$ and $\mathfrak{J}_1\phi_*(B(\xi))$ are both equal to the horizontal lift of $f(\xi)$ to the tangent space to $\mathfrak{T}_+(M)$ at J_f, and also

$$\phi_*\left[\sum_1(\sigma_f(a))\right] = \phi_*(\sigma_f(\tfrac{1}{2}[J_0, a]))$$
$$= \tfrac{1}{2}[J_f, \phi_*(\sigma_f(a))],$$

and this last expression is by definition $\mathfrak{J}_1\phi_*(\sigma_f(a))$.

The Nijenhuis tensors $\mathfrak{N}_1$ and $\mathfrak{N}_2$ of the almost-complex structures $\mathfrak{J}_1$ and $\mathfrak{J}_2$ on $\mathfrak{T}_+(M)$ are defined by

$$\mathfrak{N}_i(Y_1, Y_2) = [\mathfrak{J}_i Y_1, \mathfrak{J}_i Y_2] - [Y_1, Y_2] - \mathfrak{J}_i[\mathfrak{J}_i Y_1, Y_2] - \mathfrak{J}_i[Y_1, \mathfrak{J}_i Y_2],$$

for all vector fields Y_1 and Y_2 on $\mathfrak{T}_+(M)$. To calculate these Nijenhuis tensors we use the following theorem.

Theorem (5.17.1) Let X_1 and X_2 be vector fields on $F(M)$. Then

$$\mathfrak{N}_i(\phi_* X_1, \phi_* X_2) = \phi_*\left(\left[\sum_i X_1, \sum_i X_2\right] - [X_1, X_2]\right.$$
$$\left. - \sum_i\left[\sum_i X_1, X_2\right] - \sum_i\left[X_1, \sum_i X_2\right]\right),$$

for $i = 1$ and $i = 2$.

Proof. We claim that the right-hand side depends on the vector fields X_1 and X_2 only through their value at the element of $F(M)$ under consideration. Let us denote the right-hand side of the above equation by $B_i(X_1, X_2)$. By skew-symmetry it suffices to show that

$$B_i(X_1, gX_2) = gB_i(X_1, X_2)$$

for all smooth functions g on $F(M)$. Now

$$B_i(X_1, gX_2) = gB_i(X_1, X_2) + \left(\sum_i X_i\right)[g]\phi_*\sum_i X_2 - X_1[g]\phi_* X_2$$

$$- \left(\sum_i X_i\right)[g]\phi_*\sum_i X_2 - X_1[g]\phi_*\sum_i\sum_i X_2$$

$$= gB_i(X_1, X_2) - X_1[g](1 + \mathfrak{J}_i^2)\phi_* X_2$$

$$= gB_i(X_1, X_2)$$

as required. To evaluate $B_i(X_1, X_2)$ it suffices to consider the case when there exist vector fields Y_1 and Y_2 defined on a neighbourhood of J_f in $\mathfrak{T}_+(M)$ such that $\phi_* X_1 = Y_1$ and $\phi_* X_2 = Y_2$ in a neighbourhood of $f \in F(M)$. In this case

$$\phi_*[X_1, X_2] = [\phi_* X_1, \phi_* X_2]$$

$$= [Y_1, Y_2],$$

and thus

$$B_i(X_1, X_2) = [\mathfrak{J}_i Y_1, \mathfrak{J}_i Y_2] - [Y_1, Y_2] - \mathfrak{J}_i[\mathfrak{J}_i Y_1, Y_2] - \mathfrak{J}_i[Y_1, \mathfrak{J}_i Y_2]$$

$$= \mathfrak{N}_i(Y_1, Y_2),$$

using the fact that $\phi_*(\Sigma_i X_1) = \mathfrak{J}_i Y_1$ and $\phi_*(\Sigma_i X_2) = \mathfrak{J}_i Y_2$. This proves the theorem.

5.18 INTEGRABILITY OF THE ALMOST-COMPLEX STRUCTURES ON THE TWISTOR SPACE

We now calculate the Nijenhuis tensors of the almost-complex structures $\mathfrak{J}_1$ and $\mathfrak{J}_2$ defined on the twistor space $\mathfrak{T}_+(M)$ of an oriented even-dimensional Riemannian manifold of dimension $2n$. We have a splitting $so(n) = u(n) + w$, where $u(n)$ is the Lie algebra of $U(n)$. This consists of all elements of $so(2n)$ which commute with J_0, while w consists of all elements of $so(2n)$ which anti-commute with J_0. We define a vector space endomorphism $S: so(2n) \to so(2n)$ by $Sa = \frac{1}{2}[J_0, a]$. If $a \in u(n)$ then $Sa = 0$. If $a \in w$ then $Sa = J_0 a$. We have the following lemma.

Lemma (5.18.1) Let $f \in F(M)$ and let $a, b \in so(2n)$. Then

5.18 Integrability on the twistor space

$$\mathfrak{N}_1(\phi_*\sigma_f(a), \phi_*\sigma_f(b)) = \mathfrak{N}_2(\phi_*\sigma_f(a), \phi_*\sigma_f(b)) = 0.$$

Proof. We must show that if $a, b \in \mathrm{so}(2n)$ then $\beta(a, b) \in u(n)$ where

$$\beta(a, b) = [Sa, Sb] - [a, b] - S[Sa, b] - S[a, Sb].$$

If $a, b \in u(n)$ then $Sa = Sb = 0$, hence $\beta(a, b) = -[a, b]$ and therefore $\beta(a, b) \in u(n)$. If $a \in w$ and $b \in u(n)$ then $Sb = 0$, and

$$\begin{aligned}
\beta(a, b) &= -[a, b] - S[Sa, b] \\
&= -ab + ba - J_0(J_0 ab - bJ_0 a) \\
&= -ab + ba + ab + bJ_0^2 a = 0.
\end{aligned}$$

If $a, b \in w$ then

$$\begin{aligned}
\beta(a, b) &= [J_0 a, J_0 b] - [a, b] - S([J_0 a, b] - [a, J_0 b]) \\
&= J_0 a J_0 b - J_0 b J_0 a - ab + ba - S(J_0 ab - bJ_0 a + aJ_0 b - J_0 ba) \\
&= -J_0^2 ab + J_0^2 ba - ab + ba - S(J_0 ab + J_0 ba - J_0 ab - J_0 ba) \\
&= 0,
\end{aligned}$$

as required.

We also have the following additional result.

Lemma (5.18.2) *Let $a \in u(n)$ and $\xi \in \mathbb{R}^{2n}$. Then*

$$\mathfrak{N}_1(f_*\sigma_f(a), \phi_* B_f(\xi)) = \mathfrak{N}_2(\phi_*\sigma_f(a), \phi_* B_f(\xi)) = 0.$$

Proof. Using the fact that $\Sigma_i \sigma_f(a) = \sigma_f(Sa) = 0$ we see that

$$\begin{aligned}
\mathfrak{N}_i(\phi_*\sigma_f(a), \phi_* B_f(\xi)) &= \phi_*\left(-[\sigma(a), B(\xi)]_f - \sum_i \left[\sigma(a), \sum_i B(\xi)\right]_f\right) \\
&= \phi_*\left(-B_f(a\xi) - \sum_i B_f(aJ_0\xi)\right) \\
&= \phi^*(-B_f(a\xi) - B_f(J_0 a J_0 \xi)) \\
&= 0
\end{aligned}$$

since a commutes with J_0. This completes the proof of the lemma.

Be patient! Two more lemmas before we prove the main theorem.

Lemma (5.18.3) *Let $a \in w$ and $\xi \in \mathbb{R}^{2n}$. Then*

$$\mathfrak{N}_1(\phi_*\sigma_f(a), \phi_* B_f(\xi)) = 0,$$
$$\mathfrak{N}_2(\phi_*\sigma_f(a), \phi_* B_f(\xi)) = -4\phi_* B_f(a\xi).$$

Proof. Observe that $\Sigma_1 \sigma_f(a) = J_0 a$, $\Sigma_2 \sigma_f(a) = -J_0 a$;

194   5 Complex and almost-complex manifolds

$$\sum_1 B_f(\xi) = \sum_2 B_f(\xi) = B_f(J_0\xi).$$

Thus

$$N_1(\phi_*\sigma_f(a), \phi_*B_f(\xi)) = \phi_*\Big([\sigma(J_0a), B(J_0\xi)]_f - [\sigma(a), B(\xi)]_f$$
$$- \sum_1 [\sigma(J_0a), B(\xi)]_f - \sum_1 [\sigma(a), B(J_0\xi)]_f\Big)$$
$$= \phi_*(B_f(J_0aJ_0\xi - a\xi - J_0^2 a\xi - J_0aJ_0\xi)) = 0.$$

Moreover

$$N_2(\phi_*\sigma_f(a), \phi_*B_f(\xi)) = \phi_*\Big([-\sigma(J_0a), B(J_0\xi)]_f - [\sigma(a), B(\xi)]_f$$
$$- \sum_2 [-\sigma(J_0a), B(\xi)]_f - \sum_2 [\sigma(a), B(J_0\xi)]_f$$
$$= \phi_*(B_f(-J_0aJ_0\xi - a\xi + J_0^2 a\xi - J_0aJ_0\xi))$$
$$= -4\phi_*B_f(a\xi),$$

and the lemma is proved.

We observe from this that *the almost-complex structure $\mathfrak{J}_2$ on the twistor space $\mathfrak{T}_+(M)$ is never integrable.*

We have the following lemma.

Lemma (5.18.4) Let $\xi, \eta \in \mathbb{R}^{2n}$. Then
$$\mathfrak{N}_i(\phi_*B_f(\xi), \phi_*B_f(\eta)) = -\phi_*(\tilde{\Omega}(B_f(J_0\xi), B_f(J_0\eta)) - \tilde{\Omega}(B_f(\xi), B_f(\eta)))$$
$$+ \mathfrak{J}_i\phi_*(\tilde{\Omega}(B_f(J_0\xi), B_f(\eta))$$
$$+ \tilde{\Omega}(B_f(\xi), B_f(J_0\eta))).$$

Proof. This follows immediately from the identity
$$[B(\xi), B(\eta)] = -\tilde{\Omega}(B(\xi), B(\eta)).$$

Theorem (5.18.5) (Main theorem) *The almost-complex structure $\mathfrak{J}_2$ on the twistor space $\mathfrak{T}_+(M)$ of an oriented even-dimensional Riemannian manifold M is never integrable. The almost-complex structure $\mathfrak{J}_1$ is integrable if and only if*
$$\mathfrak{R}(JX, JY) - \mathfrak{R}(X, Y) - \tfrac{1}{2}[J, \mathfrak{R}(JX, Y) + \mathfrak{R}(X, JY)]$$
commutes with J for all $m \in M$, all $X, Y \in M_m$, and all complex structures J on M_m compatible with metric and orientation.

Proof. We have already observed that $\mathfrak{J}_2$ is never integrable. The condition for $\mathfrak{J}_1$ to be integrable is a consequence of the previous lemma and the fact that the kernel of the derivative ϕ_* of $\phi: F(M) \to \mathfrak{T}_+(M)$ corresponds to the subalgebra of $\mathrm{so}(M_m)$ consisting of all skew-symmetric endomorphisms of M_m which commute with J.

5.19 INTERPRETATION OF THE INTEGRABILITY CONDITION

In the previous section we saw that the almost-complex structure $\mathfrak{J}_1$ on the twistor space $\mathfrak{T}_1(M)$ of an oriented even-dimensional Riemannian manifold M is integrable if and only if the Riemann curvature operator $\mathcal{R}: TM \otimes TM \to \mathrm{so}(TM)$ has the property that

$$\mathcal{R}(JX, JY) - \mathcal{R}(X, Y) - \tfrac{1}{2}[J, \mathcal{R}(JX, Y) + \mathcal{R}(X, JY)]$$

commutes with J for all $m \in M$, all $X, Y \in M_m$, and all complex structures J on M_m compatible with metric and orientation. In this section we look for conditions on the curvature of M which ensure that this integrability condition is satisfied.

The Riemannian metric on M defines a natural isomorphism $\mathrm{so}(M_m) \cong \Lambda^2 M_m$. Composing the curvature operator $\mathcal{R}$ with this isomorphism we obtain a self-adjoint endomorphism $\hat{R}: \Lambda^2 M_m \to \Lambda^2 M_m$. If $(E_1, \ldots, E_n)$ is an orthonormal basis of M_m then

$$\hat{R}(X \wedge Y) = \sum_{i=1}^{n} (\mathcal{R}(X, Y)E_i) \wedge E_i.$$

The endomorphism of $\mathrm{so}(M_m)$ sending a to JaJ^{-1} corresponds to the endomorphism $\Lambda^2 J: \Lambda^2 M_m \to \Lambda^2 M_m$. The subspaces $u(M_m, J)$ and $w(M_m, J)$ of $\mathrm{so}(M_m)$ correspond to subspaces $\hat{u}(M_m, J)$ and $\hat{w}(M_m, J)$ of $\Lambda^2(M_m)$ where

$$\hat{u}(M_m, J) = \{\beta \in \Lambda^2 M_m : (\Lambda^2 J)\beta = \beta\},$$
$$\hat{w}(M_m, J) = \{\beta \in \Lambda^2 M_m : (\Lambda^2 J)\beta = -\beta\}.$$

We defined a complex structure on the space $w(M_m, J)$ of skew-symmetric endomorphisms of M_m which anti-commute with J. This complex structure sends $a \in w(M_m, J)$ to $\tfrac{1}{2}[J, a]$. There is a corresponding complex structure $\tilde{J}$ on $\hat{w}(M_m, J)$. This complex structure is the restriction to $\hat{w}(M_m, J)$ of the endomorphism of $\Lambda^2 M_m$ which sends $X \wedge Y$ to $\tfrac{1}{2}(JX \wedge Y + X \wedge JY)$.

Let $p_J: \Lambda^2 M_m \to \hat{w}(M_m, J)$ denote the orthogonal projection onto $\hat{w}(M_m, J)$. Then

$$p_J(X \wedge Y) = \tfrac{1}{2}(X \wedge Y - JX \wedge JY),$$
$$(\tilde{J} \circ p_J)(X \wedge Y) = \tfrac{1}{2}(JX \wedge Y + X \wedge JY).$$

We now prove the following

Lemma (5.19.1) The almost complex structure $\mathfrak{J}_1$ on the twistor space $\mathfrak{T}_+(M)$ of M is integrable if and only if the endomorphism $(p_J \circ \hat{R})|\hat{w}(M_m, J)$ of $\hat{w}(M_m, J)$ commutes with the complex structure $\tilde{J}$ on $\hat{w}(M_m, J)$ for all $J \in \mathfrak{T}_+(M)$.

Proof. The integrability condition states that
$$p_J \hat{R}(JX \wedge JY - X \wedge Y) - \tilde{J} p_J \hat{R}(JX \wedge Y + X \wedge JY) = 0.$$

But this is satisfied if and only if
$$(p_J \circ \hat{R}) \circ p_J = -\tilde{J} \circ (p_J \circ \hat{R}) \circ \tilde{J} \circ p_J$$

or equivalently
$$(p_J \circ \hat{R})|\hat{w}(M_m, J) = -\tilde{J} \circ ((p_J \circ \hat{R})|\hat{w}(M_m, J)) \circ \tilde{J}$$

which is precisely the condition that $(p_J \circ \hat{R})|\hat{w}(M_m, J)$ commutes with $\tilde{J}$. This proves the lemma.

Let k and l be symmetric endomorphisms of M_m. We define $k \otimes l : \Lambda^2 M_m \to \Lambda^2 M_m$ to be the symmetric endomorphism given by
$$(k \otimes l)(X \wedge Y) = 2(kX \wedge lY - kY \wedge lX)$$

for all $X, Y \in M_m$.

The Riemann curvature tensor regarded as a symmetric endomorphism of $\Lambda^2 M_m$ is of the form
$$\hat{R} = \hat{W} + \frac{1}{(\dim M - 2)}(\rho \otimes \mathrm{Id}) - \frac{\mathrm{Tr}\,\rho}{2(\dim M - 1)(\dim M - 2)} \mathrm{Id} \otimes \mathrm{Id},$$

where $\hat{W}: \Lambda^2 M_m \to \Lambda^2 M_m$ is the Weyl conformal curvature tensor, $\rho: M_m \to M_m$ is the Ricci tensor regarded as an endomorphism of M_m, $\mathrm{Tr}\,\rho$ is the scalar curvature and Id is the identity endomorphism of M_m.

Lemma (5.19.2) The almost-complex structure $\mathfrak{J}_1$ on the twistor space $\mathfrak{T}_+(M)$ of M is integrable if and only if, for all $\mathfrak{T}_+(M)$, the endomorphism $(p_J \circ \hat{W})|\hat{w}(M_m, J)$ of $\hat{w}(M_m, J)$ commutes with the complex structure $\tilde{J}$ on $\hat{w}(M_m, J)$, where $\hat{W}: \Lambda^2 M_m \to \Lambda^2 M_m$ denotes the Weyl tensor of M.

Proof. In view of the previous lemma it suffices to show that if $k: M_m \to M_m$ is a symmetric endomorphism of M_m then $(p_J \circ (k \otimes \mathrm{Id}))|\hat{w}(M_m, J)$ commutes with $\tilde{J}$. Thus it suffices to show that
$$\tilde{J} \circ p_J \circ (k \otimes \mathrm{Id}) \circ p_J = p_J \circ (k \otimes \mathrm{Id}) \circ \tilde{J} \circ p_J.$$

5.19 Interpretation of the integrability condition

Now

$$(\tilde{J} \circ p_J \circ (k \otimes (\text{Id})) \circ p_J)(X \wedge Y) = \tfrac{1}{2}(\tilde{J} \circ p_J \circ (k \otimes \text{Id}))(X \wedge Y - JX \wedge JY)$$
$$= (\hat{J} \circ p_J)(kX \wedge Y - kJX \wedge JY$$
$$\quad - kY \wedge X + kJY \wedge JX)$$
$$= \tfrac{1}{2}(JkX \wedge Y - JkJX \wedge JY - JkY \wedge X$$
$$\quad + JkJY \wedge JX$$
$$\quad + kX \wedge JY + kJX \wedge Y$$
$$\quad - kY \wedge JX - kJY \wedge X).$$

Also

$$(p_J \circ (k \otimes \text{Id}) \circ \tilde{J} \circ p_J)(X \wedge Y) = \tfrac{1}{2}(p_J \circ (k \otimes \text{Id}))(JX \wedge Y + X \wedge JY)$$
$$= p_J(kJX \wedge Y + kX \wedge JY - kY \wedge JX$$
$$\quad - kJY \wedge X)$$
$$= \tfrac{1}{2}(kJX \wedge Y + kX \wedge JY - kY \wedge JX$$
$$\quad - kJY \wedge X - JkJX \wedge JY$$
$$\quad + JkX \wedge Y - JkY \wedge X + JkJY \wedge JX).$$

Hence

$$\tilde{J} \circ p_J \circ (k \otimes \text{Id}) \circ p_J = p_J \circ (k \otimes \text{Id}) \circ \tilde{J} \circ p_J,$$

as required.

We now prove the main theorem.

Theorem (5.19.3) Let M be an oriented even-dimensional Riemannian manifold. If $\dim M = 4$, then the almost-complex structure $\mathfrak{J}_1$ on the twistor space $\mathfrak{T}_+(M)$ of M is integrable if and only if the Weyl conformal tensor of M is anti-self-dual. If $\dim M > 4$ then the almost-complex structure $\mathfrak{J}_1$ on $\mathfrak{T}_+(M)$ is integrable if and only if the Weyl conformal tensor of M vanishes, so that M is conformally flat.

Proof. Let J be a complex structure on the tangent space M_m to M at m compatible with metric and orientation. Let E_0, E_1, E_2, E_3 be orthonormal vectors with the property that

$$JE_0 = E_1, \qquad JE_2 = E_3.$$

We define 2-forms $\omega_1^\pm, \omega_2^\pm$, and $\omega_3^\pm$ by

$$\omega_1^\pm = E_0 \wedge E_1 \pm E_2 \wedge E_3,$$
$$\omega_2^\pm = E_0 \wedge E_2 \pm E_3 \wedge E_1,$$

5 Complex and almost-complex manifolds

$$\omega_3^\pm = E_0 \wedge E_3 \pm E_1 \wedge E_2.$$

Then

$$(\Lambda^2 J)\omega_1^\pm = \omega_1^\pm,$$
$$(\Lambda^2 J)\omega_2^\pm = \mp \omega_2^\pm,$$
$$(\Lambda^2 J)\omega_3^\pm = \mp \omega_3^\pm.$$

Thus

$$p_J\omega_1^- = p_J\omega_2^- = p_J\omega_3^- = 0, \qquad p_J\omega_1^+ = 0,$$

and ω_2^+ and ω_3^+ span $\hat{w}(M_m, J)$. Moreover

$$\tilde{J}\omega_2^+ = \omega_3^+,$$
$$\tilde{J}\omega_3^+ = -\omega_2^+.$$

We first prove the theorem when $\dim M = 4$. Then for all $m \in M$, $\Lambda^2 M_m$ splits as the orthogonal direct sum

$$\Lambda^2 M_m = \Lambda_+ M_m \oplus \Lambda_- M_m,$$

where $\Lambda_+ M_m$ and $\Lambda_- M_m$ denote the spaces of self-dual and anti-self-dual 2-forms. Let $p_+: \Lambda^2 M_m \to \Lambda_+ M_m$ denote the orthogonal projection onto $\Lambda_+ M_m$ and denote $(p_+ \circ R)|\Lambda_+ M_m$ by R_{++}. From the Singer–Thorpe decomposition of the curvature tensor of M (see Chapter 6 page 241) we see that the Weyl tensor of M is anti-self-dual if and only if R_{++} is diagonal.

If J is a complex structure on M_m compatible with metric and orientation and if (E_0, E_1, E_2, E_3) is an orthonormal basis of M_m such that $JE_0 = E_1$ and $JE_2 = E_3$, then this basis is positively oriented. Thus $\omega_1^+, \omega_2^+, \omega_3^+ \in \Lambda_+ M_m$ and $\omega_1^-, \omega_2^-, \omega_3^- \in \Lambda_- M_m$. We deduce that $\hat{w}(M_m, J)$ is contained in $\Lambda_+ M_m$. Thus if the Weyl tensor of M is anti-self-dual then R_{++} is diagonal so that $(p_J \circ \hat{R})|\hat{w}(M_m, J)$ commutes with $\tilde{J}$. We deduce that if the Weyl tensor of M is anti-self-dual, then $\mathfrak{J}_1$ is integrable. Conversely if $\mathfrak{J}_1$ is integrable then $(p_J \circ \hat{R})|\hat{w}(M_m, J)$ commutes with $\tilde{J}$ for all complex structures J on M_m compatible with metric and orientation. But $(p_J \circ \hat{R})|\hat{w}(M_m, J)$ is symmetric and $\hat{w}(M_m, J)$ is two-dimensional. It follows that $(p_J \circ \hat{R})|\hat{w}(M_m, J)$ is diagonal. Since this is true for all such complex structures J on M_m we deduce that R_{++} is diagonal and thus the Weyl tensor of M is anti-self-dual. This proves the theorem when $\dim M = 4$.

Finally consider the case when $\dim M > 4$. If the Weyl tensor of M vanishes then the almost-complex structure $\mathfrak{J}_1$ on $\mathfrak{T}_+(M)$ is integrable, by the previous lemma. It remains to show that if $\mathfrak{J}_1$ is integrable then the Weyl tensor of M vanishes.

5.19 Interpretation of the integrability condition

Let E_0, E_1, E_2, E_3 be orthonormal vectors in M_m. Then there exists $J \in \mathfrak{T}_+(M)$ such that $JE_0 = E_1$ and $JE_2 = E_3$. On defining $\omega_1^\pm$, $\omega_2^\pm$, and $\omega_3^\pm$ as above, we see that ω_2^+ and ω_3^+ belong to $\hat{w}(M_m, J)$ and $\tilde{J}\omega_2^+ = \omega_3^+$. If $\mathfrak{I}_1$ is integrable then $(p_J \circ \hat{W}) | \hat{w}(M_m, J)$ commutes with $\tilde{J}$. But this endomorphism is symmetric and hence

$$\langle \omega_2^+, \hat{W}\omega_2^+ \rangle = \langle \omega_3^+, \hat{W}\omega_3^+ \rangle,$$

and

$$\langle \omega_3^+, \hat{W}\omega_2^+ \rangle = 0.$$

Thus

$$\langle E_0 \wedge E_3 + E_1 \wedge E_2, \hat{W}(E_0 \wedge E_2 + E_3 \wedge E_1) \rangle = 0,$$

or

$$W(E_0, E_3, E_0, E_2) + W(E_0, E_3, E_3, E_1) + W(E_1, E_2, E_0, E_2)$$
$$+ W(E_1, E_2, E_3, E_1) = 0$$

where we define $W(X, Y, Z, V) = \langle X \wedge Y, \hat{W}(Z \wedge V) \rangle$. If we replace (E_0, E_1, E_2, E_3) by $(E_0, E_1, E_2, -E_3)$ we see that

$$-W(E_0, E_3, E_0, E_2) + W(E_0, E_3, E_3, E_1) + W(E_1, E_2, E_0, E_2)$$
$$- W(E_1, E_2, E_3, E_1) = 0.$$

On adding these two equations we see that

$$W(E_0, E_2, E_1, E_2) = W(E_0, E_3, E_1, E_3).$$

Thus we see that if $(E_0, \ldots, E_{2n-1})$ is an oriented orthonormal basis of M_m, then

$$W(E_0, E_i, E_1, E_i) = W(E_0, E_2, E_1, E_2)$$

for all $i > 2$, so that

$$(2n - 2) W(E_0, E_2, E_1, E_2) = \sum_{i=0}^{2n-1} W(E_0, E_i, E_1, E_i) = 0,$$

since all contractions of the Weyl tensor vanish. Thus

$$W(X, Z, Y, Z) = 0,$$

for all orthonormal vectors X, Y, and Z. If we replace X and Y by $1/\sqrt{2}(X + Y)$ and $1/\sqrt{2}(X - Y)$ we see that

$$W(X, Z, X, Z) = W(Y, Z, Y, Z).$$

Applying this result to an oriented orthonormal basis $(E_0, \ldots, E_{2n-1})$ of M_m we see that

$$(2n-1)W(E_1, E_0, E_1, E_0) = \sum_{i=0}^{2n-1} W(E_i, E_0, E_i, E_0) = 0.$$

A standard argument (see page 61) using the first Bianchi identity shows that the Weyl tensor of M vanishes as required.

The concept of twistor space is due to Roger Penrose who made use of this idea in his work on relativity. Since then it has had many applications in theoretical physics including the theory of fundamental particles. In pure mathematics twistors have played an important role in relating the geometry of homogeneous spaces to the theory of harmonic maps.

5.20 EXERCISES

1. Prove that every complex submanifold of a Kähler manifold is minimal.
2. Writing $z = x + iy$, prove that

$$\frac{\partial}{\partial z} = \tfrac{1}{2}\left(\frac{\partial}{\partial x} - i\frac{\partial}{\partial y}\right), \qquad \frac{\partial}{\partial \bar z} = \tfrac{1}{2}\left(\frac{\partial}{\partial x} + i\frac{\partial}{\partial y}\right).$$

Hence prove that if Π is the second fundamental form of a surface immersed in R^3, relative to isothermic coordinates x, y, with

$$L = \Pi\left(\frac{\partial}{\partial x}, \frac{\partial}{\partial x}\right), \qquad M = \Pi\left(\frac{\partial}{\partial x}, \frac{\partial}{\partial y}\right), \qquad N = \Pi\left(\frac{\partial}{\partial y}, \frac{\partial}{\partial y}\right)$$

then

$$2\Pi\left(\frac{\partial}{\partial z}, \frac{\partial}{\partial z}\right) = \tfrac{1}{2}(L - N) - iM.$$

If we denote this expression by Φ, prove that the mean curvature is constant if and only if Φ is a holomorphic function of $z = x + iy$.

3. Show that the 2-form Φdz^2 is globally defined, where Φ is holomophic. If the given compact surface has genus 0, it follows from the Riemann–Roch theorem that this 2-form must be identically zero. Hence deduce that every point is an umbilic and hence that the surface is embedded as a round sphere.

6
Special Riemannian Manifolds

The subject of special Riemannian manifolds is so rich that it is impossible to include many interesting and important examples in this chapter. For example, we have the book by Wolf (1967) of 408 pages devoted to manifolds of constant curvature. We have the book by Besse (1987) of 510 pages devoted to the study of Einstein spaces. Another book by Besse (1978) of 262 pages is devoted to the special topic of manifolds with closed geodesics. We have the book by Yano and Ishihara (1973) of 423 pages on tangent and cotangent bundles, the book by Willmore (1982) of 168 pages devoted to the special topic of total curvature. And so on, and so on ....

In the present chapter we shall report on harmonic spaces including k-stein spaces and also special Riemannian spaces associated with mean-value theorems. We choose these topics primarily because they are of special interest to the author but also because although the methods used are fairly central to the development of differential geometry, these spaces are not dealt with in most books already available.

6.1 HARMONIC AND RELATED METRICS

The theory of harmonic manifolds arose as follows. In Euclidean space $\mathbb{R}^n$ of dimension n it is well-known that a nice solution of Laplace's equation $\Delta f = 0$ is given by $f(x) = \|x\|^{2-n}$ if $n > 2$, $f(x) = \log(\|x\|)$ if $n = 2$. Observe that such a solution is a function only of the distance from the origin. In 1930 H. S. Ruse attempted to find in any Riemannian manifold (M,g) a solution of $\Delta f = 0$ which depended only on the distance $\rho(m,.)$ for each $m \in M$. This attempt failed because in general such a solution does not exist. The manifold was called *harmonic* if such a solution existed.

Then mathematicians tried to classify such manifolds for which such solutions exist. They found necessary conditions satisfied by the curvature tensor and, in particular, they found that such manifolds must be Einstein, that is, the Ricci tensor is proportional to the metric tensor. These curvature conditions were strong enough to show that two- and three-dimensional harmonic spaces must be of constant curvature, and that four-dimensional harmonic spaces with positive definite metric must be symmetric, that is, the curvature tensor has zero covariant derivative. It is known that the situation is radically different if the metric is allowed to be indefinite, certainly when $n \geq 4$. There exists an example of a four-dimensional harmonic metric of signature $(++--)$ which is not locally symmetric. On the other hand, it is known that every n-dimensional harmonic metric with Lorentzian signature is of constant curvature.

It was conjectured by Lichnerowicz in 1944 that every harmonic space with positive definite metric is symmetric. This conjecture defied resolution for over 40 years. However Szabo (1990) confirmed that the conjecture was true with the additional assumption that the manifold was compact and simply connected. What was even more surprising is that in 1991 E. Dammek and F. Ricci came up with a counterexample which showed that locally the conjecture is false. The problem remains of finding a set of necessary and sufficient conditions which locally force a harmonic space to be symmetric.

In this chapter we shall deal with harmonic metrics first from the point of view of local Riemannian geometry. We do this because the topic provides a natural opportunity to apply the techniques developed in earlier chapters. Moreover, the classification of harmonic metrics remains an outstanding problem in local Riemannian geometry. Such a treatment illustrates the value of admitting indefinite metrics often ignored in differential geometry, since in many of the theorems the signature of the metric plays an important part. Later on we shall mention global problems associated with harmonic manifolds. We shall also find that sometimes a global way of regarding a situation makes previous local results appear more natural. Indeed, sometimes by employing Jacobi vector fields, for example, the actual computations arising from local problems are substantially simplified.

The second special topic dealt with in this chapter is concerned with mean-value theorems. This originated from the study of harmonic manifolds but is of interest under much more general conditions. It has led to the study of relations between Einstein and super-Einstein manifolds, commutative metrics and their relation to D'Atri metrics, and to a characterization of some of these spaces by mean-value properties.

6.2 NORMAL COORDINATES

We have already met this topic in Chapter 3. Here we prove a theorem which is particularly useful in the study of harmonic metrics. We find first a set of necessary and sufficient conditions that a system of coordinates (y^i) shall be normal with respect to a torsion-free analytic connexion Γ defined on an analytic manifold M. In a coordinate neighbourhood U the connexion will have components Γ^i_{jk} ($=\Gamma^i_{kj}$)

6.2 Normal coordinates

which are analytic functions of the coordinates (x^i). If P_0 is a point of U with coordinates x_0^i, then for given constants a^i not all zero and for sufficiently small values of the real number $|t|$, the differential equation

$$\frac{d^2 x^i}{dt^2} + \Gamma^i_{jk} \frac{dx^j}{dt} \frac{dx^k}{dt} = 0 \tag{6.1}$$

possesses a unique solution of the form

$$x^i = \phi^i(x_0, y) \tag{6.2}$$

where

$$y^i = a^i t. \tag{6.3}$$

Here the a^i are arbitrary and the ϕ^i are analytic functions of x_0^i and y^i such that

$$\phi^i(x_0, 0) = x_0^i \tag{6.4}$$

and

$$\left(\frac{dx^i}{dt}\right)_0 = a^i, \tag{6.5}$$

the subscript denoting evaluation at P_0.

A more geometrical way of describing the above is to use the idea of the exponential map determined by Γ. This maps the tangent vector (y^i) to the point P, obtained by proceeding along the unique path determined by the direction (a^i) to the point of affine parameter t. Normally t is undetermined to within an affine transformation $t' = at + b$ but by stipulating $t = 0$ at P_0, condition (6.5) uniquely determines t.

It will be convenient to denote the components of Γ referred to normal coordinates by an asterisk, for example, $*\Gamma^i_{jk}$. Since the differential equations of the path

$$\frac{d^2 y^i}{dt^2} + *\Gamma^i_{jk}(y) \frac{dy^j}{dt} \frac{dy^k}{dt} = 0 \tag{6.6}$$

must be satisfied by $y^i = a^i t$ we have

$$*\Gamma^i_{jk}(at) a^j a^k = 0 \tag{6.7}$$

for all values of a^i. Multiply (6.7) by t^2 to get

$$*\Gamma^i_{jk}(y) y^j y^k = 0. \tag{6.8}$$

This is a *necessary* condition which must be satisfied by the normal coordinates y^i.

But clearly it is also *sufficient*. For if (6.8) is satisfied in some coordinate system (y^i), then the differential equations of the affinely parametrized paths relative to that system are satisfied by the linear equations $y^i = a^i t$.

Alternatively, the coordinates are obtained by the exponential map based at P_0.

Suppose now that the Γ^i_{jk} are Christoffel symbols obtained from the metric (g_{ij}). The paths now become geodesics with equations

$$\frac{d^2 x^i}{ds^2} + \Gamma^i_{jk} \frac{dx^j}{ds} \frac{dx^k}{ds} = 0 \qquad (6.9)$$

where s is the affine parameter. From these equations it follows that

$$g_{jk} \frac{dx^j}{ds} \frac{dx^k}{ds} = \text{constant} \qquad (6.10)$$

along any geodesic.

If the constant is non-zero we may choose the parameter s so as to be numerically equal to the arc length from a fixed point P_0 (x_0^i) on the geodesic to the point P (x^i), and the constant is then $+1$ or $+1$. When the constant is zero, the geodesic is *null*. In all cases

$$g_{ij} \frac{dx^j}{ds} \frac{dx^j}{ds} = e \qquad (6.11)$$

where the constant e, the *indicator* of the geodesic, is $+1$ or -1 if the geodesic is non-null and is equal to 0 if the geodesic is null. Null geodesics are important in relativity theory since they are precisely light paths.

As before, let P_0 (x_0^i) be a fixed point in a simple convex neighbourhood W, and let the equations of a geodesic through P_0 be

$$x^i = \phi^i(x_0, y) \qquad (6.12)$$

where

$$y^i = a^i s, \qquad (6.13)$$

$$\phi^i(x_0, 0) = x_0^i \qquad (6.14)$$

$$\left(\frac{dx^i}{ds} \right)_0 = a^i \qquad (6.15)$$

for fixed numbers a^i. The components a^i of the unit tangent vector at P_0 satisfy

$$(g_{ij})_0 a^i a^j = e. \qquad (6.16)$$

If we regard equation (6.12) as defining the transformation from the allowable coordinate system (x^i) to a normal coordinate system

6.2 Normal coordinates

(y^i), centred at P_0, then, for fixed a^i equations (6.13) are those of a geodesic through P_0 in the system (y^i). So relative to the normal coordinate system, equation (6.11) gives

$$*g_{jk}\frac{dy^j}{ds}\frac{dy^k}{ds} = e$$

and hence

$$*g_{jk}a^j a^k = e. \tag{6.17}$$

This holds along a geodesic so we have

$$*g_{jk}a^j a^k = (*g_{jk})_0 a^j a^k. \tag{6.18}$$

Multiply (6.18) by s^2 to get

$$*g_{jk}y^j y^k = (*g_{jk})_0 y^j y^k \tag{6.19}$$

which must hold for every point (y^i) in W. We can now obtain a necessary and sufficient condition that a coordinate system shall be normal.

Let us assume that (y^i) is a normal coordinate system. Then from (6.8) it follows that

$$*\{{}^l_{jk}\} y^j y^k = 0 \tag{6.20}$$

where $*\{{}^l_{jk}\}$ are now Christoffel symbols. Multiply this by $2*g_{il}$ to get

$$(\partial_k *g_{ij} + \partial_j *g_{ik} - \partial_i *g_{jk}) y^j y^k = 0.$$

Interchange of dummy suffixes in the second term gives

$$(2\partial_k *g_{ij} - \partial_i *g_{jk}) y^j y^k = 0. \tag{6.21}$$

Differentiate (6.19) with respect to y^i to get

$$\partial_i *g_{jk} y^j y^k + 2*g_{ji} y^j = 2(*g_{ji})_0 y^j,$$

and hence, by (6.21)

$$\partial_k *g_{ij} y^j y^k + *g_{ij} y^j - (*g_{ij})_0 y^j = 0$$

which may be written

$$y^k \partial_k \{*g_{ij} y^j - (*g_{ij})_0 y^j\} = 0. \tag{6.22}$$

Thus the function $f_i(y)$ defined by

$$f_i(y) = *g_{ij} y^j - (*g_{ij})_0 y^j \tag{6.23}$$

is homogeneous of degree zero in the y^i. Thus, replacing y^i by λy^i where λ is any real number, we get

$$f_i(y) = f_i(\lambda y).$$

Now let $\lambda \to 0$ to get $f_i(y) = f_i(0) = 0$. Thus a necessary condition that y^i should be a normal coordinate system is

$$*g_{ij}y^j = (*g_{ij})_0 y^j. \tag{6.24}$$

We now prove that condition (6.24) is also sufficient. Differentiate (6.24) with respect to y^k to get

$$(\partial_k *g_{ij})y^j + *g_{ik} = (*g_{ik})_0.$$

Multiply this by y^k and then by y^j; using (6.24) we get

$$(\partial_k *g_{ij})y^i y^k = 0, \qquad (\partial_k *g_{ij})y^i y^j = 0.$$

Equation (6.20) now follows immediately, and from (6.8) it follows that the coordinate system is normal.

In particular we can choose normal coordinates y^i defined in a neighbourhood U of $m \in M$ with $y^i(m) = 0$ such that

$$y^j = \left(\exp_m \sum_{k=1}^{m} t^k \mathbf{e}_k \right) = t^j$$

where $\{\mathbf{e}_k, k = 1, \ldots, n\}$ is an arbitrary orthonormal basis of M_m.

6.3 THE DISTANCE FUNCTION Ω

Let W be a simple convex neighbourhood of the manifold M, that is, we assume that any two points of W can be joined by a unique geodesic contained inside W. Let P_0, P be any two points of W and let e be the indicator of the geodesic joining them. Let r be the length of the geodesic arc joining P_0 to P. Then the *distance function* is defined by

$$\Omega = \tfrac{1}{2} er^2. \tag{6.25}$$

In terms of normal coordinates y^i we have

$$y^i = a^i s \tag{6.26}$$

where

$$*g_{ij}a^i a^j = e. \tag{6.27}$$

If the geodesic arc $P_0 P$ is non-null then $e \neq 0$ and the parameter s is equal to the length of the geodesic arc $P_0 P$. More precisely we have $|s| = r$. If the geodesic is null then $r = 0$ but the affine parameter s is not equal to zero unless $P = P_0$. Thus, multiplying (6.27) by s^2 we get

6.3 The distance function Ω

$$\Omega = \tfrac{1}{2}({}^*g_{ij})y^iy^j = \tfrac{1}{2}({}^*g_{ij})_0 y^i y^j. \tag{6.28}$$

Hence

$$\partial \Omega/\partial y^i = ({}^*g_{ij})_0 y^j = {}^*g_{ij} y^j. \tag{6.29}$$

Thus by (6.26) we get

$${}^*g^{ij} \frac{\partial \Omega}{\partial y^j} = y^i = a^i s = s \frac{dy^i}{ds}. \tag{6.30}$$

However classical tensor calculus allows us to conclude that the above equation holds in any system of coordinates, since the left-hand side and the right-hand side are components of tensors. Thus in *any allowable coordinate system* we have

$$g^{ij} \frac{\partial \Omega}{\partial x^j} = s \frac{dx^i}{ds}. \tag{6.31}$$

Similarly from (6.29) we get

$${}^*g^{ij} \frac{\partial \Omega}{\partial y^i} \frac{\partial \Omega}{\partial y^j} = ({}^*g_{ij})_0 y^i y^j = 2\Omega,$$

and, again due to the tensorial nature of the equation, we have *in any allowable coordinate system*

$$\Delta_1 \Omega = g^{ij} \frac{\partial \Omega}{\partial x^i} \frac{\partial \Omega}{\partial x^j} = 2\Omega. \tag{6.32}$$

In our derivation of (6.31) we have regarded P_0 as fixed and P as variable. If instead we had taken P as fixed and regarded P_0 as variable, since $\Omega(P_0, P) = \Omega(P, P_0)$ we would have obtained

$$(g^{ij})_0 \frac{\partial \Omega}{\partial x_0^j} = s \left(-\frac{dx^i}{ds} \right)_0 \tag{6.33}$$

The minus sign is necessary because we must take the tangent vector in the direction PP_0. From (6.33) we get, using $(dx^i/ds)_0 = a^i$,

$$y^i = -(g^{ij})_0 \frac{\partial \Omega}{\partial x_0^j}. \tag{6.34}$$

These equations give the y^i explicitly as functions of the x_0^i and x^i.

Perhaps a more natural approach is to regard Ω as a real-valued function defined on the tangent bundle TM. In fact if $v \in TM$, $v \neq 0$, and $\pi v = P_0$ where π is the bundle projection map, then the geodesic through P_0 with direction $v/\|v\|$ is given by $t \to \exp_{P_0}(vt/\|v\|)$. If P is given by $\exp_{P_0}(vs/\|v\|)$, then

$$\Omega = \tfrac{1}{2} es^2. \tag{6.35}$$

This defines Ω in a region where the exponential map is defined, irrespective of the signature of the metric.

6.4 THE DISCRIMINANT FUNCTION

As a result of considering the concept of distance in relativistic cosmology H. S. Ruse introduced a two-point invariant function called *the discriminant function* defined as follows.

Let (x^i) be an allowable coordinate system covering a simple convex neighbourhood W on M. Let the distance function Ω be expressed in terms of the coordinates x_0^i, x^i of any two points P_0, P of W. Let J denote the determinant

$$J = \det\left(\frac{\partial^2 \Omega}{\partial x_0^i \partial x^j}\right) \tag{6.36}$$

and $|J|$ denote its modulus. Then the discriminant function ρ, relative to P_0 and P is given by

$$\rho = \frac{\sqrt{(gg_0)}}{|J|}. \tag{6.37}$$

It is by no means obvious that ρ is independent of the coordinate system chosen, although this can be checked. However we shall give a neater proof of this fact by using normal coordinates y^i related to the coordinates x^i and the direction dx^i/ds of the geodesic arc $P_0 P$ at P_0.

From (6.34) we have

$$g_0^{ik} \frac{\partial^2 \Omega}{\partial x_0^j \partial x^k} = -\frac{\partial y^i}{\partial x^j}.$$

Taking the determinant of each side gives

$$\frac{J}{g_0} = (-1)^n \frac{\partial(y)}{\partial(x)}$$

However

$$g = {}^*g \left(\frac{\partial(y)}{\partial(x)}\right)^2$$

and therefore

$$\frac{gg_0}{J^2} = \frac{{}^*g}{g_0}.$$

If we take the positive square root of each side we get

6.4 The discriminant function

$$\rho = \sqrt{\left(\frac{*g}{g_0}\right)} = \sqrt{\left(\frac{*g}{*g_0}\right)}. \tag{6.38}$$

Clearly the right-hand side is independent of the particular system of orthonormal coordinates chosen at P_0, and ρ is therefore a uniquely defined function of P and P_0 with the property $\rho(P_0, P) = \rho(P, P_0)$.

A more natural definition of ρ can now be given. Let m be a fixed point of the manifold M and consider the function θ_m defined (roughly!) as

$$n \mapsto \theta_m(n) = \frac{\mu \exp_m^* g\left(\exp_m^{-1}(n)\right)}{\mu g_m}, \quad n \in M \tag{6.39}$$

where the right-hand side is the quotient of the canonical volume measure of the Riemannian metric $\exp^* g$ on M_m (the pull-back of g by the map $\exp_m$) by the Lebesgue measure of the Euclidean structure on M_m. Clearly this is consistent with (6.38) and is coordinate-free.

Calculation of $\Delta\Omega$

In this section we establish the general formula

$$\Delta\Omega = n + \Omega^k \frac{\partial}{\partial x^k} (\log \rho) \tag{6.40}$$

where

$$\Omega^k = g^{jk} \frac{\partial \Omega}{\partial x^j}. \tag{6.41}$$

In terms of normal coordinates y^i we have from (6.29)

$$\frac{\partial^* \Omega}{\partial y^i} = {}^* g_{ij} y^j$$

and hence

$$^*\Omega^i = {}^*g^{ij} \, {}^*\Omega_{,j} = y^i.$$

Thus

$$^*\Omega^i_{,j} = \delta^i_j + {}^*\{{}^i_{kj}\} y^k. \tag{6.42}$$

Contracting we have

$$\Delta^*\Omega = n + \tfrac{1}{2} y^k \frac{\partial}{\partial y^k} \log|{}^*g|, \tag{6.43}$$

since

$$*\{{}^i_{ik}\} = \tfrac{1}{2} \partial_k \log|*g|.$$

Using (6.38) we have

$$\Delta^*\Omega = n + y^k \frac{\partial}{\partial y^k} \log \rho, \tag{6.44}$$

which can be written

$$\Delta^*\Omega = n + *\Omega^k \frac{\partial}{\partial y^k} (\log \rho). \tag{6.45}$$

The tensorial nature of this equation shows that it is valid in any allowable coordinate system and we thus have the formula (6.40).

6.5 GEODESIC POLAR COORDINATES

In this section we assume that the manifold (M, g) is strictly Riemannian, that is, the metric is positive definite. The geodesic rays emerging from P_0 have equations of the type

$$y^i = a^i r, \quad r > 0, \tag{6.46}$$

r being the length of the geodesic measured from P_0, and (a^i) being the components of the unit tangent vector to the geodesic at P_0. We now introduce *geodesic polar coordinates* (r, θ^α), $\alpha = 1, \ldots, n-1$, by choosing the a^i to be functions of the θ^α in such a way that $g_{ij} a^i a^j = 1$. We than have $a^i = a^i(\theta)$ and

$$y^i = r a^i(\theta) \tag{6.47}$$

as the transformation to normal coordinates from geodesic polars. The volume element at $P(y^i)$ in normal coordinates is

$$\sqrt{|*g|} \, dy^1 \ldots dy^n$$

and hence in geodesic polars it is given by

$$dV = \sqrt{|*g|} \cdot \left| \frac{\partial(y^1, \ldots, y^n)}{\partial(r, \theta^1, \ldots, \theta^{n-1})} \right| dr \, d\theta^1 \ldots d\theta^{n-1} \tag{6.48}$$

Using the formula for ρ in terms of normal coordinates this may be written, using (6.47) and (6.48),

$$dV = r^{n-1} \rho \, \Phi(\theta) \, dr \, d\theta^1 \ldots d\theta^{n-1} \tag{6.49}$$

where $\Phi(\theta)$ depends only on the θ's and not on r. Let c be a geodesic radial coordinate at any point of a normal neighbourhood N other than P_0. Then

6.5 Geodesic polar coordinates

$$r = c \tag{6.50}$$

is the equation of a spherical surface of degree $n-1$ which we denote by $S^{n-1}(P_0, r)$. We call this the geodesic sphere of radius r centre P_0, and assume that r is sufficiently small that every geodesic ray through P_0 lies in N. From (6.49) it follows that the volume element of $S^{n-1}(P_0, r)$ is given by

$$dV_{n-1} = r^{n-1} \rho \, \Phi(\theta) \, d\theta^1 \wedge \ldots \wedge d\theta^{n-1} \tag{6.51}$$

with $r = c$. For points P on $S^{n-1}(P_0, r)$ we have

$$\Omega = \tfrac{1}{2} c^2 = k, \quad \text{say.} \tag{6.52}$$

So the equation of $S^{n-1}(P_0, r)$ in terms of normal coordinates y^i of origin P_0 is

$$\Omega = \tfrac{1}{2} (g_{ij})_0 y^i y^j = k. \tag{6.53}$$

Let $^*v^j$ denote the components of a normal vector at $Q(y^i)$ of $S^{n-1}(P_0, r)$. Then

$$^*v_i = {^*g_{ij}} \, ^*v^j = \frac{\partial \Omega}{\partial y^i} = (^*g_{ij})_0 y^j = {^*g_{ij}} y^j.$$

Thus $^*v^i = y^i = ra^i(\theta)$ where r has the value c. The *unit* normal vector at Q has components

$$^*n^i = a^i. \tag{6.54}$$

Let u be a differentiable function defined on N, and let $\partial u / \partial n$ be its derivative at Q in the direction of the normal $^*n^i$. Then

$$\frac{\partial u}{\partial n} = \frac{\partial u}{\partial y^i} \cdot {^*n^i} = a^i \frac{\partial u}{\partial y^i}.$$

But for the point P we have normal coordinates y^i and geodesic polar coordinates (r, θ^α) and so

$$\frac{\partial u}{\partial r} = \frac{\partial u}{\partial y^i} \frac{\partial y^i}{\partial r} = a^i \frac{\partial u}{\partial y^i}. \tag{6.55}$$

Hence at the point Q we have

$$\frac{\partial u}{\partial n} = \frac{\partial u}{\partial r}. \tag{6.56}$$

Relation (6.56) also follows from the Gauss lemma on geodesics.

We multiply (6.55) by r and use (6.47) to get

$$r \frac{\partial u}{\partial r} = y^i \frac{\partial u}{\partial y^i}. \tag{6.57}$$

6 Special Riemannian manifolds

Apply this with $u = \Omega$ and use (6.44) to get

$$\Delta\Omega = n + r\frac{\partial}{\partial r}(\log \rho). \tag{6.58}$$

Now

$$*g^{ij}\frac{\partial \Omega}{\partial y^i}\frac{\partial \Omega}{\partial y^j} = (*g_{ij})_0 y^i y^j = 2\Omega,$$

that is, $*g^{ij}rr_{,i}rr_{,j} = r^2$ so that

$$\Delta_1 r = *g^{ij}r_{,i}r_{,j} = 1. \tag{6.59}$$

Then $\Delta\Omega = \Omega^i_{,i} = (rr_{,i}g^{ij})_{,j} = r\Delta r + \Delta_1 r = r\Delta r + 1$. Thus we get

$$\Delta r = \frac{n-1}{r} + \frac{\partial}{\partial r}(\log \rho). \tag{6.60}$$

We shall compute the functions ρ and $\Delta\Omega$ for a space of constant sectional curvature K with a positive definite metric. Consider the metric $ds^2 = *g_{ij}dy^i dy^j$ where

$$*g_{ij} = \frac{y_i y_j}{r^2} + \frac{\sin^2\{r\sqrt{K}\}}{Kr^2}\left\{(g_{ij})_0 - \frac{y_i y_j}{r^2}\right\}, \qquad r \neq 0 \tag{6.61}$$

$$(*g_{ij})_0 = \lim_{P \to P_0} *g_{ij} = (g_{ij})_0 \qquad \text{at } P_0. \tag{6.62}$$

We leave it as an exercise to the reader to check that this metric satisfies the condition

$$R_{ijke} = K(g_{ik}g_{je} - g_{ie}g_{jk}) \tag{6.63}$$

and is therefore a space of constant curvature K. Moreover, if we multiply both sides of (6.61) by y^j we find $*g_{ij}y^j = (*g_{ij})_0 y^j$ which is sufficient to show that the coordinate system is normal. Writing

$$\lambda = \frac{\sin^2\{r\sqrt{K}\}}{Kr^2}, \qquad \mu = \frac{1-\lambda}{r^2} \tag{6.64}$$

then the metric is given by

$$*g_{ij} = y_i y_j \mu + (*g_{ij})_0 \lambda. \tag{6.65}$$

Taking determinants of each side gives

$$*g = (*g_0)\lambda^{n-1} \tag{6.66}$$

from which

$$\rho = \sqrt{\left(\frac{g^*}{g_0}\right)} = \left[\frac{\sin^2(r\sqrt{K})}{Kr^2}\right]^{\frac{n-1}{2}}. \tag{6.67}$$

6.6 *Jacobi fields*

Alternatively

$$\rho = \left[\frac{\sin^2\sqrt{(2K\Omega)}}{2K\Omega}\right]^{\frac{n-1}{2}} \quad (6.68)$$

which is valid in any coordinate system.
Moreover we have

$$\Delta\Omega = n + 2\Omega \frac{d}{d\Omega}(\log \rho)$$

and on substituting for ρ we have

$$\Delta\Omega = 1 + (n-1)\sqrt{(2K\Omega)} \cot\sqrt{(2K\Omega)}. \quad (6.69)$$

If $K = 0$ then $\rho = 1$ and $\Delta\Omega = n$. If K is negative then the same formulas hold on replacing the trigonometrical functions by the corresponding hyperbolic functions. We note that for spaces of constant curvature, ρ and $\Delta\Omega$ are functions of Ω alone. One of the main problems of this chapter is to characterize spaces which possess this property.

6.6 JACOBI FIELDS

In this section we were influenced by the book by Berger, Gauduchon and Mazet (1971), denoted by [BGM]. A *Jacobi field* Y is defined along the geodesic $\gamma: [a, b] \to (M, g)$ if it satisfies the equation

$$\nabla_{\dot\gamma}\nabla_{\dot\gamma} Y + R(Y, \dot\gamma)\dot\gamma = 0, \quad (6.70)$$

where

$$R(Y, \dot\gamma) = \nabla_Y \nabla_{\dot\gamma} - \nabla_{\dot\gamma}\nabla_Y + \nabla_{[\dot\gamma, Y]}. \quad (6.71)$$

It should be noted that in this section we do not follow the rule for the curvature tensor favoured by Milnor, Gray, and [BGM]. This uses the opposite sign convention to that favoured by Chern, Kobayashi and Nomizu, Eisenhart, Lichnerowicz, and Helgason as in (2.7). The essential difference is that some equations will contain a plus sign with one convention and a minus sign with the other. However the geometrically significant results are valid whatever convention is used.

Let $e_1, \ldots, e_n$ be n orthonormal vector fields which are parallel along the geodesic γ so that, for each $s \in [a, b]$, the vectors at $\gamma(s)$ form an orthonormal basis for $M_{\gamma(s)}$.

Write $Y(s) = Y^i(s) e_i(s)$ and

$$R(e_i(s), \dot\gamma(s))\dot\gamma(s) = a_i^j(s) e_j(s).$$

Then the Jacobi vector field equation can be written

$$\frac{d^2 Y^i}{ds^2} + a^i_j Y^j = 0, \quad i, j = 1, \ldots, n. \tag{6.72}$$

This is a differential equation of the second order. Moreover, dY^i/ds are the components of $\nabla_{\dot\gamma} Y$ in the basis $(e_1, \ldots, e_n)$. From this we get the following proposition.

Proposition (6.6.1) *For every Y_0, $Y_0' \in M$ there exists a uniquely determined Jacobi vector field Y along γ such that $Y(0) = Y_0$ and $(\nabla_{\dot\gamma} Y)_0 = Y_0'$.*

The Jacobi vector fields along γ therefore form a vector space of dimension $2n$ which will be denoted by $J_\gamma(\gamma(a), \gamma(b))$. We can put on $J_\gamma(\gamma(a), \gamma(b))$ a Euclidean inner product by writing

$$\langle Y, Z \rangle = \int_a^b \langle Y(s), Z(s) \rangle \, ds.$$

Let us now see how the projection of Y on $\dot\gamma$ varies with s.
We have

$$\frac{d}{ds} \langle \dot\gamma(s), Y(s) \rangle = \langle \nabla_{\dot\gamma}\dot\gamma(s), Y(s) \rangle + \langle \dot\gamma(s), \nabla_{\dot\gamma} Y(s) \rangle$$

$$= \langle \dot\gamma(s), \nabla_{\dot\gamma} Y(s) \rangle,$$

the first term vanishing because γ is a geodesic. Similarly

$$\frac{d^2}{ds^2} \langle \dot\gamma(s), Y(s) \rangle = \langle \dot\gamma(s), \nabla_{\dot\gamma}\nabla_{\dot\gamma} Y(s) \rangle.$$

Therefore

$$\frac{d^2}{ds^2} \langle \dot\gamma(s), Y(s) \rangle = - \langle \dot\gamma(s), R(Y(s), \dot\gamma(s))\dot\gamma(s) \rangle$$

$$= R(\dot\gamma(s), Y(s), \dot\gamma(s), \dot\gamma(s)) = 0.$$

Thus $\langle \dot\gamma(s), Y(s) \rangle$ is an affine function of s so that

$$\langle \dot\gamma(s), Y(s) \rangle = \alpha s + \beta$$

for some constants α and β. We shall be particularly interested in Jacobi vector fields which are *normal* to $\gamma(s)$, so that $\langle \dot\gamma(s), Y(s) \rangle = 0$ for all s.

Consider the field $s \to \dot\gamma(s)$. We have $\nabla_{\dot\gamma}\nabla_{\dot\gamma}\gamma = 0$ and $R(\dot\gamma, \dot\gamma)\dot\gamma = 0$. Thus this is a Jacobi field. Similarly, consider the field $s \to s\dot\gamma(s)$. We have

$$\nabla_{\dot\gamma}(s\dot\gamma(s)) = s\nabla_{\dot\gamma}\dot\gamma(s) + \dot\gamma(s) = \dot\gamma(s)$$

6.6 Jacobi fields

and so $\nabla_{\dot\gamma}\nabla_{\dot\gamma}(s\dot\gamma(s)) = 0$. Moreover $R(\dot\gamma(s), s\dot\gamma(s))\dot\gamma(s) = 0$ and hence the field $s \to s\dot\gamma(s)$ is also a Jacobi vector field.

These two vector fields are obviously linearly independent. Consider now the orthogonal complement in $J_\gamma[\gamma(a), \gamma(b)]$ of the plane which they generate. We have

$$\langle \dot\gamma, Y \rangle = \int_a^b \langle \dot\gamma(s), Y(s)\rangle \, ds = \int_a^b (\alpha s + \beta) \, ds$$

$$= \tfrac{1}{2}\alpha(b^2 - a^2) + \beta(b - a),$$

$$\langle s\dot\gamma, Y \rangle = \int_a^b \langle s\dot\gamma(s), Y(s)\rangle \, ds = \int_a^b (\alpha s^2 + \beta s) \, ds$$

$$= \tfrac{1}{3}\alpha(b^3 - a^3) + \tfrac{1}{2}\beta(b^2 - a^2).$$

The conditions for orthogonality give

$$\tfrac{1}{2}\alpha(b^2 - a^2) + \beta(b - a) = 0; \quad \tfrac{1}{3}\alpha(b^3 - a^3) + \tfrac{1}{2}\beta(b^2 - a^2) = 0,$$

which has the unique solution $\alpha = 0$, $\beta = 0$. Thus the orthogonal complement of $(\dot\gamma, s\dot\gamma)$ is precisely the set of Jacobi fields normal to γ. *Thus the set of Jacobi fields normal to γ form a vector subspace of dimension $2n - 2$.*

We now obtain a geometrical interpretation of the Jacobi field equations.

Theorem (6.6.2) *Let $K: [a, b] \times]-\varepsilon, \varepsilon[\to (M, g)$ be a variation of the geodesic γ such that, for all $t \in]-\varepsilon, \varepsilon[$, $\gamma_t(\cdot) = K(.,t)$ is a geodesic. Then $Y(s) = \partial K/\partial t\, (s, 0)$ is a Jacobi vector field along γ. Conversely every Jacobi vector field along γ arises in this way.*

To prove the first part of the theorem we have

$$\nabla_{\dot\gamma}\nabla_{\dot\gamma} Y = \nabla_{\frac{\partial}{\partial s}}\nabla_{\frac{\partial}{\partial s}} \frac{\partial K}{\partial t}$$

$$= \nabla_{\frac{\partial}{\partial s}}\nabla_{\frac{\partial}{\partial t}} \frac{\partial K}{\partial s}$$

$$= \nabla_{\frac{\partial}{\partial t}}\nabla_{\frac{\partial}{\partial s}} \frac{\partial K}{\partial s} - R\left(\frac{\partial K}{\partial t}, \frac{\partial K}{\partial s}\right)\frac{\partial K}{\partial s}$$

$$= 0 - R\left(\frac{\partial K}{\partial t}, \frac{\partial K}{\partial s}\right)\frac{\partial K}{\partial s}.$$

Putting $t = 0$ gives

$$\nabla_{\dot\gamma}\nabla_{\dot\gamma} Y + R(Y, \dot\gamma)\dot\gamma = 0$$

so that Y is a Jacobi vector field.

Conversely suppose we are given a Jacobi vector field Y. We shall construct a variation K such that the γ_t are geodesics. Let c be a geodesic issuing from $Y(0)$ defined on $]-\varepsilon, \varepsilon[$. We put

$$K(s, t) = \exp_{c(t)}[s. P_c(c(0), c(t))(\dot{\gamma}(0) + tY'(0))].$$

That is to say, we transport parallelly along $c(t)$ the tangent vector $\dot{\gamma}(0) + tY'_{(0)}$ at $c(0)$, and then take the exponential of the resulting vector at $c(t)$. Then $Z(s) = \partial K/\partial t\,(s, 0)$ is a Jacobi vector field along γ by the first part of the theorem.

It suffices to show that $Z(0) = Y(0)$ and $Z'(0) = Y'(0)$:

$$Z(0) = \left(\frac{d}{dt} \exp_{c(t)} 0\right)_{t=0} = \left(\frac{dc}{dt}\right)_{t=0} = Y(0);$$

moreover

$$Z'(0) = \left[\nabla_{\frac{\partial}{\partial s}} \frac{\partial K}{\partial t}\right](0, 0) = \left[\nabla_{\frac{\partial}{\partial t}} \frac{\partial K}{\partial s}\right](0, 0)$$

$$= \nabla_{\frac{\partial}{\partial t}}[P_c(c(0), c(t))(\dot{\gamma}(0) + tY'(0))].$$

$$= \frac{d}{dt}(\dot{\gamma}(0) + tY'(0)) = Y'(0).$$

We have established the proof of the proposition.

We shall now use Jacobi vector fields in order to calculate the differential of the exponential map. Let $m \in M$ and $x \in M_m$. Then the exponential map sends x to a point on the geodesic γ (uniquely determined by the direction of x at m) distance $|x|$ along the geodesic. The differential of this map will send a tangent vector to M_m at x to a tangent vector to M at $\exp_m x$. However we make the natural identification of the Euclidean tangent space to M_m at x with the Euclidean space M_m. With this identification, the differential of the exponential map can be regarded as mapping a tangent vector $y \in M_m$ to a tangent vector to M at $\exp_m x$. We denote this image vector by $T_x \exp_m(y)$ or by $(\exp_m*)_{|x}(y)$. We define a variation K of γ^u, where $u = x/|x|$, by putting

$$K(s, t) = \exp_m s\left(u + \frac{t}{|x|} y\right).$$

We have $T_x \exp_m(y) = \dfrac{\partial K}{\partial t}(|x|, 0)$. Write $Y(s) = \dfrac{\partial K}{\partial t}(s, 0)$.

Clearly the γ_t are geodesics and hence Y is a Jacobi vector field along γ_u. We have

6.6 Jacobi fields

$$Y(|x|) = T_x \exp_m(y), \qquad Y(0) = 0;$$

$$Y'(0) = \nabla_{\frac{\partial}{\partial s}} \frac{\partial K}{\partial t}(0,0) = \nabla_{\frac{\partial}{\partial t}} \frac{\partial K}{\partial s}(0,0) = \frac{d}{dt}\left(u + \frac{t}{|x|}y\right)_{t=0} = \frac{y}{|x|}.$$

Hence we have proved the following important result.

Proposition (6.6.3) Let u be a unit tangent vector to M at m and let r be a real number > 0. Then for every $y \in M_m$ the vector $T_{ru} \exp_m(y)$ is the value at distance r from m along the geodesic γ_u of the Jacobi vector field Y along γ_u determined by the initial conditions $Y(0) = 0$, $Y'(0) = y/r$.

We now use Jacobi vector fields to give an explicit formula for the volume density function of $\exp_m$ which we shortly define. Let (M, g) be a connected n-dimensional C^∞-Riemannian manifold, and let m be a point of M. Let r_m be the *injectivity radius* of $\exp_m: M_m \to M$, that is, the maximum value of r such that for vectors in M_m of length less than r, $\exp_m$ is a diffeomorphism. Hence $\exp_m$ induces a C^∞-diffeomorphism of the open ball $B_m := B_m(m, r_m)$ of the tangent space M_m of radius r_m about the origin o_m onto the open geodesic ball $\mathfrak{U}_m$ in M of the same radius.

Since $\mathfrak{U}_m$ is an n-cell we can choose an orientation and the corresponding volume form ω of (M, g) on $\mathfrak{U}_m$. Let ω_m denote the volume form of the coherently oriented Euclidean vector space (M_m, g_m) considered as a Riemannian manifold. Then we have

$$(\exp_m)^*(\omega) = \delta_m \omega_m \qquad (6.73)$$

for some C^∞-function δ_m on $B_m \subset M_m$. Observe that if we reverse the two coherently chosen orientations in $\mathfrak{U}_m$ and M_m, we change the sign of ω and ω_m simultaneously and hence from (6.73) it follows that δ_m is independent of the choice of orientation.

This strictly positive function δ_m is called the *volume density function* of $\exp_m$ because, for every $v \in B_m$, we have from (6.73)

$$\delta_m(v) = \lim_{\text{diam } K \to 0} \frac{\text{vol}_g(\exp_m K)}{\text{vol}_{g_m}(K)}, \qquad (6.74)$$

where K denotes any compact ball of M_m with centre v contained in B_m. Formula (6.74) is reminiscent of the classical formula of Gauss for the curvature density function of surfaces. The function δ_m gives rise to a uniquely determined C^∞-function $\theta_m: \mathfrak{U}_m \to \mathbb{R}_+$ defined on the open ball $\mathfrak{U}_m$ in M of radius r_m and centre m such that

$$\theta_m \circ \exp_m = \delta_m. \qquad (6.75)$$

Let $r \in [0, r_m[$ and let $B_m(m, r)$ denote the ball in M_m of radius r and centre o_m. Let $S_m(m, r)$ denote the sphere in M_m of radius r and centre o_m. We now obtain a formula for the n-dimensional volume of the open geodesic ball $\exp_m B_m(m, r)$ of radius r and centre m.

Consider the function $f: [0, r] \times S(m, 1) \to B(m, r)$ given by

$$(s, \xi) \to f(s, \xi) = s\xi. \tag{6.76}$$

This is a C^∞-map with the property

$$f^*(\omega_m) = s^{n-1}(\mathrm{d}s \wedge \mathrm{d}\xi)$$

where $\mathrm{d}\xi$ denotes the volume form of the sub-manifold $S(m, 1)$ of (M_m, g_m) and where for simplicity we have omitted the canonical pull-backs from the notation. From Fubini's theorem the volume of the open geodesic ball $\exp_m(B(m, r))$ in M of centre m and radius r is given by

$$V_m(r) = \int \delta_m(v) \, \mathrm{d\, vol}_{g_m}(v) = \int_0^r s^{n-1} \int_{S(m,1)} \delta_m(s\xi) \, \mathrm{d}\xi \, \mathrm{d}s \tag{6.77}$$

We recall that the domain of $\exp_m$ is an open subset E_m of M_m, star-shaped with respect to o_m. Let $v \in E_m$ be a fixed vector and express this as $v = s\xi$ where $\xi \in S(m, 1)$ and $s \in [0, \infty[$. Then for every tangent vector $a \in (M_m)_v$ there exists a unique vector $\mathbf{a} \in M_m$ such that

$$a = \text{velocity vector of the curve } t \to v + t\mathbf{a} \text{ at } t = 0. \tag{6.78}$$

The map $a \to \mathbf{a}$ describes the standard identification of $(M_m)_v$ with the vector space M_m referred to previously. Then we have

$$(\exp_m)_{*|s\xi}(a) = \begin{cases} \mathbf{a} & \text{if } s = 0, \tag{6.79} \\ \dfrac{Y(s)}{s} & \text{if } s > 0, \tag{6.80} \end{cases}$$

where Y is the uniquely determined Jacobi vector field along the unit speed geodesic curve

$$\gamma: [0, s] \to M, \quad \lambda \to \gamma(\lambda) = \exp_m(\lambda\xi) \tag{6.81}$$

satisfying the initial conditions

$$Y(0) = o_m, \quad Y'(0) = \mathbf{a}. \tag{6.82}$$

Alternatively we may say that Y is the unique solution of the Jacobi equation $Y'' + R(Y, \gamma')\gamma' = 0$ satisfying (6.82). From (6.73) and (6.79) we have immediately

$$\delta_m(o_m) = 1. \tag{6.83}$$

6.6 Jacobi fields

Moreover we have, from (6.73), (6.79), and (6.82), the following theorem.

Theorem (6.6.4) If $(e_1, \ldots, e_n)$ is any positively oriented orthonormal frame of M_m we have

$$\delta_m(s\xi) = s^{-n} \omega_{\exp_m(s\xi)}(Y_1(s), \ldots, Y_m(s)) \quad (6.84)$$

where Y_i is the Jacobi vector field along γ satisfying

$$Y_i(0) = o_m, \quad Y_i'(0) = e_i, \quad i = 1, \ldots, n. \quad (6.85)$$

We can obtain a further simplification by specializing the choice of frame $(e_1, \ldots, e_n)$ so that $e_n = \gamma'(0) = \xi$. Then

$$Y_n(\lambda) = \lambda \gamma'(\lambda) \quad \text{for all } \lambda \in [0, s]. \quad (6.86)$$

Hence we get from (6.84) and (6.86)

$$\delta(s\xi) = s^{1-n} \omega_{\exp_m(s\xi)}(Y_1(s), \ldots, Y_{n-1}(s), \gamma'(s)). \quad (6.87)$$

Let E_i denote the vector field parallel along γ with $E_i(0) = e_i$ for $i = 1, \ldots, n$. From (6.81) and (6.86), $E_n = \gamma'$, and hence $E_1, \ldots, E_{n-1}$ span the normal bundle along γ. But we know that $Y_1, \ldots, Y_{n-1}$ remain orthogonal to γ' along γ, and therefore there exists a unique C^∞-endomorphism A of the normal bundle of γ with

$$Y_i(\lambda) = A(E_i)\lambda, \quad i = 1, \ldots, n-1 \text{ and all } \lambda \in [0, s]. \quad (6.88)$$

Hence we get from (6.87), (6.88)

$$\delta(s\xi) = s^{1-n}(\det A)(s). \quad (6.89)$$

Moreover, since E_i is parallel and Y_i is a Jacobi vector field along γ satisfying (6.85) for $i = 1, \ldots, n$, then A must satisfy

$$A'' + (R(.,\gamma')\gamma')A = 0 \quad \text{with } A(0) = 0, A'(0) = \text{Id}_{\xi^\perp} \quad (6.90)$$

where $A' = \nabla_{d/d\lambda} A$ and $A'' = \nabla_{d/d\lambda} A'$. Conversely if A is an endomorphism of the normal bundle satisfying (6.90), then it is easily verified that equation (6.89) is satisfied.

We now define a real-valued function θ on the tangent bundle TM. Let v be a point in TM excluding an element of zero section. Making use of equation (6.75) we define θ as follows. Let v be a non-zero element of TM and let $\pi(v) = m$ where π is the bundle projection map, so that v is a tangent vector to M at m. Let γ denote the unit speed geodesic through m with initial direction $v/\|v\|$. Then

$$\theta(v) = \delta_m(\gamma(\|v\|)). \quad (6.91)$$

We extend the definition of $\theta(v)$ by putting $\theta(v) = 1$ whenever v is a zero vector. From (6.87) we have

$$\theta(v) = \|v\|^{1-n} \det(Y_1(\|v\|), \ldots, Y_{n-1}(\|v\|)). \tag{6.92}$$

Writing $|v| = s$ we have

$$\theta(v) = s^{1-n} \det(Y_1(s), \ldots, Y_{n-1}(s)). \tag{6.93}$$

Note that this formula differs from that given in the book by Besse (1978), page 154, (6.4), which has s^{n-1} instead of our s^{1-n}. The importance of this formula is that it makes sense wherever the exponential map is defined; for a complete manifold there is no barrier to its validity. In our earlier treatment when we regarded the volume element $\theta_m(v)$ as $\sqrt{(g)}$ in a system of normal coordinates, there was some ambiguity about the extent of the domain of $\theta(v)$.

Let γ be a geodesic parametrized by arc length and identify tangent vectors to M along γ by parallel transport relative to a fixed vector space, say $M_{\gamma(0)}$ as previously. Writing $Y(s) = Y^i(s)E_i$ the Jacobi field equation along γ gives rise to an associated endomorphism A which satisfies the equation $A'' + RA = 0$, where R is the endomorphism of $T_{\gamma(s)}M$ depending on s given by $V \to R(V, \gamma')\gamma'$.

We note that R is a symmetric endomorphism since

$$\langle R(X, \gamma')\gamma', Y \rangle = R(\gamma', X, \gamma', Y) = R(\gamma', Y, \gamma', X)$$
$$= \langle R(Y, \gamma')\gamma', X \rangle.$$

However, A is not necessarily symmetric. Suppose now that Y and Z are two Jacobi vector fields along γ with corresponding automorphisms $A(s)$ and $B(s)$. Using * to denote transposition of endomorphisms we have

$$\frac{d}{ds}[A^{*'}B - A^*B'] = A^{*''}B + A^{*'}B' - A^{*'}B' - A^*B''$$
$$= A^{*''}B - A^*B''.$$

Using the fact that $A'' + RA = 0$ we have $A^{*''} + A^*R^* = 0$; similarly $B^{*''} + B^*R^* = 0$ so that $A^{*''}B - A^*B'' = -A^*R^*B + A^*RB = 0$ since $R = R^*$. It follows that

$$A^{*'}B - A^*B' = \text{constant along } \gamma. \tag{6.94}$$

In particular if we fix s and let A (respectively B) be a solution of (6.90) such that $A(0) = 0$, $A'(0) = I$ (respectively $B(s) = 0$, $B'(s) = -I$), then from (6.94) we have

$$B(0) = A^*(s). \tag{6.95}$$

6.6 Jacobi fields

The canonical geodesic involution

This is an involution on the tangent bundle TM of a Riemannian manifold and is defined as follows. Let Ω be the domain of definition of the exponential map and let $v \in TM$ belong to Ω. Let γ be the geodesic defined by $t \to \exp(tv)$. Then i is the map $\Omega \to \Omega$ defined by

$$i(v) = -\dot{\gamma}(1). \tag{6.96}$$

In other words start with $v \in TM$; project to a point m of M such that $\pi(v) = m$. Consider the unit speed geodesic through m in the direction $v/\|v\|$ and proceed along γ a distance equal to $\|v\|$. Associate to this point the tangent vector to M in the reverse direction of length $\|v\|$. This is the image of the original vector v under the involution map i. We now prove what for us is an important proposition.

Proposition (6.6.5) For any vector $v \in \Omega$ we have

$$\theta(i(v)) = \theta(v). \tag{6.97}$$

The proof follows since, using (6.95) we get

$$\theta(s\gamma'_{(0)}) = s^{1-n} \det(A(s)) = s^{1-n} \det(B(0)) = \theta(-s\gamma'(s))$$

as required. Another way of stating this result is as follows. Let P and Q be two points of M lying in a normal coordinate neighbourhood U. Then the volume element $\sqrt{g_Q}$ at Q expressed in terms of an orthonormal coordinate system based at P is equal to the volume element $\sqrt{g_P}$ at P expressed in terms of an orthonormal coordinate system based at Q.

We now make another application of Jacobi vector fields, this time to the geometry of geodesic spheres of Riemannian manifolds. A natural research programme is to determine to what extent a knowledge of the geometry of geodesic spheres determines the geometry of the manifold M. In the following we express the second fundamental form of a geodesic sphere in terms of Jacobi vector fields.

Let $S(m, r)$ be a geodesic sphere of centre m and radius r contained inside a coordinate neighbourhood U, with orthonormal coordinates x^i centred at m. Let s denote the arc length measured along a unit speed geodesic γ from m. The unit tangent vector field along γ is given by

$$N = \frac{x^i}{s} \cdot \frac{\partial}{\partial x^i} = a^i \frac{\partial}{\partial x^i},$$

where a^i are components, relative to (x^i), of the unit tangent vector to γ at m. Consider the vector field Y defined on U, perpendicular to γ along γ, given by

6 Special Riemannian manifolds

$$Y = \lambda^i s \frac{\partial}{\partial x^i}$$

where $(g_{ij})_m \lambda^i a^j = 0$. Clearly $Y(s)$ is a Jacobi vector field defined along γ with the property that $Y(0) = 0$, and $Y'(0) = (\lambda^i \partial/\partial x^i)_m$. We now prove that $[N, Y] = 0$. Consider the geodesic sphere given by $s = r$. Then $N(s) = 1$ and $Y(s) = 0$ since Y is perpendicular to N and therefore tangential to the sphere $s = r$. We have

$$\begin{aligned}[][N, Y] &= \left[N, \lambda^i s \frac{\partial}{\partial x^i}\right] = N(\lambda^i s) \frac{\partial}{\partial x^i} + \lambda^i s \left[N, \frac{\partial}{\partial x^i}\right] \\ &= \lambda^i \frac{\partial}{\partial x^i} + \lambda^i s \left[\frac{x^j}{s} \frac{\partial}{\partial x^j}, \frac{\partial}{\partial x^i}\right] \\ &= \lambda^i \frac{\partial}{\partial x^i} - \lambda^i s \frac{\partial}{\partial x^i}\left(\frac{x^j}{s}\right) \frac{\partial}{\partial x^j} \\ &= \lambda^i \frac{\partial}{\partial x^i} - Y\left(\frac{x^j}{s}\right) \frac{\partial}{\partial x^j} \\ &= \lambda^i \frac{\partial}{\partial x^i} - \frac{1}{s} Y(x^j) \frac{\partial}{\partial x^j} \\ &= \lambda^i \frac{\partial}{\partial x^i} - \lambda^i \frac{s}{s} \frac{\partial x^j}{\partial x^i} \frac{\partial}{\partial x^j} \\ &= \lambda^i \frac{\partial}{\partial x^i} - \lambda^i \frac{\partial}{\partial x^i} = 0,\end{aligned}$$

and the claim is justified.

Let W denote the Weingarten map of the geodesic sphere $S(m, s)$ at $\gamma(s)$. Since $[N, Y] = 0$ we have

$$\nabla_N Y - \nabla_Y N - [N, Y] = 0 \Rightarrow \nabla_Y N = \nabla_N Y.$$

Thus $WY = -\nabla_Y N = -\nabla_N Y$.

Let $(e_1, \ldots, e_n)$ be a parallel orthonormal basis along γ with $e_n = N$. Let $Y_a = \Sigma A_a^b e_b$, $a, b = 1, \ldots, n-1$, be Jacobi vector fields along γ with initial conditions $Y_a(0) = 0$, $Y_a'(0) = e_a$. Then we have $WA = -A'$ from which we deduce $W = -A'A^{-1}$.

Note that all the contents of this section apply to an arbitrary Riemannian manifold. However we have selected these results because they are appropriate to the study of harmonic manifolds which we take up in the next section.

6.7 DEFINITION OF HARMONIC METRICS

At the end of Section 5 we found that in spaces of constant curvature, the two-point scalar functions ρ and $\Delta\Omega$ were functions of distance only. We now wish to determine those Riemannian metrics which have precisely this property.

Definition (6.7.1) *A Riemannian manifold (M, g) is said to be locally harmonic if for every m in M there exists a positive real number $\varepsilon(m)$ and a function $\Theta_m: [0, \varepsilon(m)[\to \mathbb{R}$ such that for all $u \in M_m$ with $\|u\| < \varepsilon(m)$, we have*

$$\theta(u) = \Theta_m(\|u\|). \tag{6.98}$$

Definition (6.7.2) *A Riemannian manifold is said to be globally harmonic if it is complete and if for every $m \in M$ there is a function $\Theta_m: \mathbb{R}_+ \to \mathbb{R}$ such that for all $u \in M_m$,*

$$\theta(u) = \Theta_m(\|u\|). \tag{6.99}$$

By using the function $\theta(u)$ rather than normal coordinates for our definition we have avoided the difficulty previously mentioned of determining where the map $(m, n) \to \theta_m(n)$ is really defined. However, for properties of harmonic manifolds which are strictly local in extent normal coordinates are a powerful and valuable tool.

In this chapter we shall be concerned primarily with analytic manifolds and we shall see that analyticity is already implied by the harmonic condition. We now have the following result.

Proposition (6.7.3) *For every real number r such that there exists m in M which satisfies $\Theta_m(r) \neq 0$, we have $\Theta_n(r) = \Theta_m(r)$ for every n in M.*

In other words, the function $\Theta_m(r)$ is the same for all points of M. To prove the result let r and m satisfy $\Theta_m(r) \neq 0$. Then from the continuity of θ there exists an open neighbourhood V of m for which $\Theta_n(r) \neq 0$ for every n in V. We define a function f on V by $n \to \Theta_n(r)$ and we shall prove that $df = 0$ so that f is constant. Since θ is continuous and M is connected we shall then have established the result.

Let $n \in V$ and let u be a unit tangent vector at n. Let γ be a geodesic such that $\gamma(r) = n$ and $\gamma'(r)$ is perpendicular to u. Let $p = \gamma(0)$. Since $\theta(r\dot\gamma(r)) = \Theta_n(r) \neq 0$, the points p and n are not conjugate along γ, that is to say that there is a unique geodesic joining p to n. Thus there exists a one-parameter family of geodesics γ_t with $\gamma_0 = \gamma$, $\gamma_t(0) = p$ for all t and $d/dt\,(t \to \gamma_t(r))|_{t=0} = u$. Now apply the invariance of θ under the canonical involution of the tangent bundle to get $\Theta_{\gamma_t(r)}(r) = \Theta_m(r)$ for every t. Thus

$$0 = \frac{\partial}{\partial t}(t \to \Theta_{\gamma_t(r)})|_{t=0} = df(u).$$

Thus f is a constant and the proof is complete.

We now prove the equivalence of three definitions of harmonicity.

(1) (M,g) is locally harmonic at m if there exists a function Θ and a positive real number ε such that $\theta(u) = \Theta(\|u\|)$ for every u in M_m with $\|u\| < \varepsilon$.

(2) (M,g) is locally harmonic if there exists a positive real number ε and a function $F: [0, \varepsilon[\to \mathbb{R}$ such that the function $\Omega = 1/2d^2(m,.)$ satisfies $\Delta\Omega = F(d(m,.))$, that is, $\Delta\Omega$ is a function of the distance from m only.

(3) (M,g) is locally harmonic if there exists a positive real number ε and a function $G: [0, \varepsilon[\to \mathbb{R}$ such that the function g defined on $B(m, \varepsilon) \setminus m$ by $n \to G(d(m,n))$ satisfies $\Delta g = 0$.

From equation (6.60) we have, for a function g which depends only on the distance from m, that is, g is of the form $G \circ d(m,.)$,

$$\Delta g = G'' + G' \left(\frac{n-1}{d(m,.)} + \frac{\theta'_m}{\theta_m} \right). \qquad (6.100)$$

Here G' and G'' are ordinary first and second derivatives of

$$G: [0, \varepsilon[\to \mathbb{R}, \qquad \theta_m = \theta \circ (\exp_m|B(m,\varepsilon))^{-1}$$

and θ'_m is the radial derivative of θ_m in M_m. We see from (6.100) that (2) implies (1) and that (3) implies (1). Moreover (1) implies (3) via the existence theorem for ordinary differential equations. We now obtain a further characterization of locally harmonic Riemannian spaces first obtained by Willmore (1950).

Theorem (6.7.4) *Let u be a harmonic function ($\Delta u = 0$) in a given neighbourhood U of a harmonic manifold. If P_0 is any point in U, then the mean value of u over every geodesic sphere of centre P_0 contained in U is equal to its value at P_0.*

To prove the theorem, let $\mu(u; P_0; r)$ denote the mean value of the harmonic function u over the sphere $S_{n-1}(P_0, r)$. Then

$$\mu(u; P_0; r) = \int_S u\, dv_{n-1} \bigg/ \int_S dv_{n-1} \qquad (6.101)$$

6.7 Definition of harmonic metrics

where the S indicates that the integral is taken over the sphere $S_{n-1}(P_0, r)$. We have already seen that the $(n-1)$-volume element of the sphere is

$$dv_{n-1} = r^{n-1} \rho \, d\omega \tag{6.102}$$

where $d\omega = \Phi(\theta) \, d\theta^1 \wedge \ldots \wedge d\theta^{n-1}$. Since the space is harmonic we may assume that $\rho = \rho(r)$, so we may write

$$dv_{n-1} = r^{n-1} \rho(r) \, d\omega.$$

Since r is constant on the sphere it follows that

$$\mu(u; P_0; r) = \frac{\int_S u \, d\omega}{\int_S d\omega}.$$

Therefore

$$\frac{\partial \mu}{\partial r}(u; P_0; r) = \frac{\int_S \frac{\partial u}{\partial r} \, d\omega}{\int_S d\omega}.$$

Multiply numerator and denominator of the right-hand side by $r^{n-1} \rho(r)$ to get

$$\frac{\partial}{\partial r} \mu(u; P_0; r) = \frac{\int_S \frac{\partial u}{\partial r} \, dv_{n-1}}{\int_S dv_{n-1}}. \tag{6.103}$$

However, by Green's theorem we have

$$\int_B \Delta u \, dv_n = \int_S \frac{\partial u}{\partial n} \, dv_{n-1} \tag{6.104}$$

where the integral on the left is taken over the closed ball $B(P_0, r)$, and $\partial u/\partial n$ denotes the derivative of u in the direction of the outward normal to $S_{n-1}(P_0, r)$. However we have seen that this derivative is equal to $\partial u/\partial r$. Hence

$$\int_B \Delta u \, dv_n = \int_S \frac{\partial u}{\partial r} \, dv_{n-1},$$

and so, since u is harmonic, we have

$$\int_S \frac{\partial u}{\partial r} dv_{n-1} = 0$$

Hence

$$\frac{\partial}{\partial r} \mu(u; P_0; r) = 0.$$

Thus the mean value is independent of the radius of the sphere, and the theorem follows by making $r \to 0$.

Theorem (6.7.5) *Let u be a C^2-function defined on a neighbourhood U of a harmonic manifold, such that its mean value over every geodesic sphere contained in U is equal to the value at the centre. Then $\Delta u = 0$.*

To prove the theorem we note that our data implies that if r is small enough so that $S_{n-1}(P_0, r)$ lies in U for every P_0, then

$$\frac{\partial}{\partial r} \mu(u; P_0; r) = 0.$$

Hence by (6.103) and (6.104)

$$\int_B \Delta u \, dv_{n-1} = 0$$

for every $B(P_0, r)$. This forces the conclusion $\Delta u = 0$. For, otherwise, there must be a point $P_1 \in U$ at which Δu is positive or negative — say positive. Since u is a C^2-function, Δu is continuous and hence is positive in some neighbourhood U_1 of P_1 where $U_1 \subset U$. Hence its integral over any geodesic *ball* of centre P_1 contained in U_1 is positive, giving a contradiction. This completes of proof.

Theorem (6.7.6) *Let (M, g) be an analytic Riemannian manifold such that the mean value over every geodesic sphere of every harmonic function in any neighbourhood containing the sphere is equal to the value of the function at the centre. Then (M, g) is a harmonic manifold.*

To prove the theorem we make use of the following lemma.

Lemma (6.7.7) *In any analytic Riemannian manifold (M, g) there exists a solution of the equation $\Delta u = 0$ which assumes prescribed continuous values $u = f$ on the surface of any given geodesic sphere $S_{n-1}(P_0, r)$ and has no singularities inside the sphere.*

6.7 Definition of harmonic metrics

The lemma says, in effect, that the Dirichlet problem for an elliptic partial differential equation of the second order with analytic coefficients has a unique solution. The proof of this standard result may be seen, for example, in the paper by Duff and Spencer (1952).

Proof Theorem (6.7.6). Let P_0 be a point of M and let $S_{n-1}(P_0, r)$ be a geodesic sphere contained in a normal coordinate neighbourhood and let u be a function harmonic in a neighbourhood containing $S_{n-1}(P_0, r)$. The element of volume of $S_{n-1}(P_0, r)$ is of the form

$$dv_{n-1} = r^{n-1} \rho(r, \theta) \, d\omega.$$

By hypothesis, the mean value of u over $S_{n-1}(P_0, r)$ is equal to the value at P_0. Hence

$$\frac{\partial}{\partial r} \mu(u; P_0; r) = 0,$$

and so by (6.101) we have

$$\frac{\partial}{\partial r} \left\{ \frac{\int_S u \, dv_{n-1}}{\int_S dv_{n-1}} \right\} = 0.$$

This can be written

$$\frac{\frac{\partial}{\partial r} \left(\int_S u \, dv_{n-1} \right)}{\int_S u \, dv_{n-1}} = \frac{\frac{\partial}{\partial r} \int_S dv_{n-1}}{\int_S dv_{n-1}}. \tag{6.105}$$

Now

$$\frac{\partial}{\partial r} \int_S u \, dv_{n-1} = \frac{\partial}{\partial r} \int_S u \, r^{n-1} \rho(r, \theta) \, d\omega$$

$$= \int_S \frac{\partial u}{\partial r} dv_{n-1} + \int_S u \frac{\partial}{\partial r} \{r^{n-1} \rho(r, \theta)\} \, d\omega. \tag{6.106}$$

Since u is harmonic, the first term on the right is zero while the second term is equal to

$$\int_S u \frac{\partial}{\partial r} \log \{r^{n-1} \rho(r, \theta)\} \, dv_{n-1}.$$

Hence using (6.60) we get from (6.106)

$$\frac{\partial}{\partial r}\int_S u\, dv_{n-1} = \int_S u\, \Delta r\, dv_{n-1}. \qquad (6.107)$$

This also holds when u is identically equal to unity. So substituting in (6.107) we get

$$\frac{\int_S u\, \Delta r\, dv_{n-1}}{\int_S u\, dv_{n-1}} = \frac{\int_S \Delta r\, dv_{n-1}}{\int_S dv_{n-1}}.$$

This relation must hold for every harmonic function u. But the right-hand side is a function independent of u, say $\zeta(r)$. Then

$$\int_S u\{\Delta r - \zeta(r)\}\, dv_{n-1} = 0.$$

Hence $\Delta r - \zeta(r) = 0$ which is equivalent to

$$\Delta\Omega = r\zeta(r) + 1$$

and hence the manifold is harmonic.

6.8 CURVATURE CONDITIONS FOR HARMONIC MANIFOLDS

If we use normal coordinates we know that the condition for harmonicity can be expressed by requiring θ to be a radial function. We define the *radial derivative* D_u as follows. Let $m \in M$ and let u be a unit tangent vector at m. Then the radial derivative of order n is given by

$$D_u^n(\theta) = \frac{d^n}{ds^n}(s \to \theta(\exp_m(su)))_0. \qquad (6.108)$$

Then a necessary condition for harmonicity is that for every integer n there exists a real number k_n such that $D_u^n(\theta) = k_n$ for every unit vector of M_m. But we know from our discussion of normal coordinates in Chapter 3 that we can expand $\theta = \sqrt{(g)}$ as a power series. In fact the paper by Gray (1973) gives the expansion for the first seven terms. Writing $p = \exp_m(r\xi)$ where ξ is a unit vector of M_m we have

6.8 Curvature conditions for harmonic manifolds

$$\theta_m(p) = 1 + \sum_{k=2}^{7} \alpha_k(m, \xi) r^k + O(r^8), \tag{6.109}$$

where the α_k are completely determined by the Riemann curvature tensor and its covariant derivatives. In particular we have

$$\alpha_2(m, \xi) = -\frac{1}{6} \rho_{\xi\xi}(m),$$

$$\alpha_3(m, \xi) = -\frac{1}{12} (\nabla_\xi \rho_{\xi\xi})(m),$$

$$\alpha_4(m, \xi) = \frac{1}{24} \left[-\frac{3}{5} \nabla^2_{\xi\xi} \rho_{\xi\xi} + \frac{1}{3} \rho^2_{\xi\xi} - \frac{2}{15} \sum_{a, b=1}^{n} R^2_{\xi a \xi b} \right] (m),$$

$$\alpha_5(m, \xi) = \frac{1}{120} \left[-\frac{2}{3} \nabla^3_{\xi\xi\xi} \rho_{\xi\xi} + \frac{5}{3} \rho_{\xi\xi} \nabla_\xi \rho_{\xi\xi} - \frac{2}{3} \sum_{a, b=1}^{n} R_{\xi a \xi b} \nabla_\xi R_{\xi a \xi b} \right] (m),$$

$$\alpha_6(m, \xi) = \frac{1}{720} \left[-\frac{5}{7} \nabla^4_{\xi\xi\xi\xi} \rho_{\xi\xi} + 3 \rho_{\xi\xi} \nabla^2_{\xi\xi} r_{\xi\xi} + \frac{5}{2} (\nabla_\xi \rho_{\xi\xi})^2 \right.$$
$$- \frac{5}{9} \rho^3_{\xi\xi} - \frac{8}{7} \sum_{a, b=1}^{n} R_{\xi a \xi b} \nabla^2_{\xi\xi} R_{\xi a \xi b} - \frac{15}{14} \sum_{a, b=1}^{n} (\nabla_\xi R_{\xi a \xi b})^2$$
$$\left. - \frac{16}{63} \sum_{a, b, c=1}^{n} R_{\xi a \xi b} R_{\xi b \xi c} R_{\xi c \xi a} + \frac{2}{3} R_{\xi\xi} \sum_{a, b=1}^{n} R^2_{\xi a \xi b} \right] (m). \tag{6.110}$$

Here ∇ denotes the Levi-Civita connexion, R is the Riemann curvature tensor, and ρ the corresponding Ricci tensor. To compute $\alpha_7(m, \xi)$ is still more complicated but it can be given by the following formula:

$$2\alpha_7(m, \xi) = \sum_{i=2}^{6} \frac{(-1)^i}{(7-i)!} (\nabla^{7-i}_{\xi \ldots \xi} \alpha_i)(m, \xi). \tag{6.111}$$

Here we have written, for example,

$$\xi = \sum_i \xi^i \frac{\partial}{\partial x^i}, \quad \rho_{\xi\xi} = \sum_{i,j} \rho_{ij} \xi^i \xi^j, \quad \nabla_\xi \rho_{\xi\xi} = \sum_{i,j,k} (\nabla_i \rho_{jk}) \xi^i \xi^j \xi^k.$$

The condition for harmonicity implies that each $\alpha_k(m, \xi)$ shall be independent of ξ. The first condition gives

$$\rho_{ij} = K g_{ij} \tag{6.112}$$

for some constant K so we see immediately that the space must be Einstein. Now we know from a paper by Deturk and Kazdan

(1981) that an Einstein manifold is always analytic in geodesic normal coordinates, so there is no real restriction in supposing that our manifold is analytic.

The second condition arising from $\alpha_3(m, \xi)$, namely $\nabla_\xi \rho_{\xi\xi} = 0$, is satisfied automatically because of the first condition. The third condition simplifies to

$$\sum_{a,b=1}^{n} R_{a\xi b\xi} R_{a\xi b\xi} = H, \quad (6.113)$$

where H is some constant, again making use of the previous conditions. It is easily verified that the fourth condition is satisfied automatically because of the previous conditions. In a similar way the fifth condition simplifies to

$$32 \sum_{a,b,c=1}^{n} R_{a\xi b\xi} R_{b\xi c\xi} R_{c\xi a\xi} - 9 \sum_{a,b=1}^{n} \nabla_\xi R_{a\xi b\xi} \nabla_\xi R_{a\xi b\xi} = L. \quad (6.114)$$

It can be verified that $\alpha_7 = 0$ because of the preceding equations. It seemed a reasonable conjecture that this was no accident, that in fact the even conditions for a harmonic space implied the odd conditions. This was proved by Vanhecke (1981). He gave a very simple proof involving the canonical geodesic involution of the tangent bundle. Using a system of normal coordinates centred at m, with ξ a unit tangent vector at m, let $p = \exp_m(r\xi)$. We get

$$\theta_m(p) = 1 + \sum_{i=2}^{\infty} \alpha_i(m, \xi) r^i.$$

We have to prove that the conditions

$$\alpha_{2i}(m, \xi) = k_{2i}, \quad i = 1, \ldots,$$

where k_{2i} is a globally defined constant function, imply $\alpha_{2i+1}(m, \xi) = 0$. We use the canonical geodesic involution on the tangent bundle to get

$$\theta_p(m) = \theta_p(-r\xi) = \theta_m(p) = \theta_m(r\xi),$$

or

$$\sum_{i=2}^{\infty} (-1)^i \alpha_i(p, \xi) r^i = \sum_{i=2}^{\infty} \alpha_i(m, \xi) r^i. \quad (6.115)$$

Also by a Taylor expansion we have

$$\alpha_i(p, \xi) = \sum_{k=0}^{\infty} \frac{1}{k!} (\nabla_{\xi\ldots\xi}^k \alpha_i)(m, \xi) r^k \quad (6.116)$$

6.8 Curvature conditions for harmonic manifolds

where ∇_ξ denotes covariant differentiation along the geodesic γ with $\gamma(0) = m$, and $\gamma'(0) = \xi$. From (6.115) and (6.116) we get

$$\sum_{i=2}^{\infty} (-1)^i \sum_{k=0}^{\infty} \frac{1}{k!} (\nabla^k_{\xi\ldots\xi} \alpha_i)(m, \xi) r^{k+i} = \sum_{s=1}^{\infty} \alpha_s(m, \xi) r^s. \qquad (6.117)$$

Writing $k + i = s$ we have

$$\sum_{k+i=s} \frac{1}{k!} (-1)^i (\nabla^k_{\xi\ldots\xi} \alpha_i)(m, \xi) = \alpha_s(m, \xi). \qquad (6.118)$$

With $s = 3$ (6.118) gives $\alpha_3(m, \xi) = \nabla_\xi \alpha_2 - \alpha_3(m, \xi)$ and hence $\alpha_3(m, \xi) = 0$. We now proceed by induction. Assume that

$$\alpha_{2q-1}(m, \xi) = 0, \quad q = 1, \ldots, k \qquad \text{and} \qquad \alpha_{2q}(m, \xi) = k_{2q}, \text{ all } q.$$

Then we have, using (6.118),

$$\alpha_{2q+1}(m, \xi) = 0 + \cdots + 0 - \alpha_{2q+1}(m, \xi)$$

giving $\alpha_{2q+1}(m, \xi) = 0$. Thus, by the principle of induction, the result is obtained.

An alternative method of procedure is due to A. J. Ledger who used it in his Ph.D. thesis at the University of Durham in 1954. Fix m in M and a unit tangent vector u in M_m. Consider the geodesic $\gamma: s \to \exp_m(su)$ through m with velocity vector u. Let $e_1, \ldots, e_n$ be an orthonormal frame at m which we suppose carried along γ by parallel transport. Using the technique of Jacobi vector fields we let $A(s)$ be the associated endomorphism-valued function defined as the solution of $A'' + RA = 0$ with initial conditions $A(0) = 0$, $A'(0) = I$. We have

$$\det A(s) = s^{n-1} \theta(s). \qquad (6.119)$$

Using the well-known rule for differentiating a determinant we have

$$\text{trace}(A'A^{-1}) = \frac{\theta'}{\theta} + \frac{(n-1)}{s}. \qquad (6.120)$$

Now set $C = sA'A^{-1}$ which is a linear map depending on s. Differentiate C to get

$$C' = A'A^{-1} + sA''A^{-1} - sA'A^{-1}A'A^{-1}$$

and hence

$$sC' = sA'A^{-1} + s^2 A''A^{-1} - s^2 A'A^{-1}A'A^{-1}$$

that is $sC' = C - s^2 R - C^2$.

6 Special Riemannian manifolds

If we differentiate this equation $(n+1)$ times using the Leibnitz rule, we get

$$nC^{n+1}(0) = -n(n+1)R^{(n-1)}(0) - \sum_{k=0}^{n+1} \binom{n+1}{k} C^{(k)}(0) C^{n+1-k}(0), \tag{6.121}$$

which is known as Ledger's formula.

This formula allows us theoretically to compute $C^{(n)}(0)$ for any n but for $n > 6$ the result is quite complicated. Here is the formula for $n = 0, \ldots, 6$.

$$C(0) = I, \tag{6.122}$$

$$C'(0) = 0, \tag{6.123}$$

$$C''(0) = -\frac{2}{3} R(0), \tag{6.124}$$

$$C'''(0) = -\frac{3}{2} R'(0), \tag{6.125}$$

$$C^{(4)}(0) = -\frac{12}{5} R''(0) - \frac{8}{15} R(0) R(0) \tag{6.126}$$

$$C^{(5)}(0) = -\frac{10}{3} R'''(0) - \frac{5}{3} (R(0) R'(0) + R'(0) R(0)) \tag{6.127}$$

$$C^{(6)}(0) = -\frac{30}{7} R^4(0) - \frac{1}{7} \Big[45 R'(0) R'(0) + 24 R(0) R''(0)$$
$$+ 24 R''(0) R(0) + \frac{32}{3} R(0) R(0) R(0) \Big] \tag{6.128}$$

where we have denoted the composition of endomorphisms by ordinary multiplication. The formula for C^7 is given by

$$C^{(7)}(0) = -\frac{21}{4} R^{(5)}(0) - \frac{35}{6} (R'''(0) R(0) + R(0) R'''(0))$$
$$- \frac{63}{4} (R''(0) R'(0) + R'(0) R''(0))$$
$$- \frac{77}{12} (R'(0) R(0) R(0) + R(0) R(0) R'(0))$$
$$- \frac{35}{6} R(0) R'(0) R(0). \tag{6.129}$$

To give some idea of the complication of the process, here is the result for $n = 8$.

6.8 Curvature conditions for harmonic manifolds

$$C^{(8)}(0) = -\frac{56}{9} R^{(6)}(0) - \frac{80}{9} (R^{(4)}(0)R(0) + R(0)R^{(4)}(0))$$

$$-\frac{260}{9} (R(0)R'(0)R'(0) + R'(0)R'(0)R(0))$$

$$-\frac{280}{9} R'(0)R(0)R'(0) - \frac{128}{9} R(0)R''(0)R(0)$$

$$-\frac{224}{5} R''(0)R''(0) - \frac{280}{9} (R'(0)R'''(0) + R'''(0)R'(0))$$

$$-\frac{256}{15} (R(0)R(0)R''(0) + R''(0)R(0)R(0))$$

$$-\frac{128}{15} R(0)R(0)R(0)R(0). \tag{6.130}$$

I have checked this formula three times at monthly intervals and believe it to be correct. Incidentally there is a disagreement between some of our formulas and those contained in Besse (1978). In particular there is an error of sign $-C$ instead of the correct $+C$ in his equation (6.34), his version of Ledger's formula in his (6.35) is incorrect, and in his (6.41) there is a sign error, namely $+5/3$ instead of the correct $-5/3$. However we agree about the final formulas!

We set

$$C_u^{(n)} = C^n(0), \quad R_u: v \mapsto R(v, u)u, \quad R_u^{(n)} = \underbrace{D_u \cdots D_u}_{n \text{ times}} (R(., u)u), \tag{6.131}$$

where the maps are endomorphisms of M_m associated with a given unit tangent vector u. Using (6.120) we get

$$\text{trace}(C) = s\frac{\theta'}{\theta} + n - 1. \tag{6.132}$$

Hence we get the following lemma.

Lemma (6.8.1) At a given point $m \in M$, if for every interger n the real number $D_u^n(\theta)$ does not depend upon the choice of unit vector u, then for every integer n the number trace $(C_u^{(m)})$ also does not depend upon u.

We now re-obtain some previous results, contained in the following theorem.

Theorem (6.8.2) If a Riemannian manifold (M, g) is harmonic, then there exist constants K, H, L such that for every unit vector u we have

$$\text{trace } (R_u) = K, \qquad \text{trace } (R_u \circ R_u) = H,$$
$$\text{trace } (32R_u \circ R_u \circ R_u - 9R'_u \circ R'_u) = L \qquad (6.133)$$

To obtain these by the present method we note that (6.124) gives trace $(R_u) = K$. Differentiate this along the geodesic determined by u at m to get trace $(R'_u) = 0$, trace $(R''_u) = 0$, trace $(R'''_u) = 0$ and trace $(R_u^{(4)}) = 0$. Then equation (6.126) yields trace $(R_u \circ R_u) = H$. Differentiate this last relation to get trace $(R_u \circ R'_u) = 0$ and trace $(R'_u \circ R'_u) = -$ trace $(R_u \circ R''_u)$. With the aid of (6.128) this completes the proof of the proposition.

By using Ledger's formula we have avoided rather tricky computations involving tensor suffixes. The method also has the advantage that it works in any coordinate system, not necessarily a system of normal coordinates.

To express these results in terms of local coordinates we note that for any tangent vector u, not necessarily unit, we have

$$\text{trace } (R_u) = K\|u\|^2, \qquad \text{trace } (R_u \circ R_u) = H\|u\|^4,$$
$$\text{trace } (32R_u \circ R_u \circ R_u - 9R'_u \circ R'_u) = L\|u\|^6. \qquad (6.134)$$

Then in any system of orthonormal coordinates we get, for every i,j,k,h:

$$\sigma \left(\sum_{a,b} R_{iajb} R_{kahb} \right) = H\sigma \left(\delta_{ij} \delta_{kh} \right), \qquad (6.135)$$

for every i,j,k,h,l,m:

$$\sigma \left(32 \sum_{a,b,c} R_{iajb} R_{kbhc} R_{lcma} - 9 \sum_{a,b} \nabla_i R_{jakb} \nabla_h R_{lamb} \right) = L\sigma \left(\delta_{ij} \delta_{kh} \delta_{lm} \right), \qquad (6.136)$$

where the sign σ means summation over the action of the symmetric group on the four letters i,j,k,h for the first formula and that for the six letters i,j,k,h,l,m for the second formula.

6.9 CONSEQUENCES OF THE CURVATURE CONDITIONS

Proposition (6.9.1). A harmonic space is an Einstein space. In particular the only harmonic spaces of dimension 2 or 3 are spaces of constant sectional curvature.

This follows immediately from the first condition and the fact that Einstein spaces of dimension 2 and 3 are necessarily of constant sectional curvature.

6.9 Consequences of the curvature conditions

We now consider those spaces for which the first two conditions of Theorem (6.8.2) are satisfied. We now introduce a symmetric bilinear differential form ξ defined in a general Riemannian manifold by

$$\xi_{ij} = \sum_{u,v,w} R_{iuvw} R_{juvw} \quad \text{for every } i,j \text{ in orthonormal coordinates.} \tag{6.137}$$

Although this form is a natural Riemannian invariant it does not seem to have received much attention in the literature. We shall prove the following result.

Proposition (6.9.2) *If an Einstein manifold (with $\rho_{ij} = K g_{ij}$) satisfies the second condition, trace $(R_u \circ R_u) = H$, for every unit vector u and some constant H then*

$$\xi = \frac{2}{3}((n+2)H - K^2)g. \tag{6.138}$$

In particular the norm of the curvature tensor is constant, that is,

$$\|R\|^2 = \frac{2n}{3}((n+2)H - K^2). \tag{6.139}$$

We follow the proof given in Besse (1978), p. 164. We work in orthonormal coordinates and put

$$\phi(ijkh) = \sum_{p,q} R_{ipjq} R_{kphq}.$$

The symmetry properties of the curvature tensor yield $\phi(ijkh) = \phi(jihk) = \phi(khij)$. Again using the symmetry of the curvature tensor together with the Einstein condition gives

$$\sum_{i,j} \phi(iijj) = K^2, \quad \sum_j \phi(ijij) = \xi_{ii}, \quad \sum_j \phi(ijji) = \frac{1}{2}\xi_{ii} \tag{6.140}$$

for every i,j. We now apply condition (6.135) with $i = j$ and $k = h$, and sum over k. Using (6.140) we get

$$\xi_{ii} = \frac{2}{3}((n+2)H - K^2).$$

We now use condition (6.135) with $i \neq j$, $k = h$ and sum over k to get $\xi_{ij} = 0$. Thus the first claim is justified. The second follows since trace $\xi = \|R\|^2$.

Exercise (6.9.3)

We leave it to the reader to prove that if, for an Einstein space, $\xi = fg$ for some function f, and if $n > 4$, then f is a constant function. However if $n = 4$, it is always the case that $\xi = fg$ but then f in general is no longer a constant function. This is reminiscent of Schur's theorem which says that if $\rho_{ij} = Kg_{ij}$ and $n > 2$ then K is a constant function. However if $n = 2$ every manifold satisfies this condition but K is no longer a constant in general.

Consequences of condition (6.136)

This is a condition of order 6. It is known from Sakai (1971) that for an Einstein space there are only three distinct orthogonal invariants of order 6, namely, when expressed in terms of orthonormal coordinates,

$$\|\nabla R\|^2 = \sum_{i,j,k,h,l} (\nabla_l R_{ijkh})^2 \qquad (6.141)$$

$$\hat{R} = \sum_{i,j,p,q,r,s} R_{ijpq} R_{pqrs} R_{rsij} \qquad (6.142)$$

$$\overset{\circ}{R} = \sum_{i,j,p,q,r,s} R_{ipjq} R_{prqs} R_{risj}. \qquad (6.143)$$

We shall find that these expressions occur several times in the subsequent analysis of this chapter. We first prove a rather remarkable theorem due to Lichnerowicz and Walker (1945).

Theorem (6.9.4) *Every harmonic pseudo-Riemannian manifold with Lorentzian signature is of constant sectional curvature.*

We remind the reader that a harmonic pseudo-Riemannian manifold is one whose curvature tensor satisfied the infinite set of conditions which we obtained in Section 6.8, except that the components of the metric tensor δ_{ij} at m are replaced by g_{ij}. In fact their proof uses only the three conditions (6.133). Their proof is based on classical tensor calculus and suffix manipulation, and shows that such methods are quite powerful.

Suppose that the manifold is harmonic and write

$$P_{hijl} = R_{hijl} + (n-1)^{-1} k_1 (g_{hj} g_{il} - g_{ij} g_{hl}). \qquad (6.144)$$

where $-\rho_{ij} = k_1 g_{ij}$ since the space is Einstein. We define (P_{ij}) by $P_{ij} = g^{hl} P_{hijl}$ so that

$$P_{ij} = -\rho_{ij} - k_1 g_{ij} = 0. \qquad (6.145)$$

6.9 Consequences of the curvature conditions

The components P_{hijl} have the same symmetry properties as the curvature tensor, and in addition they satisfy

$$P_{hijl} + P_{hjli} + P_{hlij} = 0. \tag{6.146}$$

The idea of the proof is to prove that P_{ihjl} is identically zero and hence that the manifold is of constant curvature $-k_1/(n-1)$.

Let

$$T_{ijhl} := P_{hijl} + P_{hjil}. \tag{6.147}$$

Then from (6.146) it follows that

$$P_{hijl} = \frac{1}{3}(T_{ijhl} - T_{ilhj}). \tag{6.148}$$

Also

$$T_{ijhl} = T_{jihl} = T_{ijlh} = T_{hlij}, \tag{6.149}$$

and

$$T_{ijhl} + T_{ihlj} + T_{iljh} = 0. \tag{6.150}$$

From (6.145) and (6.147) it follows that

$$T_{ijh}{}^h = 0 \tag{6.151}$$

where the indices are raised and lowered by means of the metric tensor. We now use the third condition of (6.133) which may be written in terms of the T's as

$$T_{ij}{}^{pq}T_{hlpq} + T_{ih}{}^{pq}T_{jlpq} + T_{il}{}^{pq}T_{jhpq} = k(g_{ij}g_{hl} + g_{ih}g_{jl} + g_{il}g_{jh}), \tag{6.152}$$

where k is constant. Multiplying by g^{hl} and summing we get, on using (6.149) and (6.151),

$$T_i{}^{lpq}T_{jlpq} = \frac{1}{2}k(n+2)g_{ij}, \tag{6.153}$$

from which

$$T^{jlpq}T_{jlpq} = \frac{1}{2}kn(n+2). \tag{6.154}$$

Now we choose coordinates so that at a given point P_0, we have $g_{11} = g^{11} = -1$, $g_{\alpha\alpha} = g^{\alpha\alpha} = 1$, $g_{\alpha\beta} = 0 = g^{\alpha\beta}$ ($\alpha \neq \beta$), $\alpha, \beta = 2, \ldots, n$. Then from (6.152) with the indices i, j, h, l put equal to unity we get

238   6 Special Riemannian manifolds

$$\sum_{p,q} e_p e_q (T_{11pq})^2 = k \qquad (6.155)$$

where $e_1 = -1$, $e_\alpha = 1$, $\alpha = 2, \ldots, n$.

Since T_{ijhl} is zero if any three suffixes are equal, in particular if they are equal to unity, we have

$$\sum_{\alpha,\beta=2}^{n} (T_{11\alpha\beta})^2 = k \qquad (6.156)$$

so that k must be non-negative. We now take i,j,h,l in (6.153) to be $1,\alpha,\beta,1$ and using (6.149) we get

$$T_{1\alpha\beta 1} = -\frac{1}{2} T_{11\alpha\beta}. \qquad (6.157)$$

If the i and j in (6.153) are both put equal to unity we get, on using (6.149), (6.156), and (6.157),

$$\sum (T_{1\alpha\beta\gamma})^2 = -\frac{1}{2} k(n-1), \qquad (6.158)$$

so that k is non-positive. Hence $k = 0$. Thus $T_{11\alpha\beta} = 0$ from (6.156), $T_{1\alpha\beta\gamma} = 0$ from (6.158), and $T_{1\alpha\beta 1} = 0$ from (6.157). Hence every component of T_{ijhl} with one or more subscripts equal to unity is zero. Thus from (6.154) with $k = 0$ we get $\Sigma(T_{\alpha\beta\gamma\delta})^2 = 0$ and hence $T_{\alpha\beta\gamma\delta} = 0$. It follows that $P_{hijl} = 0$ everywhere and the theorem is thus proved. This example shows the importance of signature when dealing with properties of harmonic spaces.

A Formula of Lichnerowicz

In this section we derive an important formula due to Lichnerowicz. We closely follow the treatment given by him in Lichnerowicz (1958) pp. 8–10.

Apply the Ricci identity to the curvature tensor of a Riemannnian manifold to get

$$\nabla_i \nabla_j R_{abcd} - \nabla_j \nabla_i R_{abcd} = H_{abcd,\,ij} \qquad (6.159)$$

where

$$-H_{abcd,\,ij} = R^r_{aij} R_{rbcd} + R^r_{bij} R_{arcd} + R^r_{cij} R_{abrd} + R^r_{dij} R_{abcr} \qquad (6.160)$$

We multiply by g^{ai} to get

$$\nabla_r \nabla_j R^r_{bcd} - \nabla_j \nabla_r R^r_{bcd} = H^r_{bcd,\,rj}.$$

From the Bianchi identity it follows that

6.9 Consequences of the curvature conditions

$$\nabla_r R^r_{bcd} = \nabla_c \rho_{bd} - \nabla_d \rho_{bc}.$$

Hence we get

$$\nabla_r \nabla_a R^r_{bcd} = H^r_{bcd, ra} + \nabla_a (\nabla_c \rho_{bd} - \nabla_d \rho_{bc}).$$

We take the product with R^{abcd} to get

$$R^{abcd} \nabla_r \nabla_a R^r_{bcd} = R^{abcd} \nabla_a (\nabla_c \rho_{bd} - \nabla_d \rho_{bc}) + K^* \quad (6.161)$$

where we have written

$$K^* = R^{abcd} H^r_{bcd, ra}.$$

Now the left-hand side of (6.161) can be written, using the second Bianchi identity, as

$$R^{abcd} \nabla_r \nabla_a R^r_{bcd} = \frac{1}{2} R^{abcd} \nabla_r (\nabla_a R^r_{bcd} - \nabla_b R^r_{acd}) = \frac{1}{2} R^{abcd} \nabla_r \nabla^r R_{abcd}.$$

From this we deduce

$$R^{abcd} \nabla_r \nabla_a R^r_{bcd} = \frac{1}{2} \nabla^r (R^{abcd} \nabla_r R_{abcd}) - \frac{1}{2} \nabla^r R^{abcd} \nabla_r R_{abcd}$$

$$= \frac{1}{4} \Delta \|R\|^2 - \frac{1}{2} \nabla^r R^{abcd} \nabla_r R_{abcd}$$

$$= \frac{1}{4} \Delta \|R\|^2 - \frac{1}{2} \|\nabla R\|^2.$$

We thus get

$$-\frac{1}{4} \Delta \|R\|^2 = -\frac{1}{2} \|\nabla R\|^2 - R^{abcd} \nabla_a (\nabla_c \rho_{bd} - \nabla_d \rho_{bc}) - K^*.$$

The second term of the second member can be written

$$R^{abcd} \nabla_a (\nabla_c \rho_{bd} - \nabla_d \rho_{bc}) = \nabla_a [R^{abcd} (\nabla_c \rho_{bd} - \nabla_d \rho_{bc})]$$
$$- \nabla_a R^{abcd} (\nabla_c \rho_{bd} - \nabla_d \rho_{bc}).$$

so the previous equation may be put in the form

$$-\frac{1}{4} \Delta \|R\|^2 + \nabla_a [R^{abcd} (\nabla_c \rho_{bd} - \nabla_d \rho_{bc})]$$

$$= -\frac{1}{2} \|\nabla R\|^2 + (\nabla^c \rho^{bd} - \nabla^d \rho^{bc})(\nabla_c \rho_{bd} - \nabla_d \rho_{bc}) - K^*.$$

We now apply the above formula to an Einstein space to get

$$\frac{1}{4}\Delta\|R\|^2 = \frac{1}{2}\|\nabla R\|^2 + K^*.$$

This formula holds for any Einstein space. We note that there is a difference in sign between this formula and the one in the book by Lichnerowicz but this is because he uses a different sign convention for the Laplacian. We can express this formula in terms of the invariants given in (6.141), (6.142), and (6.143). We find

$$\frac{1}{2}\Delta(\|R\|^2) = 2K\|R\|^2 - \hat{R} - 4\mathring{R} + \|\nabla R\|^2.$$

where K is the constant in (6.133).

Now the third condition (6.136) may be written, after some calculation, in the form

$$32\left(nK^3 + \frac{9}{2}K\|R\|^2 + \frac{7}{2}\hat{R} - \mathring{R}\right) - 27\|\nabla R\|^2 = n(n^2 + 6n + 8)L.$$

We thus arrive at the next proposition.

Proposition (6.9.5) *On a harmonic manifold the following three functions are constants:*

$$\hat{R} + 4\mathring{R} - \|\nabla R\|^2, \qquad 112\hat{R} - 32\mathring{R} - 27\|\nabla R\|^2, \qquad 17\hat{R} - 28\mathring{R}.$$
(6.162)

These functions are evidently constants when the manifold is an irreducible symmetric space since the covariant derivative of the curvature tensor is then identically zero.

6.10 HARMONIC MANIFOLDS OF DIMENSION 4

We have seen that harmonic manifolds of dimension 2 or 3 are necessarily spaces of constant sectional curvature, so the next case to receive detailed attention was that of dimension 4. The first result was due to A. Lichnerowicz and A. G. Walker who proved the following theorem.

Theorem (6.10.1) *A harmonic Riemannian manifold of dimension 4 is locally symmetric, that is, its curvature tensor has zero covariant derivative.*

A complete but rather complicated proof of this theorem is given in Ruse *et al.* (1961). However we give here a simpler proof which depends upon the choice of a Singer–Thorpe basis which was unknown when that book was written. A Singer–Thorpe basis has proved to be of great use in investigating properties of four-dimensional Einstein spaces so

6.10 Harmonic manifolds of dimension 4

we summarize the content of the paper Singer and Thorpe (1969) where this concept was introduced.

We summarize our description of a Singer-Thorpe basis of four-dimensional Riemannian manifolds. We fix $m \in M$ and work with $V = M_m$ with a fixed orientation. If $\beta = (e_1, e_2, e_3, e_4)$ is an orthonormal basis of V, then $\Lambda^2 V$ is endowed with a basis $\Lambda^2 \beta$ where

$$\Lambda^2 \beta = \{e_1 \wedge e_2, e_1 \wedge e_3, e_1 \wedge e_4, e_3 \wedge e_4, e_4 \wedge e_2, e_2 \wedge e_3\}.$$

The matrix of the $*$ operator with respect to this basis is

$$\begin{pmatrix} 0 & I \\ I & 0 \end{pmatrix}$$

where I is the 3×3 identity matrix. The curvature operator $\mathcal{R} : \Lambda^2 V \to \Lambda^2 V$ in the case of an Einstein space satisfies $*\mathcal{R} = \mathcal{R}*$. From this we deduce that the matrix of $\mathcal{R}$ looks like

$$\begin{pmatrix} a & 0 & 0 & \alpha & 0 & 0 \\ 0 & b & 0 & 0 & \beta & 0 \\ 0 & 0 & c & 0 & 0 & \gamma \\ \alpha & 0 & 0 & a & 0 & 0 \\ 0 & \beta & 0 & 0 & b & 0 \\ 0 & 0 & \gamma & 0 & 0 & c \end{pmatrix}, \qquad (6.163)$$

where the Bianchi identity

$$R_{1234} + R_{1342} + R_{1423} = 0$$

forces the relation $\alpha + \beta + \gamma = 0$.

It is shown further in the paper of Singer and Thorpe (although we shall not prove this) that β can be so chosen that if σ denotes the sectional curvature at m, then as a function of the Grassmannian of 2-planes we have

$$a = \sup \sigma \qquad \text{and} \qquad c = \inf \sigma.$$

We shall assume that β has been chosen so that these relations hold. We now have the following lemma.

Lemma (6.10.2) Let (M, g) be a harmonic Riemannian manifold of dimension 4. Then we may choose an orientation of the tangent space $V = M_m$ such that the curvature operator satisfies the relation

$$\left(\mathcal{R} - \frac{K}{3}I\right) \circ (I - *) = 0. \qquad (6.164)$$

Here Ricci = Kg and I is the identity on $\Lambda^2 V$. Moreover there exists a real number L'_m such that

$$\text{trace}\,(R_u \circ R_u \circ R_u) = L'_m \tag{6.165}$$

for every unit vector u in M_m, where R_u is the mapping $v \to R(u,v)u$. To prove the lemma we take a Singer–Thorpe basis. Then

$$a + b + c = K. \tag{6.166}$$

In the condition for harmonicity

$$\sigma \sum_{a,b} R_{iajb} R_{kahb} = H\sigma \delta_{ij}\delta_{kh}$$

we take $i = j = h = k = 1$ to get

$$a^2 + b^2 + c^2 = H. \tag{6.167}$$

We now take $i = j = 1$, and $k = h = 2, 3$, and 4 to get

$$a^2 + 2bc + (\beta - \gamma)^2 = b^2 + 2ca + (\gamma - \alpha)^2 = c^2 + 2ab + (\alpha - \beta)^2 = H. \tag{6.168}$$

From the same condition with $i = 1$, $j = 2$, $k = 3$, $h = 4$, we get

$$a(\beta - \gamma) + b(\gamma - \alpha) + c(\alpha - \beta) = 0. \tag{6.169}$$

This coupled with $\alpha + \beta + \gamma = 0$ gives

$$\frac{\alpha}{a - K/3} = \frac{\beta}{b - K/3} = \frac{\gamma}{c - K/3} = \lambda, \quad \text{say.}$$

Substitute in (6.168) to obtain

$$(\lambda^2 - 1)(a - b)(c - K/3) = 0. \tag{6.170}$$

The first case gives $a = b = c = K/3$ so $\alpha = \beta = \gamma = 0$. This corresponds to the case of M having constant sectional curvature. The remaining case gives $\lambda^2 = 1$. We can adjust the orientation of V to have $\lambda = 1$. Hence

$$\alpha = a - K/3, \quad \beta = b - K/3, \quad \gamma = c - K/3 \tag{6.171}$$

which is another way of writing the relation satisfied by $\mathcal{R}$. To prove the last part of the lemma we choose a unit vector $u \in M_m$ and complete it into an orthonormal basis $\beta = \{e_1 = u, e_2, e_3, e_4\}$. We write the matrix of $\mathcal{R}$ with respect to $\beta(\Lambda^2)$ as $\begin{pmatrix} A & B \\ C & D \end{pmatrix}$ where A, B, C, D are 3×3 matrices. Then using the condition that the space is Einstein together with the relation $(\mathcal{R} - K/3)(I - *) = 0$ gives

6.10 Harmonic manifolds of dimension 4

$$\begin{pmatrix} A & B \\ C & D \end{pmatrix} = \begin{pmatrix} A & A - K/3 \\ A - K/3 & A \end{pmatrix}.$$

The matrix A is none other than the matrix of R_u with respect to the basis β. Hence

$$\text{trace } (R_u \circ R_u \circ R_u) = \text{trace } (A \circ A \circ A) = \frac{1}{8} \text{trace } (\mathfrak{R} \circ \mathfrak{R} \circ \mathfrak{R}).$$

This completes the proof of the lemma.

Lemma (6.10.3) *In a harmonic manifold the numbers a, b, c in a Singer–Thorpe basis are constant on M. Moreover the number L'_m in lemma (6.10.2) is constant.*

To prove this, a direct computation from (6.163) gives

$$\hat{R} = 16(a^3 + b^3 + c^3 + 3a\alpha^2 + 3b\beta^2 + 3c\gamma^2), \qquad (6.172)$$

$$\overset{\circ}{R} = 24(abc + a\beta\gamma + b\gamma\alpha + c\alpha\beta). \qquad (6.173)$$

Substituting $\alpha = a - K/3$, and so on, we find that

$$\hat{R} = 192abc + \text{constant}$$

$$\overset{\circ}{R} = 96abc + \text{constant}.$$

Then relation (6.162) implies that abc is constant. Using (6.166) and (6.167) it follows that a, b, c are constant. Moreover $L'_m = 2(a^3 + b^3 + c^3) = $ constant, and the lemma is proved.

We are now finally in a position to prove our main result.

Theorem (6.10.4) *A harmonic four-dimensional Riemannian manifold is locally symmetric.*

In doing this we follow closely the argument in Besse (1978) pages 170–171. We fix a point $m \in M$ and a Singer–Thorpe basis such that $a = \sup \sigma$, $c = \inf \sigma$. We extend the basis at m into a family of orthonormal frames in some neighbourhood V of m. Since σ is continuous it follows that in some neighbourhood V' of m we have $R_{1212}(n) \leq R_{1212}(m)$ for every n in V'.

Hence we have

$$\nabla_i R_{1212}(m) = 0 \qquad \text{for any } i.$$

Similarly since $R_{1414}(m) = c$ we have

$$\nabla_i R_{1414}(m) = 0 \qquad \text{for every } i.$$

We now prove that

$$\nabla_i R_{1313}(m) = \nabla_i R_{2323}(m) = \nabla_i R_{2424}(m) = \nabla_i R_{3434}(m) = 0.$$

6 Special Riemannian manifolds

We note, using the Einstein condition, that

$$\nabla_i(R_{1313} + R_{1212} + R_{1414}) = \nabla_i(Kg_{11}) = 0.$$

Since we have already proved $\nabla_i R_{1212}(m) = 0$ and $\nabla_i R_{1414}(m) = 0$ we have $\nabla_i R_{1313} = 0$. Similarly we have

$$\nabla_i(R_{2121} + R_{2323} + R_{2424}) = 0,$$

giving $\nabla_i R_{2323} + \nabla_i R_{2424} = 0$.

Similarly we have $\nabla_i R_{2424} + \nabla_i R_{3434} = 0$ and $\nabla_i R_{3434} + \nabla_i R_{3232} = 0$. These are three linear equations iin three unknowns with a unique solution (all zero). Thus we have proved $\nabla R_{jkjk} = 0$, for every i,j,k.

Since $\alpha = a - K/3$ we have $R_{1212} - R_{1234} = K/3$ and hence $\nabla_i R_{jkhl} = 0$ for every i with every j,k,h,l all distinct. So now we just have to deal with components of types $\nabla_i R_{ijik}(m)$ and $\nabla_i R_{ijkj}(m)$ with i,j,k all distinct. To do this we set $(ijk) = \nabla_i R_{ijik}$ not summed for i. Then an argument similar to that above, this time using the Bianchi identity and the Einstein condition, shows that

$$(ijk) = (jki) = (kij), \quad i,j,k \text{ all distinct}.$$

Following the notation of Besse we denote by ⓗ their common value, where h is the index such that i,j,k,h are all distinct. A similar argument proves that

$$\nabla_i R_{ijkj}(m) = \pm ⓗ.$$

Now, by (6.133) and (6.164) we know that there exists a constant L''_m such that trace$(\nabla_u R_u \circ \nabla_u R_u) = L''_m$ for every unit vector u of M_m. Taking $u = e_1$ we get

$$L''_m = \sum_{a,b} (\nabla_1 R_{1a1b(m)})^2 = ②^2 + ③^2 + ④^4.$$

In the same manner with $u = e_2, e_3, e_4$ we get

$$②^2 + ③^2 + ④^2 = ③^2 + ④^2 + ①^2 = ④^2 + ①^2 + ②^2 = ①^2 + ②^2 + ③^2.$$

Hence $① = \pm ② = \pm ③ = \pm ④$.

Now use (6.136) with $i = j = k = l = 1$ and $m = 2$. We find, on using previously obtained equalities relating $\nabla_i R_{ijik}(m)$ and $\nabla_i R_{ijkj}(m)$, we get $①② = 0$. Therefore $① = ② = ③ = ④ = 0$ and so $\nabla R = 0$. The metric is therefore locally symmetric and this completes the proof.

It could be argued that our proof is much longer than the exceedingly clever but *ad hoc* proof of Lichnerowicz and Walker. However we have introduced tools which have been used extensively by other mathematicians in proving results about more general spaces than harmonic

6.10 Harmonic manifolds of dimension 4

spaces. Whether the theorem is true for harmonic spaces of dimension 5 or 6 remains an open question. But as we have remarked earlier there exists a counterexample for dimension 7. It should also be remarked that both the original proof of the theorem of Lichnerowicz and Walker and also in the Besse proof given above, all the three conditions of (6.133) have been used, and the proofs rely heavily upon the constancy of K, H, and L. However, in a remarkable paper by Saloman (1984), using extensively the theory of group representations, an alternative proof is obtained making use of just the first and third condition of (6.133) and moreover without assuming that H, K, and L are constants.

A simplification of the condition $\nabla R = 0$

We now give a result, due to A. Gray, valid in a pseudo-Riemannian manifold of arbitrary signature, which gives a sufficient condition for the conclusion $\nabla R = 0$.

Lemma (6.10.5)

In a pseudo-Riemannian manifold the following conditions are equivalent:

(i) $\nabla R = 0$;
(ii) $\nabla_X R_{XYXY} = 0$;
(iii) $\nabla_V R_{WXYZ} + \nabla_X R_{WZYV} + \nabla_Z R_{WVYX} = 0$.

Clearly (i) $\Rightarrow$ (ii). To prove that (ii) $\Rightarrow$ (iii), replace X in (ii) by $aU + bV + cW$ and expand. This gives a polynomial in a, b, c, which vanishes identically. In particular the coefficient of abc must vanish giving

$$\underset{U,V,W}{\mathscr{O}} \nabla_U R_{VYWY} = 0$$

where $\mathscr{O}$ denotes the symmetric sum. We polarize this equation with respect to Y to get

$$0 = \underset{U,V,W}{\mathscr{O}} \{\nabla_U R_{VYWZ} + \nabla_U R_{VZWY}\}$$

$$= \underset{UVW}{\mathscr{O}} \{\nabla_U R_{VYWZ} - \nabla_U R_{VZYW}\}$$

$$= \underset{UVW}{\mathscr{O}} \{2\nabla_U R_{VYWZ} + \nabla_U R_{VWZY}\} \quad \text{(by the first Bianchi identity)}$$

$$= 2 \underset{UVW}{\mathscr{O}} \nabla_U R_{VYWZ} \quad \text{(by the second Bianchi identity)}.$$

Thus (ii) $\Rightarrow$ (iii).

We write out (iii) in the form
$$\nabla_V R_{WXYZ} + \nabla_Z R_{WVYX} + \nabla_X R_{WZYV} = 0.$$
Interchange X and Y and subtract the resulting equation to get
$$0 = \nabla_V(R_{WXYZ} - R_{WYXZ}) + (\nabla_Z R_{WVYX} - \nabla_Z R_{WVXY}) + (\nabla_X R_{WZYV} - \nabla_Y R_{WZXY}).$$

We use the first Bianchi identity on the first term and the second Bianchi identity on the third term to get
$$0 = -\nabla_V R_{WZXY} - 2\nabla_Z R_{WVXY} - \nabla_V R_{WZXY},$$
that is,
$$\nabla_V R_{WZXY} + \nabla_Z R_{WVXY} = 0.$$

Now interchange V and W in the above equation and subtract. We get
$$\nabla_V R_{WZXY} - \nabla_W R_{VZXY} + 2\nabla_Z R_{WVXY} = 0,$$
which we may rewrite as
$$3\nabla_Z R_{WVXY} = \nabla_Z R_{WVXY} + \nabla_W R_{VZXY} + \nabla_V R_{ZWXY} = 0$$
by the second Bianchi identity and the lemma is proved.

It seems possible that our proof of the previous theorem may be simplifed by using this lemma but attempts so far have been unsuccessful.

6.11 MEAN-VALUE PROPERTIES

We have seen that a harmonic Riemannian metric is completely characterized by the property
$$M_m(r, f) = f(m) \tag{6.174}$$
which must hold for harmonic functions f. Here the left-hand side is defined by
$$M_m(r, f) = \frac{\int_{\exp_m(S^{n-1}(r))} f * dr}{V(\exp_m(S^{n-1}(r)))} \tag{6.175}$$
where $\exp_m$ is the exponential map which maps a small neighbourhood of the origin in M_m onto a neighbourhood of $m \in M$, and $*dr$ is the volume element of $\exp_m(S^{n-1}(r))$. Here $\exp_m(S^{n-1}(r))$ is the set of points distant r from m, and $V(\exp_m(S^{n-1}(r)))$ is its volume.

6.11 Mean-value properties

It was Alfred Gray who suggested to me that we should obtain a formula analogous to (6.174) for a smooth function f in an arbitrary Riemannian manifold. This resulted in the paper Gray and Willmore (1982). Such a generalization for an analytic function in Euclidean space $\mathbb{R}^n$ had been found by Pizetti (see, for example, Courant and Hilbert (1962), page 287). Pizetti's formula is

$$M_m(r, f) = \Gamma\left(\frac{1}{2}n\right) \sum_0^\infty \left(\frac{r}{2}\right)^{2k} \frac{1}{k!\,\Gamma(\frac{1}{2}n + k)} (\Delta^k f)_m. \qquad (6.176)$$

A slick demonstration of this formula is given by Zalcman (1973). It is interesting to note that the formula can be more compactly written as a Bessel function in $(\sqrt{(-\Delta)})$, namely

$$M_m(r, f) = [j_{(n/2)-1}(r\sqrt{(-\Delta)})f](m) \qquad (6.177)$$

where

$$j_l\{z\} = 2^l \Gamma \frac{(l+1)J_l(z)}{z^l} \qquad (6.178)$$

and J_l is a Bessel function of the first kind.

We shall show that for a given Riemannnian manifold there is an analogous function defined at each point. The power series can be regarded as a generalization of the Bessel function. The underlying idea in obtaining the generalized formula is to use the exponential map $\exp_m$ to transfer the formula between M and the Euclidean space M_m. In this way we obtain a formula similar to Pizetti's but unfortunately the right-hand side is not expressed in terms of the Riemannian Laplacian Δ but in terms of another operator $\tilde{\Delta}_m$ which we call the *Euclidean Laplacian*.

Let $(x_1, \ldots, x_n)$ be any system of orthonormal coordinates at m. Then $\tilde{\Delta}_m$ is given by

$$\tilde{\Delta}_m = \sum_{i=1}^n \frac{\partial^2}{\partial x_i^2}. \qquad (6.179)$$

This is to be compared with the ordinary Laplacian Δ of the Riemannian manifold M given by

$$\Delta = \frac{1}{\sqrt{g}} \sum_{i,j=1}^n \frac{\partial}{\partial y_i}\left(g^{ij}\sqrt{g}\,\frac{\partial}{\partial y^j}\right) \qquad (6.180)$$

where $y_1, \ldots, y_n$ is any coordinate system. Although $(\Delta f)_m = (\tilde{\Delta}_m f)_m$, it is in general false that $(\Delta^k f)_m = (\tilde{\Delta}_m^k f)_m$ for $k > 1$.

What is important about $\tilde{\Delta}_m^k$ is that, for each k, there is a globally defined differential operator L^k of degree $2k$ which coincides with $\tilde{\Delta}_m^k$ at m, where

6 Special Riemannian manifolds

$$L^k f = \frac{1}{1.3 \cdots (2k-1)} \sum_{i_1 \cdots i_k = 1}^{n} \{ \nabla^{2k}_{i_1 i_1 \cdots i_k i_k} + \cdots + \nabla^{2k}_{i_1 \cdots i_k i_1 \cdots i_k} \} f. \quad (6.181)$$

It is not too difficult to express $\tilde{\Delta}^k_m$ in terms of Δ^k and other terms involving the curvature tensor. In principle this could be done for arbitrary k, but the computations become very complicated quickly and only the terms corresponding to $k = 1, 2, 3$ have been computed. We define the following fourteen functions:

$$\langle \nabla^2(\Delta f), \rho \rangle = \Sigma \rho_{ij} \nabla^4_{ijkk} f,$$

$$\langle \nabla(\Delta f), \nabla \tau \rangle = \Sigma (\nabla_i \tau)(\nabla^3_{ijj} f),$$

$$\langle \nabla^3 f, \nabla \rho \rangle = \Sigma (\nabla_i \rho_{jk})(\nabla^3_{ijk} f),$$

$$\langle \nabla^2 f, \nabla^2 \tau \rangle = \Sigma (\nabla^2_{ij} \tau)(\nabla^2_{ij} f),$$

$$\langle \nabla^2 f, \Delta \rho \rangle = \Sigma (\nabla^2_{ii} \rho_{jk})(\nabla^2_{jk} f),$$

$$\langle \nabla^2 f \otimes \rho, \bar{R} \rangle = \Sigma R_{ikjl} \rho_{ij} \nabla^2_{kl} f,$$

$$\sigma(\nabla^2 f \otimes \rho \otimes \rho) = \Sigma \rho_{ij} \rho_{jk} \nabla^2_{ki} f,$$

$$\langle \nabla^2 f, \dot{R} \rangle = \Sigma R_{ijkl} R_{ijkh} \nabla^2_{lh} f,$$

$$\langle \nabla f, \nabla(\Delta \tau) \rangle = \Sigma (\nabla^3_{ijj} \tau)(\nabla_j f),$$

$$\rho(\nabla f, \nabla \tau) = \Sigma \rho_{ij} (\nabla_i \tau)(\nabla_i f),$$

$$\langle \rho \otimes \nabla f, \nabla \rho \rangle = \Sigma (\nabla_i \rho_{jk}) \rho_{ij} (\nabla_k f),$$

$$\langle \nabla f \otimes \rho, \nabla \rho \rangle = \Sigma (\nabla_i \rho_{jk})(\nabla_i f) \rho_{jk},$$

$$\langle \nabla f \otimes \nabla \rho, \bar{R} \rangle = \Sigma (\nabla_i f)(\nabla_j \rho_{kl}) R_{ikjl},$$

$$\langle \nabla f \otimes R, \nabla R \rangle = \Sigma (\nabla_i f) R_{jklh} (\nabla_i R_{jklh}).$$

We find that

$$(\tilde{\Delta}^2_m f)_m = \left[\Delta^2 f + \frac{1}{3} \langle \nabla f, \nabla \tau \rangle + \frac{2}{3} \langle \nabla^2 f, \rho \rangle \right]_m. \quad (6.182)$$

Also we find that

$$(\tilde{\Delta}^3_m f)_m = \Big\{ \Delta^3 f + 2 \langle \nabla^2(\Delta f), \rho \rangle + \langle \nabla(\Delta f), \nabla \tau \rangle + 2 \langle \nabla^3 f, \nabla \rho \rangle$$

$$+ \frac{6}{5} \langle \nabla^2 f, \nabla^2 \tau \rangle + \frac{2}{5} \langle \nabla^2 f, \Delta \rho \rangle - \frac{12}{5} \langle \nabla^2 f \otimes \rho, \bar{R} \rangle$$

6.11 Mean-value properties

$$+ \frac{52}{15}\sigma(\nabla^2 f \otimes \rho \otimes \rho) + \frac{4}{15}\langle \nabla^2 f, \dot{R}\rangle + \frac{2}{5}\langle \nabla f, \nabla(\Delta \tau)\rangle$$

$$+ \frac{4}{3}\rho(\nabla f, \nabla \tau) + \frac{8}{3}\langle \rho \otimes \nabla f, \nabla \rho\rangle - \frac{4}{5}\langle \nabla f \otimes \rho, \nabla \rho\rangle$$

$$+ \frac{4}{3}\langle \nabla f \otimes \nabla \rho, \bar{R}\rangle + \frac{2}{15}\langle \nabla f \otimes R, \nabla R\rangle \bigg\}_m. \quad (6.183)$$

The mean value for a Riemannian space is given by the analogue of Pizetti's formula

$$M_m(r, f) = \frac{j_{(n/2)-1}(r\sqrt{(-\tilde{\Delta}_m)})[f\theta](m)}{j_{(n/2)-1}(r\sqrt{(-\tilde{\Delta}_m)})[\theta](m)}$$

$$= \frac{\sum_{k=0}^{\infty}\left(\frac{r}{2}\right)^{2k}\frac{1}{k!\Gamma(\frac{1}{2}n+k)}\tilde{\Delta}_m^k[f\theta](m)}{\sum_{k=0}^{\infty}\left(\frac{r}{2}\right)^{2k}\frac{1}{k!\Gamma(\frac{1}{2}n+k)}\tilde{\Delta}_m^k[\theta](m)}. \quad (6.184)$$

On substituting for $\tilde{\Delta}_m^k$ in terms of Δ^k we find

$$M_m(r, f) = f(m) + A(n)_m r^2 + B(n)_m r^4 + C(n)_m r^6 + O(r^8) \quad (6.185)$$

where

$$\begin{cases} A(n) = \frac{1}{2n}\Delta f, \\ B(n) = \frac{1}{24n(n+2)}\left[3\Delta^2 f - 2\langle \nabla^2 f, \rho\rangle - 3\langle \nabla f, \nabla \tau\rangle + \frac{4}{n}\tau \Delta f\right], \\ C(n) = \frac{1}{720n(n+2)(n+4)}\bigg[15\Delta^3 f - 30\langle \nabla^2(\Delta f), \rho\rangle \\ \quad - 45\langle \nabla(\Delta f), \nabla \tau\rangle - 30\langle \nabla^3 f, \nabla \rho\rangle - 36\langle \nabla^2 f, \nabla^2 \tau\rangle \\ \quad - 12\langle \nabla^2 f, \Delta \rho\rangle + 32\langle \nabla^2 f \otimes \rho, \bar{R}\rangle - 24\sigma(\nabla^2 f \otimes \rho \otimes \rho) \\ \quad - 8\langle \nabla^2 f, \dot{R}\rangle - 30\langle \nabla f, \nabla(\Delta \tau)\rangle - 20\rho(\nabla f, \nabla \tau) \\ \quad - 20\langle \rho \otimes \nabla f, \nabla \rho\rangle + 30\langle \nabla f \otimes \rho, \nabla \rho\rangle - 20\langle \nabla f \otimes \nabla \rho, \bar{R}\rangle \\ \quad - 10\langle \nabla f \otimes R, \nabla R\rangle + \frac{20}{n}(3\Delta^2 f - 2\langle \nabla^2 f, \rho\rangle - 3\langle \nabla f, \nabla \tau\rangle)\tau \\ \quad + \frac{4}{n}(18\Delta \tau - 8\|\rho\|^2 + 3\|R\|^2)\Delta f + \frac{80}{n^2}\tau^2 \Delta f\bigg]. \quad (6.186) \end{cases}$$

Applications

Theorem (6.11.1) *Let M be any Riemannian manifold and let $m \in M$. Then:*

(i) *any function on M harmonic near m satisfied $M_m(r, f) = f(m) + O(r^4)$ as $r \to \infty$;*

(ii) *if M is an Einstein manifold and f is harmonic near m, then*

$$M_m(r, f) = f(m) + O(r^6) \quad \text{as } r \to \infty; \quad (6.187)$$

(iii) *if M is an irreducible symmetric space and f is harmonic near m then*

$$M_m(r, f) = f(m) + O(r^8) \quad \text{as } r \to \infty. \quad (6.188)$$

In a similar way there are mean-value properties for eigenfunctions of the Laplacian Δ.

Theorem (6.11.2) *Let M be any Riemannian manifold and let $m \in M$, and let f be an eigenfunction of Δ defined near m corresponding to the eigenvalue λ.*

(i) *Then as $r \to 0$*

$$M_m(r, f) = f(m)\left\{1 + \frac{\lambda r^2}{2n}\right\} + O(r^4). \quad (6.189)$$

(ii) *If M is an Einstein space then as $r \to 0$*

$$M_m(r, f) = f(m)\left\{1 + \frac{\lambda r^2}{2n} + \frac{(3n\lambda^2 + 2\lambda\tau)}{24n^2(n+2)}r^4\right\}$$
$$+ \frac{1}{720n(n+2)(n+4)}\left[\left(15\lambda^3 + \frac{30\tau\lambda^2}{n}\right.\right.$$
$$+ \frac{1}{n^2}(16\tau^2 + 12n\|R\|^2)\lambda\right)f - 8\langle\nabla^2 f, \dot{R}\rangle$$
$$\left.\left. - 10\langle\nabla f \otimes R, \nabla R\rangle\right]_m r^6 + O(r^8). \quad (6.190)$$

(iii) *If M is an irreducible symmetric space then as $r \to 0$*

$$M_m(r, f) = f(m)\left\{1 + \frac{\lambda r^2}{2n} + \frac{(3n\lambda^2 + 2\lambda\tau_m)r^4}{24n^2(n+2)}\right.$$
$$+ \frac{1}{720n(n+2)(n+4)}\left[15\lambda^3 + \frac{30\tau\lambda^2}{n}\right.$$

6.11 Mean-value properties

$$+ \frac{1}{n^2}(16\tau^2 + 4n\|R\|^2)\lambda\Big]_m r^6 \Big\} + O(r^8)$$
(6.191)

To prove these results we note that the first two follow immediately from (6.185). To prove the third result we note that irreducible symmetric spaces are Einstein. Moreover, consider the symmetric tensor field $\dot{R}$ given by

$$\dot{R}(x, y) = \sum_{i,j,k=1}^{n} R(e_i, e_j, e_k, x) R(e_i, e_j, e_k, y) \quad (6.192)$$

for $x, y \in M_m$, where $\{e_1, \ldots, e_n\}$ is an orthonormal basis of M_m. It follows that $\dot{R}$ must be a scalar multiple of the metric tensor, otherwise M would be reducible. In fact we have

$$\dot{R}(x, y) = \frac{1}{n}\|R\|^2 \langle x, y \rangle \quad (6.193)$$

for $x, y \in M_m$. Then (6.193) implies that

$$\langle \nabla^2 f, \dot{R} \rangle = \frac{1}{n}\|R\|^2 \Delta f \quad (6.194)$$

and (iii) follows from the mean-value expansion and from equations (6.190) and (6.194).

It has been suggested in the book by Besse (1978) that an interesting class of manifolds consists of those Einstein manifolds which satisfy the additional condition (6.193). We call such manifolds *super-Einstein* when $n > 4$. Of course, every irreducible symmetric space is super-Einstein.

In fact, an inspection of the proof shows that in order that (6.188) or (6.191) hold it is only necessary to assume that M is super-Einstein if $n > 4$.

It follows from the Bianchi identity that when $n > 4$ a super-Einstein manifold has the property that $\|R\|^2$ is constant. However condition (6.193) is satisfied identically in an Einstein manifold when $n = 4$. For this reason a four-dimensional Einstein manifold will be called super-Einstein when $\|R\|^2$ is constant.

It was shown by Gray and Vanhecke that there exist metrics on spheres of dimension $4n+3$ which are Einstein but not super-Einstein. These spheres occur as geodesic spheres in quaternionic projective space of dimension $4n+4$.

We can use our mean-value expansion formula to give an immediate proof of the following theorem.

Theorem (6.11.3) Let $f: M \to R$ be an analytic function, where M is an analytic Riemannian manifold, and suppose for each $m \in M$ we have

$$M_m(r, f) = f(m)\left\{1 + \frac{\lambda r^2}{2n}\right\} + O(r^4) \qquad \text{as } r \to 0 \qquad (6.195)$$

Then $\Delta f = \lambda f$.

This theorem generalizes a result due to Walker (1947) who used a different method to obtain the same result in the particular case when M is a three-dimensional manifold of constant curvature.

We remind the reader that a totally geodesic submanifold is characterized by the condition that a geodesic tangent to the submanifold at a point lies entirely in the submanifold. Let M be a globally symmetric space. Then the *rank* of M is the maximal dimension of a flat totally geodesic submanifold of M. It is known that symmetric spaces of rank 1 are precisely the two-point homogeneous spaces considered by Wang (1952). These have the property that for any two pairs of equidistant points (A, B), (C, D), there is an isometry of the space which sends A to C and B to D. If we apply this to a geodesic sphere of radius r centred at m it follows that the volume element is constant at all points on the geodesic sphere, and hence the space is harmonic. Moreover, it follows from Carpenter *et al.* (1982), denoted by [CGW], that if a symmetric space is harmonic, its rank is necessarily 1. However for irreducible symmetric spaces of rank greater than 1, all we know is that the weaker mean-value property (6.188) holds for harmonic functions. So an interesting research problem is the following: Let M be an irreducible symmetric space of rank l. If f is a harmonic function we may write

$$M_m(r, f) = f(m) + A_k r^{2k} + O(r^{2k+2}), \qquad (6.196)$$

where $k \geqslant 4$. What is the relation between k, A_k, and the rank l?

It is not difficult to prove that if the curvature tensor of a symmetric Riemannian manifold satisfies the first two non-trivial conditions for a harmonic space, then the coefficients $A(n)$, $B(n)$, and $C(n)$ all vanish. Also it is shown in [CGW] that the exceptional Lie group E_8 with its natural metric satisfies the first three non-trivial conditions but not the fourth. Thus E_8 is a non-harmonic symmetric space of rank 8 which, for a harmonic function f, satisfies

$$M_m(r, r) = f(m) + O(r^{10}). \qquad (6.197)$$

On the other hand, it is shown in [CGW] that if a symmetric space satisfies the first four conditions for a harmonic space then it is necessarily harmonic. We are thus led to the problem of determining the smallest k such that the coefficients of $r^2, r^4, \ldots, r^{2k}$ in the mean-value expansion vanish only for flat spaces or for symmetric spaces of rank 1.

We finally note that by applying the Cauchy-Kowalewski existence theorem the inequalities satisfied by Einstein and super-Einstein spaces can be shown to characterize such spaces.

6.12 COMMUTATIVE METRICS

An alternative (but in general different) way of defining a mean-value over a geodesic sphere is to use the formula

$$\hat{M}_m(r,f) = \frac{\int \exp_m(S^{n-1}(r)) f* \, dr}{V(S^{n-1}(r))}. \tag{6.198}$$

Here we are averaging over what is effectively the sphere of directions in the tangent space M_m. Suppose now that we are given two values of r, namely r_1 and r_2. From a given function f we can define a new function denoted by f_{r_1} where

$$f_{r_1}(m) = \hat{M}_m(r_1, f). \tag{6.199}$$

From f_{r_1} we define a new function denoted by $(f_{r_1})_{r_2}$ where

$$(f_{r_1})_{r_2}(m) = \hat{M}_m(r_2, f_{r_1}). \tag{6.200}$$

We now seek the condition that

$$(f_{r_1})_{r_2} = (f_{r_2})_{r_1}, \tag{6.201}$$

holds for all smooth functions f and for all sufficiently small real numbers r_1, r_2. It turns out that the above condition is equivalent to

$$L^k(L^l f) = L^l(L^k f), \tag{6.202}$$

for all integers k and l, where L^k is the globally defined differential operator defined in terms of the Euclidean Laplacian. We take equation (6.202) as the defining condition for a *commutative Riemannian metric*.

The idea of a commutative Riemannian metric first appeared in the paper by Roberts and Ursell (1960). These authors were concerned with the problem of random walks on the sphere and with the possibility of extending their work to consider random walks on more general

Riemannian manifolds. They restricted their attention to a special type of random walk in which the probability arising from two consecutive steps is independent of the order in which the steps are taken. They called such Riemannian metrics *commutative*. They were primarily concerned with the propagation of earthquake waves and, although with hindsight, their paper is seen to contain ideas subsequently rediscovered and elucidated by Allamigeon (1965), their non-invariant approach makes it difficult for a mathematician to understand what is really going on. Nevertheless they showed that a *necessary* condition for a commutative metric is that its curvature tensor satisfies the infinite set of conditions that we have previously denoted by $\alpha_{2k+1} = 0$, so that a commutative space is necessarily a D'Atri space. Whether every D'Atri space is commutative is possibly true but the problem remains open. Certainly all symmetric spaces have commutative metrics as is implied by the following theorem due to Lichnerowicz.

Theorem (6.12.1) In a compact symmetric space the algebra of invariant differential operators is commutative.

An outline of the proof is as follows. Let D_1, D_2 be invariant differential operators on a symmetric space. Let S be a symmetry associated with the space, and let D_1^*, D_2^* be operators defined by

$$D_i^* = SD_iS^{-1}, \qquad i = 1, 2. \tag{6.203}$$

Lichnerowicz proved that D_1^*, D_2^* are duals of the operators $D_1 D_2$. Then

$$D_1^*D_2^* = (SD_1S^{-1})(SD_2S^{-1}) = SD_1D_2S^{-1} = (D_1D_2)^* = D_2^*D_1^*, \tag{6.204}$$

and this implies that $D_1D_2 = D_2D_1$, and the theorem is proved. Since L^k, L^l are invariant differential operators, it follows that all compact symmetric spaces have commutative metrics.

Conclusion

We started this chapter by considering metrics for which Laplace's equation had a simple solution. This led to an infinite set of conditions on the curvature tensor and its covariant derivatives. The relation between harmonic spaces and symmetric spaces led to the idea of a k-stein space. An infinite subset of conditions gave rise to D'Atri spaces (see section 6.14) and indeed to commutative spaces. The generalization of the mean-value property of harmonic spaces led to an analytic characterization of Einstein and super-Einstein spaces. Mean-value considerations gave rise to a precise definition of a commutative space.

Two outstanding results have been achieved during the last two or three years. The first is due to Szabo (1990) who proved that every compact simply connected harmonic manifold is a symmetric space. The second result which I find even more surprising, due to Damek and Ricci (1991), is that locally the conjecture of Lichnerowicz is false, that is, there exists an example of a seven-dimensional manifold with a positive definite metric which is harmonic but not symmetric. What additional restrictions must be added to local harmonicity to force symmetry is likely to remain a fruitful field of research for several years to come.

6.13 THE CURVATURE OF EINSTEIN SYMMETRIC SPACES

The content of this section is based on the paper [CGW]. We recall that there is an infinite sequence $\{H_k\}$ of conditions that the curvature tensor of a harmonic space must satisfy. Using a slightly different notation from that used previously, the first four of these are

$$H_1: \sum_{a=1}^{n} R_{axax} = \lambda_1 \|x\|^2,$$

$$H_2: \sum_{a,b=1}^{n} R_{axbx} R_{axbx} = \lambda_2 \|x\|^4,$$

$$H_3: 32 \sum_{a,b,c=1}^{n} R_{axbx} R_{bxcx} R_{cxax} - 9 \sum_{a,b=1}^{n} (\nabla_x R_{axbx})^2 = \lambda_3 \|x\|^6,$$

$$H_4: 72 \sum_{a,b,c,d=1}^{n} R_{axbx} R_{bxcx} R_{cxdx} - 50 \sum_{a,b,c=1}^{n} \nabla_x R_{axbx} \nabla_x R_{bxcx} R_{cxax}$$
$$+ 8 \sum_{a,b,c=1}^{n} \nabla_{xx}^2 R_{axbx} R_{bxcx} + 3 \sum_{a,b=1}^{n} (\nabla_{xx}^2 R_{axbx})^2 = \lambda_4 \|x\|^8,$$

for any tangent vector x to M. Hence each λ_k ($k = 1, \ldots$) is a constant. If $x = \sum_{i=1}^{n} x^i e_i$, by R_{axbx} we mean $\sum_{a,b,i,j=1}^{n} R_{aibj} x^i x^j$.

The question arises naturally whether or not there exist Riemannian manifolds M which satisfy some but not all of the conditions $\{H_k\}$. We have already seen that condition H_1 is that M be an Einstein manifold. However little seems to be known about Einstein spaces which satisfy some condition H_k for $k > 1$. The following result was obtained in that paper.

Theorem (6.13.1) Let M be a non-flat locally symmetric space which satisfies H_1 and some other H_k. Then

(i) *M is irreducible*;
(ii) *either M is harmonic or $k \leq 3$*;
(iii) *M is harmonic if and only if M is locally isometric to one of the eight simply connected rank 1 symmetric spaces shown in Table 6.2*;
(iv) *M satisfies H_1, H_2, and H_3 but no other H_k if and only if M is locally isometric to the Lie group E_8, the symmetric space E_8/D_8, the symmetric space $E_{8(-24)}$, or the non-compact dual of these spaces*;
(v) *M satisfies H_1 and H_2 but no other H_k if and only if M is locally isometric to one of the twenty compact symmetric spaces listed in Tables 6.1 and 6.3 or to its non-compact dual*;
(vi) *M satisfies H_1 and H_3, but no other H_k, if and only if M is locally isometric to the symmetric space $SO(12)/SO(10) \times SO(2)$ or to its non-compact dual.*

Corollary (6.13.2) Let M be a harmonic symmetric space. Then either M is flat or has rank 1.

We shall not prove the theorem here — the method of proof is to rewrite the conditions H_k for a symmetric space M in terms of the $2k$th powers of the roots of M. Explicit knowledge of the root systems of all the irreducible symmetric spaces allows one to determine which symmetric spaces satisfy each H_k. However in the course of doing this the ideas of generalized Ricci curvatures and k-stein spaces were introduced, and since these concepts have proved useful in other contexts, we describe them here.

Let $x \in M_m$ and let R_x be the linear operator defined by $\langle R_x u, v \rangle = R_{xuxv}$ for $u, v \in M_m$.

Definition (6.13.3) The kth Ricci curvature $\rho[k]$ or $(\rho_m^{[k]})$ of M is the symmetric covariant tensor field of degree 2k given by

$$\rho^{[k]}(x, \ldots, x) = \mathrm{tr}\,(R_x^k)$$

for $x \in M_m$ and $m \in M$.

We write $\rho^{[k]}(x) = \rho^{[k]}(x, \ldots, x)$. Evidently $\rho^{[1]}$ is the usual Ricci curvature. For $k \geq 2$ we have

$$\rho^{[k]}(x) = \sum_{a_1, \ldots, a_k = 1}^{n} R_{a_1 x a_2 x} R_{a_2 x a_3 x} \cdots R_{a_k x a_1 x}$$

for $x \in M_m$ and $m \in M$. Here $\{1, \ldots, n\}$ is an arbitrary orthonormal basis of M_m.

Definition (6.13.4) We say that M is a k-stein space provided that there are real-valued functions μ_l on M such that

$$\rho^{[l]}(x) = \mu_l \|x\|^{2l}$$

for all $x \in M_m$ and $m \in M$ for $1 \leq l \leq k$.

The definition is sensible because a 1-stein space is Einstein!

We now prove that k-stein spaces are irreducible. More precisely we have the next theorem.

Theorem (6.13.5) Let M be a Riemannian manifold for which, for some $k \geq 2$, $\rho^{[k]}(z) = \mu_k \|z\|^{2k}$ for all $z \in M_m$ and for all $m \in M$, where μ_k is a non-zero function. Then M is an irreducible manifold.

To prove this, suppose that $M = M_1 \times M_2$ locally, where M_1 and M_2 have positive dimension. Let x, y be tangent vectors at M such that x is tangent to M_1 and y is tangent to M_2. Then, for all real numbers a and b, we have

$$\rho^{[k]}(ax + by) = a^{2k}\rho^{[k]}(x) + b^{2k}\rho^{[k]}(y)$$
$$= \mu_k \{a^{2k}\|x\|^{2k} + b^{2k}\|y\|^{2k}\}.$$

On the other hand

$$\mu_k \|ax + by\|^{2k} = \mu_k \sum_{p=0}^{k} \binom{k}{p} a^{2ap} b^{2k-2p} \|x\|^{2p} \|y\|^{2k-2p}.$$

This is incompatible with the previous equation when $k \geq 2$ giving a contradiction. This completes the proof.

Observe that $\rho^{[2]} \equiv 0$ if and only if M is flat.

The various 2-stein and 3-stein symmetric spaces are listed in Tables 6.1 and 6.3. Evidently a harmonic space is k-stein for all values of k.

6.14 D'ATRI SPACES

In their paper D'Atri and Nickerson (1969) proposed the study of a new class of Riemannian manifolds. As we shall see, these can be regarded as the class of spaces whose curvature tensor satisfies all the odd conditions $\alpha_{2k+1} = 0$ for a harmonic space — in fact precisely those conditions which in a harmonic space are satisfied as a consequence of the even conditions.

258   6 Special Riemannian manifolds

Table 6.1 Harmonic, 2-stein, and 3-stein compact Lie groups and their non-compact dual spaces

Compact	Non-compact	Dynkin diagram	Type	$\dfrac{(\rho[1])^k}{\rho^{[k]}}$		Orders of the primitive invariants	Dimension
$SU(2)$	$\dfrac{SU(2,\mathbb{C})}{SU(2)}$	○	harmonic	2^{k-1}	$(k=1,2,\ldots)$	2	3
$SU(3)$	$\dfrac{SU(3,\mathbb{C})}{SU(3)}$	○—○	2-stein	4	$(k=2)$	2, 3	8
$SO(8)$	$\dfrac{SO(8,\mathbb{C})}{SO(8)}$	○—○⟨○○	2-stein	12	$(k=2)$	2, 4, 4, 6	28
G_2	$\dfrac{G_2^{\mathbb{C}}}{G_2}$	○⇛○	2-stein	$\dfrac{32}{5}$	$(k=2)$	2, 6	14
F_4	$\dfrac{F_4^{\mathbb{C}}}{F_4}$	○—○⇒○—○	2-stein	$\dfrac{108}{5}$	$(k=2)$	2, 6, 8, 12	52
E_6	$\dfrac{E_6^{\mathbb{C}}}{E_6}$	○—○—○—○—○ with ○	2-stein	32	$(k=2)$	2, 5, 6, 8, 9, 12	78
E_7	$\dfrac{E_7^{\mathbb{C}}}{E_7}$	○—○—○—○—○—○ with ○	2-stein	54	$(k=2)$	2, 6, 8, 10, 12, 14, 18	133
E_8	$\dfrac{E_8^{\mathbb{C}}}{E_8}$	○—○—○—○—○—○—○ with ○	3-stein	100 7,200	$(k=2)$ $(k=3)$	2, 8, 12, 14, 18, 20, 24, 30	248

6.14 D'Atri spaces

Table 6.2 Simply connected symmetric harmonic spaces

Compact	Non-compact	Root system	Dynkin diagram	Dimension
—	$\mathbb{R}^n$	—	—	n
$S^n = \dfrac{SO(n+1)}{SO(n)}$	$H^n = \dfrac{SO(n,1)}{SO(n)}$	A_1	$\underset{\circledcirc}{n-1}$	n
$\mathbb{C}P^n = \dfrac{U(n+1)}{U(n) \times U(1)}$	$\mathbb{C}H^n = \dfrac{U(n,1)}{U(n) \times U(1)}$	BC_1	$\underset{\circledcirc}{2(n-1)[1]}$	$2n$
$\mathbb{Q}P^n = \dfrac{Sp(n+1)}{Sp(n) \times Sp(1)}$	$\mathbb{Q}H^n = \dfrac{Sp(n,1)}{Sp(n) \times Sp(1)}$	BC_1	$\underset{\circledcirc}{4(n-1)[3]}$	$4n$
$CayP^2 = \dfrac{F_4}{Spin(9)}$	$CayH^2 = \dfrac{F_4^*}{Spin(9)}$	BC_1	$\underset{\circledcirc}{8[7]}$	16

260   6 Special Riemannian manifolds

Table 6.3 2-stein and 3-stein spaces of type I

Space	Root system	Dynkin diagram	$\dfrac{(\rho^{[1]})^k}{\rho[k]}$		Type	Dimension
$A_2^R = \dfrac{SU(3)}{SO(3)}$	A_2	∘—∘	2	$(k=2)$	2-stein	5
$D_4^{R,4} = \dfrac{SO(8)}{SO(4)\times SO(4)}$	D_4		6	$(k=2)$	2-stein	16
$D_4^{R,3} = \dfrac{SO(8)}{SO(5)\times SO(3)}$	B_3		6	$(k=2)$	2-stein	15
					2-stein	
$D_4^{R,2} = \dfrac{SO(8)}{SO(6)\times SO(2)}$	B_2		6	$(k=2)$	Hermitian Symmetric	12
$A_5^H = \dfrac{SU(6)}{Sp(3)}$	A_2	∘—∘	8	$(k=2)$	2-stein	14
$G_{2(2)} = \dfrac{G_2}{SO(4)}$	G_2		$\dfrac{16}{5}$	$(k=2)$	2-stein	8
$F_{4(4)} = \dfrac{F_4}{A_1 \times C_3}$	F_4		$\dfrac{54}{5}$	$(k=2)$	2-stein	28
$E_{6(-26)} = \dfrac{E_6}{F_4}$	A_2	∘—∘	16	$(k=2)$	2-stein	26

6.14 D'Atri spaces

$E_{6(-14)} = \dfrac{E_6}{D_5 \times SO(2)}$	BC_2		16	$(k=2)$	2-stein Hermitian Symmetric	32
$E_{6(2)} = \dfrac{E_6}{A_1 \times A_5}$	F_4		16	$(k=2)$	2-stein	40
$E_{6(6)} = \dfrac{E_6}{C_4}$	E_6		16	$(k=2)$	2-stein	42
$E_{7(-25)} = \dfrac{E_7}{E_6 \times SO(2)}$	C_3		27	$(k=2)$	2-stein Hermitian Symmetric	54
$E_{7(-5)} = \dfrac{E_7}{D_6 \times A_1}$	F_4		27	$(k=2)$	2-stein	64
$E_{7(7)} = \dfrac{A_7}{A_7}$	E_7		27	$(k=2)$	2-stein	70
$E_{8(-24)} = \dfrac{E_8}{E_7 \times A_1}$	F_4		50, 1,800	$(k=2)$, $(k=3)$	3-stein	112
$E_{8(8)} = \dfrac{E_8}{D_8}$	E_8		50, 1,800	$(k=2)$, $(k=3)$	3-stein exceptional	128
$D_6^{R,2} = \dfrac{SO(12)}{SO(10) \times SO(2)}$	B_2		100	$(k=3)$	Hermitian Symmetric	20

6 Special Riemannian manifolds

The local geodesic symmetry ϕ_m at m is defined by

$$\phi_m: \exp_m(r\xi) \to \exp_m(-r\xi) \qquad (6.205)$$

where ξ is a unit tangent vector to M at m. This map is a local diffeomorphism if r is sufficiently small. It preserves volume (up to sign) if and only if

$$\theta_m(\exp_m(r\xi)) = \theta_m(\exp_m(-r\xi)) \qquad (6.206)$$

for all unit vectors $\xi \in M_m$. However, in Section 6.8 we obtained the expression of $\theta_m(p)$ as a power series given by (6.109), (6.110), and (6.111). From this we obtain the following lemma.

Lemma (6.14.1) *Let (M, g) be a Riemannian manifold such that all local geodesic symmetries are volume preserving (up to sign). Then we have for each point m and each unit vector $\xi \in M_m$,*

$$\alpha_{2k+1}(m, \xi) = 0 \qquad (6.207)$$

for all integers $k \geq 0$. These conditions are also sufficient if M is analytic.

The first condition is equivalent to

$$\nabla_\xi \rho_{\xi\xi} = 0 \quad \text{or} \quad \nabla_i \rho_{jk} + \nabla_j \rho_{ki} + \nabla_k \rho_{ij} = 0. \qquad (6.208)$$

From the work of Kazdan referred to earlier in connection with analyticity of Einstein spaces, it follows that a C^∞-Riemannian manifold which satisfies (6.208) is always analytic in geodesic polar coordinates, so that there is no real restriction in assuming that our manifolds are analytic.

The next condition $\alpha_5(m, \xi) = 0$ becomes, on taking account of (6.208),

$$\sum_{a,b=1}^n R_{\xi a \xi b} \nabla_\xi R_{\xi a \xi b} = 0. \qquad (6.209)$$

The condition $\alpha_7(m, \xi) = 0$ becomes, on taking account of (6.208) and (6.209),

$$3 \sum_{a,b=1}^n \nabla_\xi R_{\xi a \xi b} \nabla_{\xi\xi}^2 R_{\xi a \xi b} = 16 \sum_{a,b,c=1}^n R_{\xi b \xi c} R_{\xi c \xi a} \nabla_\xi R_{\xi a \xi b}. \qquad (6.210)$$

However it seems difficult to extract geometric information from the set of equations (6.208), (6.209), and (6.210). One such result was obtained by Segigawa and Vanhecke. (1986). They proved the next theorem.

Theorem (6.14.2) *Let (M, g) be a connected four-dimensional 2-stein space with volume preserving local geodesic symmetries. Then either*

6.14 D'Atri spaces

(M, g) is locally flat or else it is locally isometric to a rank 1 symmetric space.

However, in spite of the simplifying assumptions of dimension = 4, conditions involving α_2, α_4, as well as $\alpha_3 = \alpha_5 = \alpha_7 = 0$, their proof involves long and tedious calculations. For example, they require to establish that the hypotheses of their theorem imply the following identities:

$$\sum \nabla_j R_{iakb} \nabla^2_{jk} R_{ia\xi b} = \frac{1}{4} \xi \|\nabla R\|^2 + \frac{1}{6} \xi \hat{R} = \frac{11}{36} \xi \|\nabla R\|^2,$$

$$\sum \nabla_j R_{iakb} \nabla^2_{j\xi} R_{iakb} = \frac{1}{2} \xi \|\nabla R\|^2 + \frac{1}{3} \xi \hat{R} = \frac{11}{18} \xi \|\nabla R\|^2,$$

$$\sum \nabla_j R_{iakb} \nabla^2_{kj} R_{ia\xi b} = \frac{1}{4} \xi \|\nabla R\|^2 + \frac{1}{4} \xi \hat{R} = \frac{1}{3} \xi \|\nabla R\|^2,$$

$$\sum \nabla_j R_{ia\xi b} \nabla^2_{kj} R_{iakb} = -\frac{1}{12} \xi \hat{R} - \frac{1}{3} \xi \mathring{R} = -\frac{1}{12} \xi \|\nabla R\|^2,$$

$$\sum \nabla_\xi R_{kbia} \nabla^2_{jj} R_{kbia} = -\frac{1}{3} \xi \hat{R} - \frac{4}{3} \xi \mathring{R} = -\frac{1}{3} \xi \|\nabla R\|^2,$$

$$\sum \nabla_j R_{ka\xi b} \nabla^2_{ij} R_{iakb} = -\frac{1}{24} \xi \|\nabla R\|^2,$$

$$\sum \nabla_j R_{ka\xi b} \nabla^2_{ik} R_{iajb} = -\frac{1}{12} \xi \|\nabla R\|^2,$$

$$\sum \nabla_j R_{\xi akb} \nabla^2_{ik} R_{iajb} = -\frac{1}{24} \xi \|\nabla R\|^2,$$

$$\sum \nabla_\xi R_{jakb} \nabla^2_{ij} R_{iakb} = -\frac{1}{6} \xi \|\nabla R\|^2,$$

$$\sum \nabla_\xi R_{jakb} \nabla^2_{ik} R_{iajb} = -\frac{1}{12} \xi \|\nabla R\|^2$$

for all $\xi \in M_m$ and all $m \in M$.

The proof also relies upon the following result due to Derdzinski.

Theorem (6.14.3) *Let $W \in (\text{End } \Lambda^2 M)$ be the Weyl conformal tensor. Let (M, g) be a four-dimensional Einstein manifold such that W has constant eigenvalues. Then (M, g) is locally symmetric.*

It seems desirable to obtain a simpler proof of the theorem of Segigawa and Vanhecke, preferably with some relaxation of some of the hypotheses. However the following example suggests that this may prove difficult.

6 Special Riemannian manifolds

The so-called Heisenberg group consists of real 3×3 matrices of the form

$$\begin{bmatrix} 1 & x & y \\ 0 & 1 & z \\ 0 & 0 & 1 \end{bmatrix}. \tag{6.211}$$

It is known that this group manifold can be given the following metric which remains invariant under left translations of the group:

$$\mathrm{d}s^2 = \mathrm{d}x^2 + \mathrm{d}z^2 + (\mathrm{d}y - z\mathrm{d}x)^2. \tag{6.212}$$

Direct calculation shows that this is a D'Atri metric but it is not that of a symmetric space.

An alternative approach appeared in a paper by Gray (1978b). Gray considered the two classes of Riemannian manifolds $\mathcal{A}$ and $\mathcal{B}$ defined by

$$\mathcal{A}: \nabla_i \rho_{jk} + \nabla_k \rho_{ij} + \nabla_j \rho_{ki} = 0, \tag{6.213}$$

$$\mathcal{B}: \nabla_i \rho_{jk} = \nabla_j \rho_{ik}, \tag{6.214}$$

where ρ_{ij} denotes the Ricci tensor. In our terminology a metric belongs to $\mathcal{A}$ if it satisfies the first condition for a D'Atri space. In investigating class $\mathcal{B}$, Gray found it convenient to consider 'Codazzi tensors', that is, the class of symmetric covariant second-order tensors which satisfy equation (6.214).

Let V be an n-dimensional vector space with inner product $\langle \ , \ \rangle$ acted on by the orthogonal group $O(n)$. Let V^* denote the dual space and consider the space $V^* \otimes V^* \otimes V^*$, which is naturally isomorphic to the space of trilinear covariant tensors on V. Let W be the subspace of $V^* \otimes V^* \otimes V^*$ defined by

$$W = \left\{ \phi \in V^* \otimes V^* \otimes V^* \,|\, \phi(x, y, z) = \phi(x, z, y) \text{ and} \right.$$

$$\left. \sum_{i=1}^n \phi(x, e_i, e_i) = 2 \sum_{i=1}^n \phi(e_i, e_i, x) \text{ for all } x, y, z \in V \right\}, \tag{6.215}$$

where $\{e_1, \ldots, e_n\}$ is an arbitrary orthonormal basis of V. We choose these relations because we want W to be the space of tensors which satisfy all the identities of the covariant derivative of the Ricci tensor and no others. The second condition follows from the identity

$$X\tau = 2 \sum_{i=1}^n \nabla_{e_i}(\rho)(x, e_i) \tag{6.216}$$

6.14 D'Atri spaces

where τ is the scalar curvature. This condition is itself a consequence of the second Bianchi identity. There is a naturally defined inner product on W given by

$$\langle \phi, \psi \rangle = \sum_{i,j,k=1}^{n} \phi(e_i, e_j, e_k) \psi(e_i, e_j, e_k). \tag{6.217}$$

We define three subspaces of W as follows:

$$A = \{\phi \in W | \phi(x, y, z) + \phi(z, x, y) + \phi(y, z, x) = 0$$
$$\text{for all } x, y, z \in V\}, \tag{6.218}$$

$$B = \{\phi \in W | \phi(x, y, z) = \phi(y, z, x) \text{ for all } x, y, z \in V\}, \tag{6.219}$$

$$C = \left\{ \phi \in W \,\middle|\, \sum_{i=1}^{n} \phi(x, e_i, e_i) = 0 \text{ for all } x \in V \right\}. \tag{6.220}$$

Consider now the induced representation of $O(n)$ on W. The following theorem describes the decomposition of W into subspaces on which the induced representation of $O(n)$ is irreducible.

Theorem (6.14.4) **We have**

$$C = A \oplus B, \tag{6.221}$$

$$W = A \oplus B \oplus C^{\perp}. \tag{6.222}$$

These direct sums are orthogonal. Moreover the induced representations of O(n) on A, B, and C are irreducible.

The proof that $A \cap B = \{0\}$ is straightforward. For $\phi \in C$ we define ϕ_A and ϕ_B by

$$\phi_A(x, y, z) = \frac{1}{3}\{2\phi(x, y, z) - \phi(z, x, y) - \phi(y, z, x)\}, \tag{6.223}$$

$$\phi_B(x, y, z) = \frac{1}{3}\{\phi(x, y, z) + \phi(z, x, y) + \phi(y, z, x)\}. \tag{6.224}$$

Then

$$\phi_A \in A, \quad \phi_B \in B, \quad \text{and} \quad \phi = \phi_A + \phi_B. \tag{6.225}$$

Thus conditions (6.221) and (6.222) hold. The real difficulty consists of proving the irreducibility of the representations, and this involves a rather deep theorem of Weyl on the invariants of the orthogonal group. We assume that this is indeed the case.

For a Riemannian manifold there is a representation of the orthogonal group $O(n)$ on each tangent space M_m. Let

266   6 Special Riemannian manifolds

$$W_m = \left\{ \phi \in M_m^* \otimes M_m^* \otimes M_m^* \,|\, \phi(x,y,z) = \phi(x,z,y) \text{ and} \right.$$

$$\left. \sum_{i=1}^n \phi(x,e_i,e_i) = 2 \sum_{i=1}^n \phi(e_i,e_i,x) \text{ for all } x,y,z \in M_m \right\}.$$

Then the induced representation of $O(n)$ on W_m has three components A_m, B_m, $C_m^\perp$ as described above. Thus alternative descriptions for the spaces $\mathcal{A}$ and $\mathcal{B}$ are $M \in \mathcal{A}$ if and only if for each $m \in M$, $(\nabla \rho)_m$ lies in the subspace A_m; $M \in \mathcal{B}$ if and only if for each $m \in M$, $(\nabla \rho)_m$ lies in the subspace B_m.

Note that this analysis leads to a consideration of the class $\mathcal{C}^\perp$ of manifolds corresponding to $C^\perp$. The defining condition for this class is

$$\nabla_X(\rho)(Y,Z) = \frac{1}{(n+2)(n-1)} \left\{ n(X\tau)\langle Y,Z\rangle \right.$$
$$\left. + \frac{1}{2}(n-2)[(Y\tau)\langle X,Z\rangle + (Z\tau)\langle X,Y\rangle] \right\}.$$
(6.226)

This condition is satisfied by any two-dimensional Riemannian manifold, but it is unknown whether there are other interesting manifolds in $\mathcal{C}^\perp$.

Similarly it is possible to consider the classes $\mathcal{A} \oplus \mathcal{C}^\perp$ and $\mathcal{B} \oplus \mathcal{C}^\perp$ corresponding to $A \oplus C^\perp$ and $B \oplus C^\perp$. The defining relations are

$$\mathcal{A} \oplus \mathcal{C}^\perp : \underset{X,Y,Z}{\mathfrak{S}} \left\{ \nabla_X(\rho)(Y,Z) + \frac{2}{n+2}(X\tau)\langle Y,Z\rangle \right\} = 0, \quad (6.227)$$

$$\mathcal{B} \oplus \mathcal{C}^\perp : \nabla_X(\rho)(Y,Z) - \nabla_Y(\rho)(X,Z)$$
$$= \frac{1}{(n-1)} \{(X\tau)\langle Y,Z\rangle - (Y\tau)\langle X,Z\rangle\}. \quad (6.228)$$

In his paper Gray gives examples of homogeneous metrics on the three-dimensional sphere S^3 in $\mathbb{R}^4$ which are in $\mathcal{A}$ but do not satisfy $\nabla \rho = 0$. Let N denote the unit outward normal to the unit sphere S^3 in $\mathbb{R}^4$. If we regard $\mathbb{R}^4$ as the space of quaternions we obtain vector fields IN, JN, KN tangent to S^3. Let ϕ_I, ϕ_J, ϕ_K be the 1-forms on S^3 given by

$$\phi_I(X) = \langle X, IN \rangle,$$
$$\phi_J(X) = \langle X, JN \rangle,$$

$$\phi_K(X) = \langle X, KN \rangle.$$

We consider metrics of the form

$$ds^2 = \alpha^2 \Phi_i^2 + \beta^2 \Phi_j^2 + \gamma^2 \Phi_K^2.$$

Denote by $\mathcal{P}$ the class of Riemannian metrics such that $\nabla_X(\rho)(Y, Z) = 0$. Then Gray proved the next theorem.

Theorem (6.14.5)

(i) $S^3(\alpha, \beta, \gamma) \in \mathcal{P} \Leftrightarrow \alpha^2 = \beta^2 = \gamma^2$;
(ii) $S^3(\alpha, \beta, \gamma) \in \mathcal{B} \Leftrightarrow S^3(\alpha, \beta, \gamma) \in \mathcal{P}$;
(iii) $S^3(\alpha, \beta, \gamma) \in \mathcal{A} \Leftrightarrow$ precisely two of α^2, β^2, γ^2 are equal.

No doubt similar group representation arguments could be used to obtain information about metrics which satisfy $\alpha_5 = 0$, $\alpha_7 = 0$, and so on, but the calculations would become progressively more and more complicated.

6.15 EXERCISES

1. The metric of a four-dimensional manifold is given in a coordinate neighbourhood by

   $$ds^2 = f(x, y) dx^2 + 2dxdt + 2dydz,$$

   where f is any smooth function of x and y. Prove that the space is symmetric or recurrent. Moreover it is symmetric if and only if $\partial^2 f / \partial y^2$ is constant. Prove also that the metric is harmonic.

2. Prove that the space with metric

   $$ds^2 = yzdx^2 - xtdy^2 + 2dxdz + 2dydt$$

   is harmonic but is neither symmetric nor recurrent.

3. We recall that if $A(s)$ is the $(n-1) \times (n-1)$ matrix obtained from a consideration of Jacobi vector fields, then the space is harmonic if $\det A(s)$ is a radial function. However, instead of taking the determinant we could have taken the kth invariant of $A(s)$, the first being the trace and the last the determinant. Thus we can define a space to be k-harmonic if the kth invariant is a radial function. Prove that a 1-harmonic space is harmonic and conversely.

6 Special Riemannian manifolds

4. Show that if a symmetric harmonic space is k-harmonic for some k then it is k-harmonic for $k = 1, \ldots, n-1$.

5. Investigate whether the result of Exercise 4 is still true if the requirement of symmetry is omitted.

6. Find a set of necessary and sufficient conditions for a harmonic Riemannian space to be symmetric. If you succeed you should publish your results in a mathematical research journal!

7
Special Riemannian submanifolds

7.1 INTRODUCTION

Again we find ourselves with the problem of selection. One of the most important classes of submanifolds corresponds to the set of immersions for which the mean curvature vector vanishes. These are the minimal submanifolds. In this case there is a fundamental difference between immersions where the receiving space is Euclidean space $\mathbb{R}^n$ or the sphere S^n contained in $\mathbb{R}^{n+1}$. For example, compact minimal surfaces do not exist in $\mathbb{R}^3$ but they certainly do in S^3 contained in $\mathbb{R}^4$. A more general class is obtained by considering immersions for which the mean curvature vector is parallel with respect to the connexion in the normal bundle. In this chapter we consider immersions which are generalizations of minimal submanifolds known in the literature as 'Willmore surfaces', although the topic was mentioned in Blaschke (1929) and by Thomsen (1923). We make this selection because the topic has proved a very fruitful field of research, especially during the last fifteen years. Moreover, most of the techniques involved have already appeared in the previous chapters. Finally we remark that the subject matter has received scant attention in previous texts on Riemannian geometry.

Our treatment broadly follows the historical development of the subject rather than what might be considered, with hindsight, as a logical development from a general viewpoint. We do this because we believe that most readers will prefer to see how the subject actually developed over the years. Of course, an axiomatic treatment beginning with the most general case of an isometric immersion of a Riemannian manifold (M, g) into a Riemannian manifold (M', g') would include all the special cases. But often it is the properties of special cases which motivates study of the more general case. Moreover, my experience of teaching suggests that most students find the understanding of abstract mathematics more natural after they have studied particular cases in some detail. I must admit, however, that one of the most gifted students whom I have supervised had no difficulty in assimilating abstract ideas immediately, and for him the examples were self-evident (and rather boring!) special cases.

We shall begin by studying the very special case of Willmore surfaces immersed in Euclidean 3-space. Even here there remain many unsolved problems. My interest in the subject dates from the early 1960s when

I gave several talks on the topic at the Mathematical Institute, Oberwolfach, Germany. This was followed by my first paper on the topic, Willmore (1965).

7.2 SURFACES IN E^3

Let $f: M^2 \to \mathbb{R}^3$ be an isometric immersion of a smooth compact orientable surface into three-dimensional Euclidean space. At each point $p \in M^2$ there corresponds a shape operator $A_\xi: M_p^2 \to M_p^2$ whose eigenvalues are the principal curvatures k_1, k_2 at p. We define

$$\mathcal{W}(f) = \int_{M^2} H^2 \, dA \tag{7.1}$$

where $H = \tfrac{1}{2}(k_1 + k_2)$.

A related functional is $C(f)$ where

$$C(f) = \frac{1}{4\pi} \int_{N(f)} \left(\frac{\kappa_1 - \kappa_2}{2}\right)^2 dA$$

where $N(f)$ is the unit normal bundle (7.2)

$$= \frac{1}{2\pi} \int_{M^2} \left\{ \left(\frac{\kappa_1 + \kappa_2}{2}\right)^2 - \kappa_1 \kappa_2 \right\} dA.$$

Hence

$$C(f) = \frac{1}{2\pi} \mathcal{W}(f) - \chi(M^2) \tag{7.3}$$

where we have used the Gauss–Bonnet theorem. It follows that the study of $C(f)$ is equivalent to the study of $\mathcal{W}(f)$.

We note that we can easily extend our study to include non-orientable surfaces by defining $\mathcal{W}(f)$ by

$$\mathcal{W}(f) = \tfrac{1}{2} \int_{N(f)} H^2 \, dA \tag{7.4}$$

where we now integrate over the unit normal bundle $N(f)$ which is effectively two copies of M^2 when the latter is orientable.

Let S be a compact surface immersed in $\mathbb{R}^3$. Let $K^+ = \max\{K, 0\}$, that is, K^+ is the function on the surface whose value equals the Gaussian curvature K when K is positive and is zero when K is negative or zero. We have the following lemma.

Lemma (7.2.1) Let S be a compact orientable surface in $\mathbb{R}^3$. Then

7.2 Surfaces in E^3

$$\int_S K^+ \, dA \geq 4\pi. \tag{7.5}$$

The left hand side of (5) represents the area of the image of the Gauss normal map (counting multiplicities) of that part of S where $K \geq 0$. Hence it will be sufficient to show that this image covers the whole unit sphere $S^2(1)$. The Heinz Hopf recipe for this is to 'set the surface down on the floor'. The first point of S where it touches the floor has a normal vector pointing downwards and here the Gauss curvature is non-negative, because at a point where $K < 0$ the surface is saddle shaped and crosses itself. But the floor can be in any direction and we get a point where $K \geq 0$ in every normal direction. Since the area of the unit sphere is 4π, equation (7.5) is established.

With the aid of this lemma we obtain the first result.

Theorem (7.2.2) Let M^2 be a closed orientable surface and let $f: M^2 \to \mathbb{R}^3$ be a smooth immersion of M^2 into Euclidean 3-space. Then

$$\mathcal{W}(f) = \int_{M^2} H^2 \, dS \geq 4\pi. \tag{7.6}$$

Moreover $\mathcal{W}(f) = 4\pi$ if and only if M^2 is embedded as the round sphere.

Suppose we divide the surface into the part where K is positive and its complement as in the lemma. Since

$$H^2 - K = [\tfrac{1}{2}(\kappa_1 - \kappa_2)]^2 \geq 0 \tag{7.7}$$

we have

$$\int_S H^2 \, dA \geq \int_{K^+} H^2 \, dA \geq \int_{K^+} K \, dA \geq 4\pi$$

using the lemma. Thus (7.6) is established. Moreover equality holds if and only if $\kappa_1 = \kappa_2$ at every point. Thus every point is an umbilic. It follows (see, for example, Willmore (1959), page 128) that $f(M^2)$ is embedded as a round sphere.

We now wish to prove the following theorem due to W. Fenchel. Our proof follows the ideas of Do Carmo (1976) p. 399.

Theorem (7.2.3) Let C be a simple closed curve in Euclidean 3-space. Then

$$\int_C |\kappa| \, ds \geq 2\pi \tag{7.8}$$

with equality if and only if C is a plane convex curve.

7 Special Riemannian submanifolds

This theorem is interesting in its own right but we shall make use of it in calculating $\mathcal{W}(f)$ for a torus.

Since the curve lies in $\mathbb{R}^3$ it seems natural to use classical vector methods. The Serret–Frenet formulas in terms of the unit tangent vector **t**, the unit principal normal vector **n**, and the unit binormal $\mathbf{b} = \mathbf{t} \times \mathbf{n}$, are given by

$$\mathbf{t} = \mathbf{r}', \qquad \mathbf{t}' = \kappa \mathbf{n}, \qquad \mathbf{n}' = \tau \mathbf{b} - \kappa \mathbf{t}, \qquad \mathbf{b}' = -\tau \mathbf{n},$$

where **r** is the position vector, dash denotes differentiation with respect to the arc length s, and κ and τ are respectively, the curvature and the torsion of C.

We consider a thin tube round C. The proof consists in showing that the left-hand side of (7.5) is equal to twice the left-hand side of (7.8).

The tube of radius r about C is given by

$$\mathbf{r}(s, t) = \mathbf{r}(s) + r(\cos t \, \mathbf{n} + \sin t \, \mathbf{b})$$

where $0 \leqslant t \leqslant 2\pi$. We calculate the Gaussian curvature of this surface using the classical formula

$$K = \frac{LN - M^2}{EG - F^2}$$

where the coefficients of the first fundamental form E, F, G are given by

$$E = \mathbf{r}_1 \cdot \mathbf{r}_1, \qquad F = \mathbf{r}_1 \cdot \mathbf{r}_2, \qquad G = \mathbf{r}_2 \cdot \mathbf{r}_2,$$

and the coefficients of the second fundamental form L, M, N are given by

$$L = -\mathbf{N}_1 \cdot \mathbf{r}_1, \qquad M = -\mathbf{N}_1 \cdot \mathbf{r}_2 = -\mathbf{N}_2 \cdot \mathbf{r}_1, \qquad N = -\mathbf{N}_2 \cdot \mathbf{r}_2.$$

Here the suffix 1 denotes partial differentiation with respect to s, and the suffix 2 denotes partial differentiation with respect to t. We find

$$\mathbf{r}_1 = (1 - r\kappa \cos t)\mathbf{t} - r\tau \sin t \, \mathbf{n} + r\tau \cos t \, \mathbf{b},$$

$$\mathbf{r}_2 = -r \sin t \, \mathbf{n} + r \cos t \, \mathbf{b},$$

$$\mathbf{r}_1 \times \mathbf{r}_2 = -r(1 - r\kappa \cos t)(\cos t \, \mathbf{n} + \sin t \, \mathbf{b}).$$

Thus the unit normal **N** is given by

$$\mathbf{N} = \cos t \, \mathbf{n} + \sin t \, \mathbf{b}.$$

$$E = (1 - r\kappa \cos t)^2 + r^2 \tau^2; \qquad F = r^2 \tau; \qquad G = r^2.$$

$$EG - F^2 = r^2 (1 - r\kappa \cos t)^2.$$

$$\mathbf{N}_1 = -\kappa \cos t \, \mathbf{t} - \tau \sin t \, \mathbf{n} + \tau \cos t \, \mathbf{b},$$

$$\mathbf{N}_2 = -\sin t \, \mathbf{n} + \cos t \, \mathbf{b},$$

7.2 Surfaces in E^3

$$L = -r\tau^2 + \kappa \cos t(1 - r\kappa \cos t),$$
$$M = -r\tau,$$
$$N = -r.$$

By substitution we find that $K = -\kappa \cos t/r(1 - r\kappa \cos t)$. Hence

$$K\,dA = -\kappa \cos t\, dt\, ds.$$

Thus $k = 0$ implies that $K(s, t) = 0$ for all $0 \leqslant t \leqslant 2\pi$. We see that K is non-negative when $|\kappa| \cos t$ is non-positive, that is when $\pi/2 \leqslant t \leqslant 3\pi/2$. Hence

$$\int_{K^+} K\,dA = \int_C |\kappa|\,ds \int_{\pi/2}^{3\pi/2} -\cos t\,dt = 2\int |\kappa|\,ds.$$

Using the previous lemma we have

$$\int_C |\kappa|\,ds \geqslant 2\pi.$$

Note that the image of the Gauss normal map when restricted to each circle $s = $ constant is bijective and that its image is a great circle $\Gamma_s \subset S^2$. We denote by Γ_s^+ the closed semicircle corresponding to points where $K \geqslant 0$. Assume now that C is a plane convex curve. Then all the semicircles Γ_s^+ have the same end points p, q. Since C is convex it follows that $\Gamma_{s_1} \cap \Gamma_{s_2} = \{p\} + \{q\}$ for $s_1 \neq s_2$. Hence $\int_{K^+} K\,dA = 4\pi$ so that $\int_C |\kappa|\,ds = 2\pi$ as required.

Conversely let us assume that $\int_C |\kappa|\,ds = 2\pi$. This implies $\int_{K^+} K\,dA = 4\pi$. We claim that all the semicircles Γ_s^+ must have the same end points p and q. Otherwise there are two distinct great circles $\Gamma_{s_1}, \Gamma_{s_2}$, with s_2 arbitrarily close to s_1, which intersect in two antipodal points. These points are not in the image of the Gauss map restricted to $K^+ \cap Q$ where Q is the set of points in S with non-positive curvature. It follows that there are two points of positive curvature which are mapped by the Gauss map into a single point of S^2. Since the Gauss map is a local diffeomorphism at such points and since each point of S^2 is the image of at least one point of K_+, we conclude that $\int_{K_+} K\,dA > 4\pi$ giving a contradiction. Thus all the Γ_s^+ have the same end points p and q.

Moreover since the points of zero Gaussian curvature are the intersections of the binormal of C with S, we see that the binormal of C is parallel to the line pq. Thus C is contained in a plane normal to this line.

We finally prove that C is convex. Since we assume $\int |\kappa|\,ds = 2\pi$ we have

$$2\pi = \int_0^l |\kappa|\,ds \geqslant \int_0^l \kappa\,ds.$$

However

$$\int_J \kappa \, ds \geq 2\pi$$

holds for any closed plane curve with positive orientation number so we conclude

$$\int_0^l \kappa \, ds = \int_0^l |\kappa| \, ds = 2\pi.$$

It follows that $\kappa \geq 0$ and the curve is therefore convex. This completes the proof of the theorem.

We now make an application of Fenchel's theorem. From our calculations of the fundamental coefficients for the tube of radius r about the closed curve C of length l we find that the mean curvature, given by the formula $H = (EN + GL - 2FM)/2(EG - F^2)$, is given by $H = -(1 - 2r\kappa \cos t)/2r(1 - r\kappa \cos t)$. Thus

$$\int_{M^2} H^2 \, dA = \int_0^l \int_0^\pi \frac{(1 - 2r\kappa \cos t)^2}{2r(1 - r\kappa \cos t)} \, dt \, ds.$$

We evaluate one of the integrals (using *Mathematica* it took just 10 seconds!). We find

$$\int_{M^2} H^2 \, dA = \pi \int_0^l \frac{ds}{2r\sqrt{(1 - r^2\kappa^2)}}.$$

It is interesting to note that this expression is independent of the torsion of the curve C. We may rewrite this expression as

$$\frac{\pi}{2} \int_0^l \frac{|\kappa| \, ds}{|\kappa r|\sqrt{(1 - \kappa^2 r^2)}}.$$

But for any real number y the expression $y\sqrt{(1 - y^2)}$ takes its maximum value $\frac{1}{2}$ when $y = 1/\sqrt{2}$. So we get

$$\int_{M^2} H^2 \, dA \geq \pi \int_0^l |\kappa| \, ds.$$

Making use of Fenchel's theorem gives the relation

$$\int_{M^2} H^2 \, dA \geq 2\pi^2.$$

Moreover, equality holds if and only if $\kappa r = 1/\sqrt{2}$. This agrees with the result obtained in Willmore (1965) considering the anchor ring

$$x = (a + b \cos u)\cos v, \qquad y = (a + b \cos u)\sin v, \qquad z = b \sin u.$$

We found the result

7.2 Surfaces in E^3

$$\int H^2 \, dA = \frac{\pi^2}{c\sqrt{(1-c^2)}}$$

where $c = b/a$ and $c < 1$. The minimum value occurs when $c = 1/\sqrt{2}$, giving

$$\int H^2 \, dA = 2\pi^2.$$

A picture of this minimizing torus of revolution is given in Fig. 7.1.

Summarizing the result of this section, we have proved the following theorem.

Theorem (7.2.4) Let M^2 be a torus embedded in E^3 as a tube of constant circular cross-section. More precisely, let the embedded surface be formed by carrying a (small) circle round a closed space curve C (for which $\kappa \neq 0$) such that the centre moves along the curve and the plane of the circle is in the normal plane of the curve at each point. Then

$$\int H^2 \, dA \geq 2\pi^2,$$

equality holding if and only if the generating curve is a circle and the ratio of the radii is $1/\sqrt{2}$.

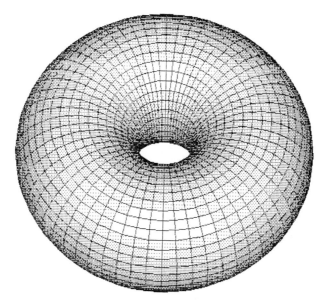

Fig. 7.1 Torus of revolution with $\mathcal{W} = 2\pi^2$.

This result adds weight to the conjecture that $2\pi^2$ is the infimum value for any torus. The theorem was obtained independently by Shiohama and Takagi (1970), and by Willmore (1971).

7.3 CONFORMAL INVARIANCE

We now prove a result explicitly due to White (1973). However the result was already known to Blaschke (1929).

Theorem (7.3.1) Let $f: M^2 \to E^3$ be a smooth immersion of a compact orientable surface into Euclidean 3-space. Let

$$\mathcal{W}(M^2, f) = \int_{M^2} H^2 \, dS. \tag{7.9}$$

Then $\mathcal{W}(M^2, f)$ is invariant under conformal transformations of E^3.

To prove the theorem we make use of the fact that any conformal transformation of E^3 can be decomposed into a product of similarity transformations and inversions. Clearly (7.9) is invariant under similarity transformations so it will be sufficient to consider invariance under inversions.

We take the centre of inversion to be the origin of E^3 which we can assume does not lie on $f(M^2)$. To a point x on $f(M^2)$ we associate the inverse point $\bar{\mathbf{x}}$ where

$$\bar{\mathbf{x}} = c^2 \frac{\mathbf{x}}{r^2} \quad \text{and} \quad r^2 = \langle \mathbf{x}, \mathbf{x} \rangle. \tag{7.10}$$

Then

$$d\bar{\mathbf{x}} = c^2 \frac{d\mathbf{x}}{r^2} - 2c^2 dr \frac{\mathbf{x}}{r^3}, \tag{7.11}$$

with

$$r\,dr = \langle \mathbf{x}, d\mathbf{x} \rangle \tag{7.12}$$

Hence

$$\langle d\bar{\mathbf{x}}, d\bar{\mathbf{x}} \rangle = c^4 \frac{\langle d\mathbf{x}, d\mathbf{x} \rangle}{r^4} - 4c^4 \frac{\langle d\mathbf{x}, dr\,\mathbf{x} \rangle}{r^5} + 4c^4 r^2 \frac{dr^2}{r^6}$$

$$= c^4 \frac{\langle d\mathbf{x}, d\mathbf{x} \rangle}{r^4}, \tag{7.13}$$

since the last two terms cancel.

Let $dS, d\bar{S}$ denote the corresponding volume elements of M^2 and $\bar{M}^2$. (Here for simplicity we have identified M^2 with $f(M^2)$.) Then

7.3 Conformal invariance

$$d\bar{S} = \frac{c^4}{r^4} dS. \tag{7.14}$$

Let e_3 be the unit normal vector to M^2 at x, and let $\bar{e}_3$ be the corresponding normal to $\bar{M}^2$ at $\bar{x}$. We claim that

$$\bar{e}_3 = \frac{2}{r^2}\langle x, e_3\rangle x - e_3 \tag{7.15}$$

is a unit vector perpendicular to $d\bar{x}$. To prove this we have

$$\langle \bar{e}_3, d\bar{x}\rangle = \left\langle \frac{2}{r^2}\langle x, e_3\rangle x - e_3, \frac{c^2}{r^2} dx - \frac{2c^2}{r^3} dr\, x \right\rangle$$

$$= \frac{2}{r^3}\langle x, e_3\rangle\, dr - \frac{4c^2}{r^3}\langle x, e_3\rangle\, dr + \frac{2c^2}{r^3}\langle x, e_3\rangle\, dr$$

$$= 0.$$

Also

$$\langle \bar{e}_3, \bar{e}_3\rangle = \left\{\frac{2}{r^2}\langle x, e_3\rangle - e_3\right\}^2 = \frac{4}{r^2}\langle x, e_3\rangle\langle x, e_3\rangle$$

$$+ 1 - \frac{4}{r^2}\langle x, e_3\rangle\langle x, e_3\rangle = 1,$$

so $\bar{e}_3$ really is a unit normal vector to $\bar{M}$. We now calculate the second fundamental form of $\bar{M}$. We have

$$d\bar{e}_3 = -\frac{4}{r^3} dr\langle x, e_3\rangle x + \frac{2}{r^2}\langle x, de_3\rangle x + \frac{2}{r^2}\langle x, e_3\rangle\, dx - de_3, \tag{7.16}$$

$$d\bar{x} = \frac{c^2}{r^2} dx - \frac{2c^2}{r^3} dr\, x. \tag{7.17}$$

The expression $d\bar{e}_3 \cdot d\bar{x}$ consists of eight terms. Four of these are

$$-\frac{4c^2}{r^4} dr^2\langle x, e_3\rangle + \frac{2c^2}{r^3}\langle x, de_3\rangle\, dr + \frac{2c^2}{r^4}\langle x, e_3\rangle\langle dx, dx\rangle - \frac{c^2}{r^2}\langle dx, de_3\rangle.$$

The remaining four terms are

$$\frac{8c^2}{r^4} dr^2\langle x, e_3\rangle - \frac{4c^2}{r^3}\langle x, de_3\rangle\, dr - \frac{4c^2}{r^4}\langle x, e_3\rangle\, dr^2 + \frac{2c^2}{r^2} dr\langle x, de_3\rangle,$$

and the sum simplifies to

$$\frac{2c^2}{r^4}\langle x, e_3\rangle - \frac{c^2}{r^4}\langle dx, de_3\rangle.$$

Thus the second fundamental form is given by

$$-d\bar{\mathbf{e}}_3 \cdot d\bar{\mathbf{x}} = \frac{c^2}{r^2}\langle d\mathbf{x}, d\mathbf{e}_3\rangle - \frac{2c^2}{r^4}\langle \mathbf{x}, \mathbf{e}_3\rangle\langle d\mathbf{x}, d\mathbf{x}\rangle. \tag{7.18}$$

Using (7.13) we get

$$-\frac{c^2}{r^2}\kappa_i\langle d\mathbf{x}, d\mathbf{x}\rangle - \frac{2c^2}{r^4}\langle \mathbf{x}, \mathbf{e}_3\rangle\langle d\mathbf{x}, d\mathbf{x}\rangle = \bar{\kappa}_i\frac{c^4}{r^4}\langle d\mathbf{x}, d\mathbf{x}\rangle, \tag{7.19}$$

so that

$$\bar{\kappa}_i = -\frac{r^2}{c^2}\kappa_i - \frac{2}{c^2}\langle \mathbf{x}, \mathbf{e}_3\rangle, \qquad i = 1, 2. \tag{7.20}$$

Thus

$$\bar{\kappa}_1 - \bar{\kappa}_2 = -\frac{r^2}{c^2}(\kappa_1 - \kappa_2), \tag{7.21}$$

$$\bar{\kappa}_1^2 + \bar{\kappa}_2^2 - 2\bar{\kappa}_1\bar{\kappa}_2 = \frac{r^4}{c^4}(\kappa_1^2 + \kappa_2^2 - 2\kappa_1\kappa_2), \tag{7.22}$$

$$\bar{H}^2 - \bar{K} = \frac{r^4}{c^4}(H^2 - K), \tag{7.23}$$

$$(\bar{H}^2 - \bar{K})\,d\bar{S} = (H^2 - K)\,ds, \tag{7.24}$$

$$\int \bar{H}^2\,d\bar{S} - 2\pi\chi(M^2) = \int H^2\,dS - 2\pi\chi(M^2). \tag{7.25}$$

It follows that $\int H^2\,dS$ is a conformal invariant.

We may note that a similar proof applies when M^2 is immersed in Euclidean space E^{2+p} where $p \geq 2$. Instead of the uniquely determined vector $\mathbf{e}_3$, we take any p mutually orthogonal unit normal vectors to M^2, $\mathbf{e}_\alpha$, $\alpha = 3,\ldots,p+2$. This time we find

$$\bar{\mathbf{e}}_\alpha = \frac{2}{r^2}\langle \mathbf{x}, \mathbf{e}_\alpha\rangle\mathbf{x} - \mathbf{e}_\alpha, \tag{7.26}$$

and

$$-d\bar{\mathbf{e}}_\alpha\,d\bar{\mathbf{x}} = \frac{c^2}{r^2}\langle d\mathbf{x}, \mathbf{e}_\alpha\rangle - \frac{2c^2}{r^4}\langle \mathbf{x}, \mathbf{e}_\alpha\rangle\langle d\mathbf{x}, d\mathbf{x}\rangle. \tag{7.27}$$

Now let $\mathbf{e}$ be any unit vector in the normal space to M^2 at $\mathbf{x}$. Then, just as before, we find that the principal curvatures satisfy

$$(\bar{\kappa}_1(\bar{\mathbf{e}}) + \bar{\kappa}_2(\bar{\mathbf{e}}))^2 - 4\bar{\kappa}_1(\bar{\mathbf{e}})\bar{\kappa}_2(\bar{\mathbf{e}}) = \frac{r^4}{c^4}\{(\kappa_1(\mathbf{e}) + \kappa_2(\mathbf{e}))^2 - 4\kappa_1(\mathbf{e})\kappa_2(\mathbf{e})\}. \tag{7.28}$$

We now average over the unit normal bundles of M^2 and $\bar{M}^2$ to get

$$\bar{H}^2 - \bar{K} = \frac{r^4}{c^4}(H^2 - K) \tag{7.29}$$

as before.

It follows that $\int H^2 \, dS$ is a conformal invariant for immersions of M^2 in E^{2+p}. This was first proved explicitly by Chen (1973a).

7.4 THE EULER EQUATION

One method of considering the infimum of $\int H^2 \, dS$ over the space of immersions of M^2 into E^3 is to apply standard techniques of the calculus of variations. Let M^2 be a compact oriented surface in E^3 given by the vector-valued function $\mathbf{x}(u_1, u_2)$. The outer unit normal $\mathbf{N}$ at $\mathbf{x}$ is given by $\mathbf{x}_1 \times \mathbf{x}_2 / |\mathbf{x}_1 \times \mathbf{x}_2|$, where $\mathbf{x}_i = \partial \mathbf{x}/\partial u^i$, $i = 1, 2$. Let $g_{ij} = \langle \mathbf{x}_i, \mathbf{x}_j \rangle$. Then the first fundamental form on M^2 is given by

$$I = \sum_{i,j} g_{ij} \, du^i \, du^j. \tag{7.30}$$

The second fundamental form is given by

$$II = -\langle d\mathbf{N}, d\mathbf{x} \rangle = \sum_{i,j} h_{ij} \, du^i \, du^j, \tag{7.31}$$

where

$$h_{ij} = -\langle \mathbf{N}_i, \mathbf{x}_j \rangle = h_{ji}, \quad \mathbf{N}_i = \frac{\partial \mathbf{N}}{\partial u^i}. \tag{7.32}$$

If we denote the inverse matrix of (g_{ij}) by (g^{ij}) then the mean curvature vector is given by

$$\mathbf{H} = \frac{1}{2} \sum_{i,j} g^{ij} h_{ij} \mathbf{N}. \tag{7.33}$$

The volume element of the surface is given by

$$dS = \sqrt{(\det(g_{ij}))} \, du^1 \wedge du^2. \tag{7.34}$$

The equations of Gauss may be written

$$\mathbf{x}_{ij} = \sum \Gamma^k_{ij} \mathbf{x}_k + h_{ij} \mathbf{N} \tag{7.35}$$

where $\mathbf{x}_{ij} = \partial^2 \mathbf{x}/\partial u^i \partial u^j$ and the Γ^k_{ij} are Christoffel symbols given by

$$\Gamma^k_{ij} = \tfrac{1}{2} \sum g^{hk} \left(\frac{\partial g_{ih}}{\partial u^j} + \frac{\partial g_{ij}}{\partial u^k} \right). \tag{7.36}$$

The Weingarten equations are

$$\mathbf{N}_i = -\sum h^j_i \mathbf{x}_j \tag{7.37}$$

where $h^j_i = g^{jk} h_{ki}$.

Consider now a normal variation of the immersion given by

$$\bar{\mathbf{x}}(u^1, u^2, t) = \mathbf{x}(u^1, u^2) + t\phi(u^1, u^2)\mathbf{N} \tag{7.38}$$

where ϕ is a smooth real-valued function and t is a real number such that $-\tfrac{1}{2} < t < \tfrac{1}{2}$. We denote by δ the operator $\partial/\partial t|_{t=0}$. According to the literature we say that M is a Willmore surface if

$$\delta \int H^2 \, dS = 0. \tag{7.39}$$

We have $\delta \mathbf{x} = \phi \mathbf{N}$, $\delta \mathbf{x}_i = \phi_i \mathbf{N} + \phi \mathbf{N}_i$;

$$\bar{g}_{ij} = \langle \mathbf{x}_i + t\phi_i \mathbf{N} + t\phi \mathbf{N}_i, \mathbf{x}_j + t\phi_j \mathbf{N} + t\phi \mathbf{N}_j \rangle$$
$$= g_{ij} + t\phi \langle \mathbf{x}_i, \mathbf{N}_j \rangle + t^2 \phi_i \phi_j + t\phi \langle \mathbf{N}_i, \mathbf{x}_j \rangle + t^2 \langle \mathbf{N}_i, \mathbf{N}_j \rangle.$$

Thus $\delta g_{ij} = -\phi \Sigma g_{ik} h^k_j - \phi \Sigma g_{jk} h^k_i = -2\phi h_{ij}$.

Using the relation $\Sigma g_{ij} g^{jk} = \delta^k_i$ we have

$$\sum \delta g_{ij} g^{jk} + \sum g_{ij} \delta g^{jk} = 0.$$

Hence

$$\delta g^{ij} = 2\phi \sum g^{jk} h^i_k.$$

Let $W = \sqrt{(\det(g_{ij}))}$, so $W^2 = \det(g_{ij})$. Then

$$2W \frac{\partial W}{\partial t} = \sum \frac{\partial g_{ij}}{\partial t} W^2 g^{ij}.$$

Hence $\delta W = \Sigma \tfrac{1}{2} \delta g_{ij} W g^{ij} = \Sigma \tfrac{1}{2} \cdot -2\phi W h_{ij} g^{ij}$, so $\delta W = -2\phi H W$.

We now compute $\delta \mathbf{N}$. Since $\langle \mathbf{N}, \mathbf{x}_i \rangle = 0$ we have $\langle \delta \mathbf{N}, \mathbf{x}_i \rangle + \langle \mathbf{N}, \delta \mathbf{x}_i \rangle = 0$ and hence

$$\langle \delta \mathbf{N}, \mathbf{x}_i \rangle = -\langle \mathbf{N}, \phi_i \mathbf{N} + \phi \mathbf{N}_i \rangle = -\phi_i.$$

Since $\langle \mathbf{N}, \mathbf{N} \rangle = 1$ we have $\langle \delta \mathbf{N}, \mathbf{N} \rangle = 0$. We write $\delta \mathbf{N} = \Sigma b^j \mathbf{x}_j$. Then $b^i = -g^{ij} \phi_j$ and hence $\delta \mathbf{N} = -g^{ij} \phi_j \mathbf{x}_i$.

From the Gauss equation we have

$$\langle \delta \mathbf{N}, \mathbf{x}_{ij} \rangle = -\langle g^{pq} \phi_q \mathbf{x}_p, \Gamma^k_{ij} \mathbf{x}_k \rangle = -\sum \phi_k \Gamma^k_{ij}.$$

7.4 The Euler equation

We now wish to obtain the formula

$$\langle \mathbf{N}, \delta \mathbf{x}_{ij} \rangle = \phi_{ij} - \phi \sum h_i^k h_{jk}.$$

We have

$$\bar{\mathbf{x}} = \mathbf{x} + t\phi \mathbf{N},$$
$$\bar{\mathbf{x}}_i = \mathbf{x}_i + t(\phi_i \mathbf{N} + \phi \mathbf{N}_i),$$
$$\bar{\mathbf{x}}_{ij} = \mathbf{x}_{ij} + t(\phi_{ij}\mathbf{N} + \phi_i \mathbf{N}_j + \phi_j \mathbf{N}_i + \phi \mathbf{N}_{ij}).$$

Hence

$$\delta \mathbf{x}_{ij} = \phi_{ij}\mathbf{N} + \phi_i \mathbf{N}_j + \phi_j \mathbf{N}_i + \phi \mathbf{N}_{ij},$$

giving

$$\langle \mathbf{N}, \delta \mathbf{x}_{ij} \rangle = \phi_{ij} + \phi \langle \mathbf{N}, \mathbf{N}_{ij} \rangle = \phi_{ij} - \phi \langle \mathbf{N}_i, \mathbf{N}_j \rangle$$
$$= \phi_{ij} - \phi \langle h_i^p \mathbf{x}_p, h_j^q \mathbf{x}_q \rangle = \phi_{ij} - \phi h_{ik} h_j^k,$$

as required. We have $\langle \mathbf{N}, \mathbf{x}_{ij} \rangle = h_{ij}$. Hence

$$\delta h_{ij} = \langle \delta \mathbf{N}, \mathbf{x}_{ij} \rangle + \langle \mathbf{N}, \delta \mathbf{x}_{ij} \rangle = -\phi_k \Gamma_{ij}^k + \phi_{ij} - \phi h_i^k h_{jk},$$
$$\delta h_{ij} = \nabla_i \nabla_j \phi - \phi h_i^k h_{jk}.$$

We can now (at last!) apply δ to our integral. We have

$$\delta H = \delta \left(\frac{1}{2} g^{ij} h_{ij} \right) = \frac{1}{2} \left(\delta g^{ij} \right) h_{ij} + \frac{1}{2} g^{ij} \delta h_{ij}$$
$$= \frac{1}{2} \cdot 2\phi g^{jk} h_k^i h_{ij} + \frac{1}{2} g^{ij}(\nabla_i \nabla_j \phi - \phi h_i^k h_{kj})$$
$$= \frac{1}{2}(\Delta \phi + \phi h_k^i h_i^k).$$

The matrix (h_k^i) has as eigenvalues the principal curvatures κ_1, κ_2. Hence $h_k^i h_i^k = \text{trace } h^2 = \kappa_1^2 + \kappa_2^2 = 4H^2 - 2K$. Thus we get

$$2\delta H = \Delta \phi + \phi(4H^2 - 2K).$$

Hence

$$\delta \int H^2 \, dS = \int 2H \delta H \, dS + \int H^2 \delta \, dS$$
$$= \int H\{\Delta \phi + \phi(4H^2 - 2K)\} \, dS - 2\int H^2 \phi H \, dS.$$

Since M^2 is a closed surface, it follows from Green's theorem that

$$\int H\Delta\phi \, dS = \int \phi \Delta H \, dS.$$

Thus we get

$$\delta \int H^2 \, dS = \int \phi(\Delta H + 2H(H^2 - K)) \, dS.$$

Thus the condition that the integral is stationary for all smooth functions ϕ is

$$\Delta H + 2H(H^2 - K) = 0. \tag{7.40}$$

Equation (7.40) is now taken as the defining property of a Willmore surface, whether or not the surface is compact or even orientable.

The history of equation (7.40) is interesting. I lectured on this topic at Oberwolfach in the early 1960s and thought that the equation was new. However Konrad Voss told me then that he had already obtained this equation in the 1950s but he had never published it. Both Voss and I were amused to find subsequently that this precise equation is mentioned specifically in Blaschke's volume III in 1929, where it is attributed to Thomsen and Schadow in 1923.

The formula was generalized by Chen (1973b). Chen considered an m-dimensional closed oriented hypersurface immersed in Euclidean space E^{m+1}. Let α denote the mean curvature. Then Chen proved the inequality

$$\int_{M^m} |\alpha|^m \, dV \geqslant C_m \tag{7.41}$$

where C_m denotes the surface area of the unit m-sphere in E^{m+1}. He also proved that the variational problem associated with this integral has Euler equation

$$\Delta \alpha^m + m(m-1)\alpha^{m+1} - \alpha^{m-1}\tau = 0, \tag{7.42}$$

where the scalar curvature τ satisfies

$$\tau = m^2 \alpha^2 - h^i_j h^j_i. \tag{7.43}$$

Equation (7.42) reduces to our equation (7.40) when $m = 2$. Then $\tau = 2K$. The method used by Chen is very similar to that given here for the case $m = 2$. However, although his choice of generalizing the surface integral of H^2 leads to a nice Euler equation, the integral is no longer a conformal invariant except when $m = 2$. We shall see later that conformal invariance is an essential property of an integrand which leads to a suitable definition of a general Willmore submanifold of an arbitrary Riemannian manifold.

7.5 WILLMORE SURFACES AND MINIMAL SUBMANIFOLDS OF S^3

An important paper by Weiner (1978) considered immersions of an orientable surface, with or without boundary, into a Riemannian manifold $\bar{M}$ of constant sectional curvature $\bar{K}$. Instead of considering $\int H^2 \, dA$ he considered the integral

$$\int (|H|^2 + \bar{K}) \, dA. \tag{7.44}$$

He used an argument similar to that in Section 7.4 but using covariant differentiation with respect to the Riemannian manifold. In the particular case when $\bar{M}$ is $S^3(1)$, the unit sphere contained in $\mathbb{R}^4$, the stationary values of the integral occur when the following Euler equation is satisfied:

$$\Delta H - 2H^3 + HS = 0 \tag{7.45}$$

where S is the square of the length of the second fundamental form A. An alternative form of the Euler equation is

$$\Delta H + 2H^3 - 2HG = 0 \tag{7.46}$$

where $G = \det(A)$. In particular for immersions in E^3 we have $G = K$ and the equation reduces to

$$\Delta H + 2H(H^2 - K) = 0 \tag{7.47}$$

as we have found previously.

Weiner's paper is more general than earlier papers because it applies to surfaces with a boundary, and moreover the receiving space is an n-dimensional space-form which includes $\mathbb{R}^n$ as a very special case. From our point of view, equation (7.45) shows that a closed surface immersed in S^3 as a minimal surface gives a stationary value to the integral.

Let $f: M^2 \to S^3$ be an immersion of the closed orientable surface M^2 into the unit sphere $S^3(1)$. Let $\sigma: S^3(1) \setminus \text{point} \to \mathbb{R}^3$ be a stereographic projection of the sphere onto $\mathbb{R}^3$. Since the integral is a conformal invariant it follows that f satisfies (7.46) on S^3 if and only if $\sigma \circ f$ satisfies (7.47) in $\mathbb{R}^3$.

Briefly we may say that stereographic projection of a minimal surface in S^3 gives rise to a Willmore surface in $\mathbb{R}^3$. Now it was proved by Lawson (1970) that there are minimal embeddings into S^3 of surfaces of arbitrary genus. From this we deduce that there exist in $\mathbb{R}^3$ Willmore surfaces of arbitrary genus.

It is interesting to note that the Clifford torus given as a submanifold of $S^3(1)$ contained in $\mathbb{R}^4$ is embedded as a minimal surface.

Stereographic projection of this minimal surface in $\mathbb{R}^3$ gives rise to the '$\sqrt{2}$ anchor ring' mentioned in Section 7.3.

A deeper result in Weiner's paper is that the '$\sqrt{2}$ anchor ring' and surfaces differing from this by a conformal transformation of $\mathbb{R}^3$ form a *stable* Willmore surface with respect to the second-order variation of the integral.

A natural question arises: Are all Willmore surfaces in $\mathbb{R}^3$ obtainable as stereographic projections of minimal surfaces in S^3? We shall see later, thanks to the work of Ulrich Pinkall, that the answer is 'No!'.

7.6 PINKALL'S INEQUALITY

In his paper *Inequalities of Willmore Type for Submanifolds*, Max Planck Institut für Mathematik, 85-24, subsequently published as Pinkall (1986), Pinkall expressed the inequality

$$\int H^2 \, dA \geq 4\pi$$

in the form

$$C(f) \geq \beta_1 \qquad (7.48)$$

where β_1 is the first Z_2-Betti number of M^2 and

$$C(f) = \frac{1}{2\pi} \int (H^2 - K) \, dA.$$

This follows trivially since

$$\begin{aligned} C(f) &= \frac{1}{2\pi} \int H^2 \, dA - \chi(M^2) \\ &= \frac{1}{2\pi} \int H^2 \, dA - (1 - \beta_1 + 1) \\ &\geq 2 - (2 - \beta_1) = \beta_1. \end{aligned}$$

In his paper Pinkall defines $C(f)$ for all immersions $f: M^n \to \mathbb{R}^m$ where M^n is an arbitrary compact manifold, and he obtains a generalization of (7.48).

Let M^n be a compact smooth manifold, $f: M^n \to \mathbb{R}^m$ an immersion, and $N(f)$ the unit normal bundle of f. Corresponding to each $\xi \in N_p(f)$ there is a shape operator $A_\xi: M_p^n \to M_p^n$ whose eigenvalues $\kappa_1(\xi), \ldots, \kappa_n(\xi)$ are the principal curvatures at ξ. Let $\sigma(A_\xi)$ denote the dispersion of the principal curvatures in the sense of probability theory:

7.6 Pinkall's inequality

$$[\sigma(A_\xi)]^2 = \frac{1}{n^2}\sum_{i<j}(\kappa_i - \kappa_j)^2. \tag{7.49}$$

Then the total conformal curvature $C(f)$ of the immersion f is defined by

$$C(f) = \frac{1}{\operatorname{vol}(S^m(1))}\cdot\int_{N(f)}[\sigma(A_\xi)]^n\,d\xi. \tag{7.50}$$

Here $d\xi$ denotes the natural volume element on $N(f)$. As pointed out by Pinkall, the functional $C(f)$ has the following remarkable properties:

(1) $n = 2 \Rightarrow C(f) = \frac{1}{2\pi}\int_{M^2}(|H|^2 - K)\,dA$;
(2) $C(f) \geq 0$, $C(f) = 0 \Leftrightarrow f$ is totally umbilical;
(3) if $i: \mathbb{R}^m \to \mathbb{R}^p, p \geq m$ denotes the canonical inclusion, then $C(i\circ f) = C(f)$;
(4) if $\sigma: \mathbb{R}^m \cup \{\infty\} \to \mathbb{R}^m \cup \{\infty\}$ is conformal, then $C(\phi\circ f) = C(f)$.

The proof of (4) is straightforward – see, for example, Abe (1982).

In his paper Pinkall proves the following inequality.

Theorem (7.6.1) *Let M^n be compact, F a field, $\beta_1,\ldots,\beta_n$ the Betti numbers of M^n with respect to F, and $f: M^n \to \mathbb{R}^n$ an immersion. Then*

$$C(f) \geq \sum_{k=1}^{n-1} a_k\beta_k \tag{7.51}$$

where

$$a_k = [k/(n-k)]^{\frac{n}{2}-k}.$$

To prove the theorem, we observe that a study of the Gauss normal map $N(f) \to S^{m-1}(1) \subset \mathbb{R}^m$ shows that

$$\int_{N_k}|\det A_\xi|\,d\xi \geq \beta_k \operatorname{vol}(S^{m-1}(1)), \tag{7.52}$$

where $N_k = \{N(f)$ has exactly k negative eigenvalues$\}$. The underlying reason for this is as follows. Let p be any point on $S^{m-1}(1)$. We wish to calculate the number of points of $N(f)$ which map to p under the Gauss map. These are readily seen to be the points on M where the normal unit vector is parallel to the radial vector through p, that is, the points where the tangent plane to M is perpendicular to the radial vector, that is, where $\langle\xi, df(x)\rangle = 0$. But the points $x \in M$ where this is satisfied are precisely the critical points of the function $\langle\xi, f(x)\rangle$. From

286   7 Special Riemannian submanifolds

Morse theory the number of such points is equal to β_k. Actually we can prove that the set of such critical points is non-degenerate except for a set of measure zero which contributes zero to the integral. This explains equation (7.52).

However we also have

$$C(f) = \frac{1}{\text{vol}(S^{m-1}(1))} \sum_{k=0}^{n} \int_{N_k} [\sigma(A_\xi)]^n \, d\xi. \tag{7.53}$$

So the theorem follows if we can prove the following lemma.

Lemma (7.6.2) *Let $\kappa_1, \ldots, \kappa_n$ be real numbers, let r be an integer such that $0 < r < n$, and $\kappa_1, \ldots, \kappa_r < 0$; $\kappa_{r+1}, \ldots, \kappa_n \geq 0$. Then*

$$\left[\frac{1}{n^2} \sum_{i<j} (\kappa_i - \kappa_j)^2\right]^{\frac{n}{2}} \geq a_r |\kappa_1, \ldots, \kappa_n| \text{ where } a_r = \left(\frac{r}{n-r}\right)^{\frac{n}{2} - r}.$$

The following proof of this lemma is due to P. J. Higgins. I think this is an improvement of the proof given by Pinkall and I thank him for his permission to use it here. We first prove a preliminary result.

Lemma (7.6.3) *Let $\kappa_1, \ldots, \kappa_n$ be arbitrary real numbers, let $r + s = n$, and let a be the arithmetic mean of $\kappa_1, \ldots, \kappa_r$ and let b be the arithmetic mean of $\kappa_{r+1}, \ldots, \kappa_n$. Then*

$$\sum_{i<j} (\kappa_i - \kappa_j)^2 \geq rs(a-b)^2.$$

Note that the left-hand side is equal to the right-hand side when the first r κ's are each equal to a, and the remaining s κ's are each equal to b. To prove the lemma put $\kappa_i = a + x_i (i \leq r)$, $\kappa_i = b + x_i (i > r)$. Then $\sum_1^r x_i = 0$ and $\sum_{r+1}^n x_i = 0$. Hence

$$\sum_{i<j} (\kappa_i - \kappa_j)^2 = \sum_{\substack{1 \leq r < j}} [(a-b) + (x_i - x_j)]^2 + \sum_{i<j \leq r} (x_i - x_j)^2$$

$$+ \sum_{r<i<j} (x_i - x_j)^2$$

$$= rs(a-b)^2 + 2(a-b) \sum_{i \leq r < j} (x_i - x_j) + \sum_{i<j} (x_i - x_j)^2.$$

The second term vanishes so we have

$$\sum_{i<j} (\kappa_i - \kappa_j)^2 \geq rs(a-b)^2.$$

as required.

Application of Lemma (7.6.3) to the proof Lemma (7.6.2)

Suppose now that $\kappa_1, \ldots, \kappa_r$ are negative and $\kappa_{r+1}, \ldots, \kappa_n$ are positive. Then $a < 0$ and $b > 0$. Consider now the set of positive numbers

$$\frac{|\kappa_1|}{r}, \frac{|\kappa_2|}{r}, \ldots, \frac{|\kappa_r|}{r}, \frac{|\kappa_{r+1}|}{s}, \ldots, \frac{|\kappa_n|}{s}.$$

Their arithmetic mean is $(b-a)/n$ and their geometric mean is

$$\left(\frac{|\kappa_1 \cdots \kappa_n|}{r^r s^s}\right)^{\frac{1}{n}}.$$

So we get

$$\left[\frac{1}{n^2}\sum_{i<j}(\kappa_i - \kappa_j)^2\right]^{\frac{n}{2}} \geq (rs)^{\frac{r+s}{2}}\left(\frac{b-a}{n}\right)^n \quad \text{(from Lemma (7.6.3))}$$

$$\geq \frac{(rs)^{\frac{r+s}{2}}}{r^r s^s}|\kappa_1 \cdots \kappa_n| \quad \text{(from the standard arithmetic/geometric inequality),}$$

$$= \left(\frac{r}{s}\right)^{\frac{s-r}{2}}|\kappa_1 \cdots \kappa_n|$$

and this proves the lemma and also Pinkall's inequality.

Another surprising result from this paper is the following. For $n \geq 2$ let

$$C(n) := \inf\{C(M^n)\,|\,M^n \text{ not homeomorphic to } S^n\}.$$

Theorem (7.6.4) $C(2) = C(\mathbb{R}P^2) = 2$.

That is to say that the real projective plane is the unique surface (non-orientable!), other than S^2, for which $C(M^2)$ attains its infimum value, namely 2. The proof follows immediately from a lemma in an important paper by Li and Yau (1982). We shall refer to this paper in some detail later in the chapter.

7.7 RELATED WORK BY ROBERT L. BRYANT

An important contribution to the subject appeared from Bryant (1984). This paper will repay detailed study but the reader is warned that it is not suitable for bed-time reading!

Bryant uses conformal invariance from the beginning. He obtains the structure equations for S^3 considered as conformal 3-space. Then, by the method of moving frames, he studies immersed surfaces

$X: M^2 \to S^3$. From this he constructs a conformally invariant 2-form Ω_X on M^2 and shows that for any stereographic projection $\sigma: S^3 \setminus \{y_0\} \to \mathbb{R}^3$, the equation

$$\Omega_X = (H^2 - K)\, dA \tag{7.54}$$

holds, where the quantities on the right are computed for the immersion $\sigma \circ X: M^2 \to \mathbb{R}^3$. In this way he demonstrates the conformal invariance of the Willmore integrand, and the conformal invariance of the umbilic locus $\mathfrak{U}_X = \{m \in M \mid \Omega_X(m) = 0\}$.

He then constructs on the complement of $\mathfrak{U}_X$ a smooth map $X^*: M - \mathfrak{U}_X \to S^3$ with the defining property that if $m_0 \notin \mathfrak{U}_X$, then $X^*(m_0)$ is the unique point such that the mean curvature of $\sigma \circ X$ vanishes to the second order at m_0 for any stereographic projection $\sigma: S^3 - \{X^*(m_0)\} \to \mathbb{R}^3$. He calls X^* the *conformal transform* of X.

He computes the Euler equation for the functional $\mathcal{W}$. The derivation is conformally invariant and this leads to a consideration of the complex structure on M^2 induced by the conformal structure and the choice of orientation on M^2.

Bryant defines $X: M^2 \to S^3$ to be a *Willmore immersion* if it is a critical point of the Willmore functional. He then relates a Willmore immersion to the complex structure by means of two theorems. The first theorem constructs a holomorphic quartic differential on M from the Willmore immersion X, denoted by Q_x. The second theorem states that if a Willmore immersion is not totally umbilical, then the conformal transform completes smoothly to a conformal branched immersion $X^*: M \to S^3$. If $Q_x \equiv 0$ then X^* is a constant map. If $Q_x \not\equiv 0$ then X^* is also a Willmore (branched) immersion and $(X^*)^* = X$. He defines X^* as the *Willinore dual* of X.

By choosing $M = S^2$ he guarantees that $Q_x \equiv 0$ [this follows from the Riemann–Roch theorem of algebraic geometry]. He thus obtains a complete classification of Willmore immersions $X: S^2 \to S^3$ in terms of a special family of minimal surfaces of finite total curvature in $\mathbb{R}^3$. Bryant then shows his skill as a geometer in reducing the problem to one of algebraic geometry concerning the zeros and poles of meromorphic functions on $\mathbb{C}P^1$, the complex projective line. One of the main results of his paper is the following theorem. Its understanding requires a knowledge of algebraic geometry from the standpoint of a differential geometer—a good reference is the book by Griffiths and Harris (1978).

Theorem (7.7.1) Let $X: S^2 \to S^3$ be a $\mathcal{W}$-critical immersion. Endow S^2 with the induced conformal structure. Then there exists a point $y_0 \in X(S^2) \subseteq S^3$ (unique if X is not totally umbilic) so that $D = X^{-1}(y_0)$ is a divisor of S^2 with $d > 0$ distinct points, a stereographic projection $\sigma: S^3 - \{y_0\} \to \mathbb{R}^3$, and a meromorphic curve $f: S^2 \to \mathbb{C}^3$

7.7 Related work by Robert L. Bryant

whose polar divisor is D so that $\sigma \circ X = \text{Re}(f)$. Moreover, f is an immersion with null tangents (that is, $(df, df) = 0$). Conversely if $f: S^2 \to \mathbb{C}^3$ is a meromorphic immersion with simple poles along D and null tangents, then, regarding $\mathbb{R}^3 = S^3 - \{y_0\}$, we have $\text{Re}(f): S^2 - D \to \mathbb{R}^3$ completes as a smooth immersion across D to be a conformal $\mathcal{W}$-critical immersion $X: S^2 \to S^3$.

In this theorem S^2 is identified with $\mathbb{C}P^1$, the unique Riemann surface of genus 0. If z is a rationalizing parameter on S^2 with its pole at $\infty \in S^2$, then a meromorphic function $f: S^2 \to \mathbb{C}^3$ with simple poles at $z = \lambda_i (i = 1, \ldots, d)$ can be written

$$f(z) = v_0 + \frac{v_1}{z - \lambda_1} + \cdots + \frac{v_d}{z - \lambda_d} \qquad (7.55)$$

where $v_0, \ldots, v_d \in \mathbb{C}^3$.

There is no loss of generality in setting $v_0 = 0$ since this may be achieved by a conformal transformation of S^3. To satisfy the condition $(df, df) = 0$ requires the imposition of a large number of algebraic constraints on the complex parameters $\{\lambda_1, \ldots, \lambda_d; v_1, \ldots, v_d\}$. The map f is also required to be an immersion.

The special case $d = 1$ corresponds to the umbilical 2-spheres in S^3. Bryant shows that the cases $d = 2$ and $d = 3$ cannot occur. However the space of solutions for $d = 4$ is completely determined. In fact he shows that the moduli space of $\mathcal{W}$-critical immersions is non-empty for all even degrees except 2. In a pre-print dated 1986 M.S.R.I. 'Some new examples of minimal surfaces in $\mathbb{R}^3$ and its applications', C. K. Peng explicitly computed examples for all even degrees $\geqslant 4$ and for all odd degrees $d = 2k + 1 \geqslant 9$.

In a subsequent paper Bryant (1988) again shows geometric insight by using a 'twistor' formulation of the problem to replace $\mathcal{W}$-critical immersions $S^2 \to S^3$ by holomorphic rational null curves $C^1 \to Q^3$, the complex quadric. In this way he shows that there are no solutions corresponding to $d = 5$ or $d = 7$. This completes the classification.

In a similar manner he considers maps of $\mathbb{R}P^2$ into S^3, but in this case there arises a technical difficulty due to non-compactness of one of the spaces involved. He constructs an explicit immersion ϕ of $\mathbb{R}P^2$ into S^3 for which $\mathcal{W}(\phi) = 12\pi$, which, as we shall see in the next section, is the infimum value. However, in a letter recently received by I. Sterling, Bryant claims that he has overcome this difficulty and has now a complete classification of the $\mathcal{W}$-critical values for immersions of $\mathbb{R}P^3$ into S^3.

Figure 7.2 shows the sphere S^2 immersed as a Willmore surface in $\mathbb{R}^3$ with $d = 4$ so that the value of the integral is 16π. Figure 7.3 shows

290   7 Special Riemannian submanifolds

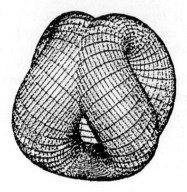

Fig. 7.2   Topological sphere as Willmore surface with $\mathcal{W} = 16\pi$.

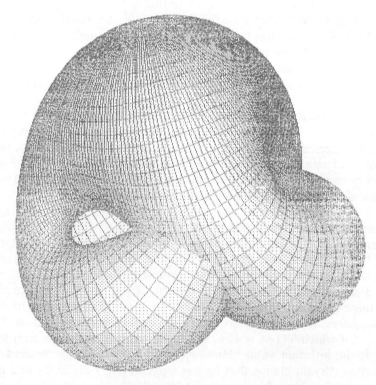

Fig. 7.3 $\mathbb{R}P^2$ as Willmore surface with $\mathcal{W} = 12\pi$.

7.8 The conformal invariants of Li and Yau, and of Gromov

$\mathbb{R}P^2$ immersed as a Willmore surface in $\mathbb{R}^3$, the value of the integral being 12π.

7.8 THE CONFORMAL INVARIANTS OF LI AND YAU, AND OF GROMOV

The concept of conformal volume was introduced by Li and Yau (1982). About the same time a related conformal invariant was considered by Gromov (1983). The following introduction to the Li–Yau invariant follows the description by Bryant.

Let G denote the conformal group acting on $S^n = \mathbb{R}^n + \infty$. The group G acts smoothly on S^n and is isomorphic to $SO(n+1,1)$. We regard S^n as the boundary of hyperbolic $(n+1)$-space $H^{n+1} = \{w \in \mathbb{R}^{n+1} \mid |w| < 1\}$ equipped with the hyperbolic metric

$$ds^2 = \frac{4|dw|^2}{(1-|w|^2)^2}. \tag{7.56}$$

Then the group G extends to an isometric action on H^{n+1}. For each $u \in H^{n+1}$, let $S_u \subset H_u^{n+1}$ denote the unit sphere in the tangent space. Then the exponential map $\exp_u: H_u^{n+1} \to H^{n+1}$ gives rise to a diffeomorphism $E_u: S_u \to S^n = \partial_\infty H^{n+1}$ by setting

$$E_u(v) = \lim_{t \to \infty} \exp_u(tv). \tag{7.57}$$

Now the map E_u is conformal and we deduce that there exists a conformal metric g_u on S^n for which E_u is an isometry. However, a standard result shows that the set $\{g_u \mid u \in H^{n+1}\}$ is the full set of conformal metrics on S^n with sectional curvature $+1$. Explicitly

$$g_u = \frac{(1 - u \cdot u)}{(1 - u \cdot x)^2} |dx|^2. \tag{7.58}$$

Now, for any immersion $\phi: M^k \to S^n = \mathbb{R}^n + \infty$ we can define a function A_ϕ on H^{n+1} by the formula

$$A_\phi(u) = \int_{M^k} d\,\mathrm{vol}_{g_u}(M^k) = \int_{M^k} \frac{(1 - u \cdot u)^{k/2}}{(1 - u \cdot \phi)^k} \Omega_0, \tag{7.59}$$

where Ω_0 is the volume form on M^k induced by $\phi: M^k \to S^n \subset \mathbb{R}^{n+1}$.

The *visual volume* of $\phi(M^k)$ at $u \in H^{n+1}$ was defined by Gromov to be given by

$$\frac{A_\phi(u)}{V(S^n)}. \tag{7.60}$$

292   7 *Special Riemannian submanifolds*

The *conformal volume* of $\phi(M^k)$ was defined by Li and Yau to be given by

$$V_c(\phi) = \sup \{A_\phi(u) | u \in H^{n+1}\}. \tag{7.61}$$

In their deep but rather difficult paper Li and Yau prove a fundamental inequality relating $\mathscr{W}(\phi)$ and $V_c(\phi)$ when $k = 2$, namely

$$\mathscr{W}(\phi) \geq V_c(\phi). \tag{7.62}$$

Moreover they show that for conformal structures near that of the '$\sqrt{2}$ torus' we have

$$V_c(\phi) = 2\pi^2. \tag{7.63}$$

This supports the Willmore conjecture that

$$\inf \int H^2 \, dA = 2\pi^2$$

for *any* torus, but proves only for a special class of tori that it is valid.

This fundamental inequality also yields another result. Since any immersion $\phi: \mathbb{R}P^2 \to \mathbb{R}^3 \subset S^3$ has a triple point, it follows that $V_c(\phi) \geq 12\pi$. However Bryant has constructed an explicit immersion

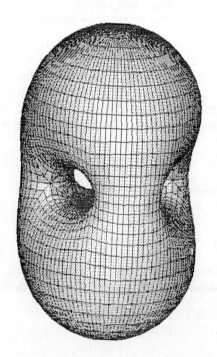

Fig. 7.4 Two-holed torus, genus 2.

7.9 Contributions from the German school

of $\mathbb{R}P^2$ into S^3 with $\mathcal{W}(\phi) = 12\pi$. It follows from the inequality that $V_c(\mathbb{R}P^2) = 12\pi$.

The value of $V_c(M^k)$ for a general M^k is unknown even in the case $k = 2$. For a general surface M^2 the moduli space of $\mathcal{W}$-critical immersions of M^2 into S^3 is also quite unknown.

7.9 CONTRIBUTIONS FROM THE GERMAN SCHOOL AND THEIR COLLEAGUES

We have already referred to the first paper by Pinkall (1986) on this topic. This was followed by a paper by Pinkall and Sterling, Max Planck Institut für Mathematik, 86-29. In this they remarked that as a result of Lawson's work on minimal surfaces in S^3, it followed that for any genus g there are compact orientable surfaces in R^3 for which

$$\int H^2 \, dA < 8\pi. \tag{7.64}$$

However, in the paper by Li and Yau, it was proved that for any immersed surface in $\mathbb{R}^3$ with self-intersection, one has

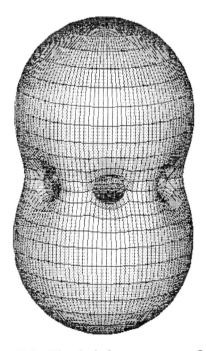

Fig. 7.5 Five-holed torus, genus 5.

$$\int H^2 \, dA \geq 8\pi. \tag{7.65}$$

Putting these two results together produces the result that all surfaces in R^3 of genus g which are absolute minima of $\mathcal{W}$ (if they exist!) are necessarily *embedded* as Willmore surfaces.

Examples of minimal surfaces in S^3 which when projected into $\mathbb{R}^3$ give Willmore surfaces are shown in Figs. 7.4 and 7.5. However new examples of compact embedded minimal surfaces in S^3 and hence embedded Willmore surfaces in $\mathbb{R}^3$ were discovered by Karcher *et al.* (1988). These are shown in Figs. 7.6–7.11. Numerical computation shows, however, that these new Willmore surfaces have a higher value of $\mathcal{W}$ than those obtained via the Lawson immersions. This suggests the probability that the Lawson examples do in fact produce examples of surfaces of genus g for which $\mathcal{W}$ takes absolute minimal values.

A significant breakthrough was made by Pinkall (1985). Following Pinkall, we identify S^3 with the set of unit quaternions $\{q \in \mathbb{H}, q\bar{q} = 1\}$ and S^2 with the unit sphere in the subspace of $\mathbb{H}$ spanned by 1, j, and k. Let $q \to \tilde{q}$ denote the anti-automorphism of $\mathbb{H}$ which fixes 1, j, and k but sends i to $-$i. Define the map $\pi \colon S^3 \to \mathbb{H}$ by

$$\pi(q) = \tilde{q}q. \tag{7.66}$$

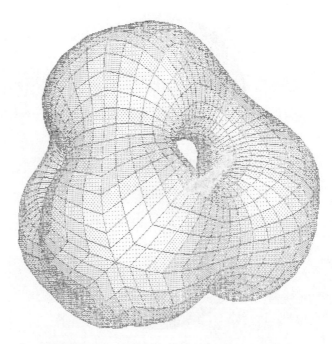

Fig. 7.6 Two tetrahedral bones, genus 3.

7.9 Contributions from the German school 295

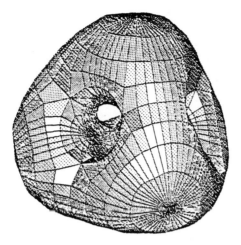

Fig. 7.7 Five tetrahedral bones, genus 6.

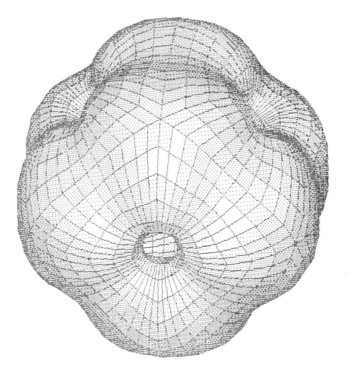

Fig. 7.8 Two cubical bones, genus 5.

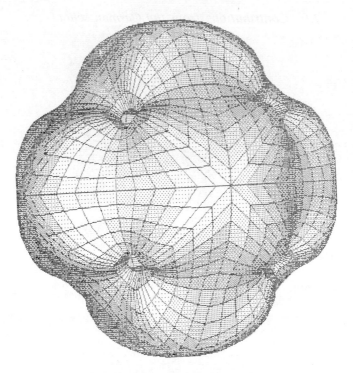

Fig. 7.9 Two octahedral bones, genus 7.

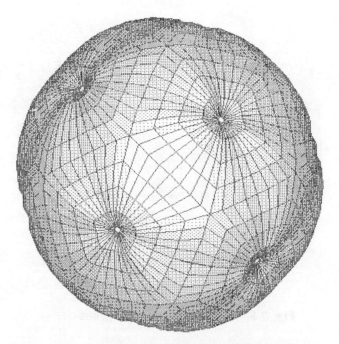

Fig. 7.10 Two dodecahedral bones, genus 11.

7.9 Contributions from the German school

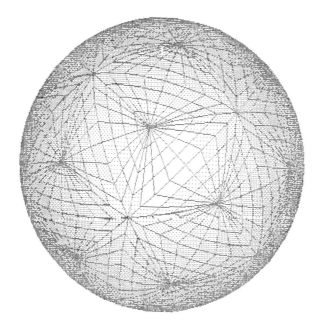

Fig. 7.11 Two icosahedral bones, genus 19.

Then it is easy to check that π has the following properties.

(1) $\pi(S^3) = S^2$.
(2) $\pi(e^{i\phi}q) = \pi(q)$ for all $q \in S^3$, $\phi \in \mathbb{R}$.
(3) The group S^3 acts isometrically on S^3 by right multiplication and on S^2 by

$$q \to \tilde{r}qr, \quad r \in S^3. \tag{7.67}$$

These actions are related as follows: for all $q, r \in S^3$ we have

$$\pi(qr) = \tilde{r}\pi(q)r.$$

Let $\mathfrak{p}: [a, b] \to S^2$ be a curve immersed in S^2. Choose $\mathfrak{y}: [a, b] \to S^3$ such that $\pi \circ \mathfrak{y} = \mathfrak{p}$. Then by regarding $S^1 = \mathbb{R}/2\pi\mathbb{Z}$ we define an immersion $\mathfrak{r}$ of the cylinder $[a, b] \times S^1$ into S^3 by

$$\mathfrak{r}(t, \phi) = e^{i\phi}\mathfrak{y}(t). \tag{7.68}$$

This surface is called the *Hopf cylinder* corresponding to the curve $\mathfrak{p}$. We assume that $\mathfrak{p}$ is parametrized according to arc length and that it cuts the fibres of π orthogonally. This implies that $\mathfrak{y}'(t)$ has unit norm for all t and is orthogonal to

298   7 *Special Riemannian submanifolds*

$$\mathfrak{x}_\phi(t,\phi) = ie^{i\phi}\mathfrak{y}(t), \tag{7.69}$$

where lower indices denote partial differentiation.

Since $\langle \mathfrak{y}(t), \mathfrak{y}'(t) \rangle = 0$, it follows that there is a function $u: [a,b] \to \text{span}(j,k)$ such that $|u| = 1$ and

$$\mathfrak{y}' = u\mathfrak{y}. \tag{7.70}$$

For the curve $\mathfrak{p} = \pi \circ \mathfrak{y}$ on S^2 we have

$$\mathfrak{p} = \tilde{\mathfrak{y}}\mathfrak{y}, \tag{7.71}$$

$$\mathfrak{p}' = 2\tilde{\mathfrak{y}}u\mathfrak{y}, \tag{7.72}$$

$$|\mathfrak{p}'| = 2. \tag{7.73}$$

Moreover

$$\mathfrak{x}_\phi = e^{i\phi}i\mathfrak{y}, \tag{7.74}$$

$$\mathfrak{x}_s = e^{i\phi}u\mathfrak{y}, \tag{7.75}$$

are unit vectors which are orthogonal. Equations (7.71), (7.72), and (7.73) imply the following conclusion.

Lemma (7.9.1) Let $\mathfrak{p}$ be a curvilinear arc on S^2 of length L. Then the corresponding Hopf cylinder is isometric to $[0, L/2] \times S^1$.

If $\mathfrak{p}$ is a closed curve so that $\mathfrak{p}(t + L/2) = \mathfrak{p}(t)$ for all t, then equation (7.68) defines a covering of the (t, ϕ)-plane onto an immersed torus. This torus will be called the *Hopf torus* corresponding to $\mathfrak{p}$. We now determine the mean curvature of a Hopf torus. From (7.68) and (7.70) we have

$$\mathfrak{x}_t(t,\phi) = e^{i\phi}u(t)\mathfrak{y} = u(t)e^{-i\phi}\mathfrak{y}(t). \tag{7.76}$$

Consider the vector field given by

$$\mathfrak{n}(t,\phi) = iu(t)e^{-i\phi}\mathfrak{y}(t). \tag{7.77}$$

From (7.74), (7.75), and (7.77) we see that $\mathfrak{n}$ is a normal vector field for the immersion. Moreover, it is a unit vector field. Taking derivatives of (7.77) gives

$$\mathfrak{n}_t = iu'(t)e^{-i\phi}\mathfrak{y}(t) + iu(t)e^{-i\phi}\mathfrak{y}'(t),$$

that is,

$$\begin{aligned} \mathfrak{n}_t &= -2\kappa\mathfrak{x}_t - \mathfrak{x}_\phi, \\ \mathfrak{n}_\phi &= -\mathfrak{x}_t, \end{aligned} \tag{7.78}$$

where we have written

$$u' = 2i\kappa u. \tag{7.79}$$

7.9 Contributions from the German school

Since

$$(\tilde{\mathfrak{y}}\mathfrak{y})' = 2\tilde{\mathfrak{y}}u\mathfrak{y} \tag{7.80}$$

$$(\tilde{\mathfrak{y}}\mathfrak{y})'' = 2\mathfrak{y}(u' - 2)\mathfrak{y}$$

it follows that $t \to \kappa(t)$ gives the curvature of the spherical curve $\mathfrak{p} = \tilde{\mathfrak{y}}\mathfrak{y}$. However, since the mean curvature of the surface $\mathfrak{r}$ is given by $-\frac{1}{2} \times$ trace of the matrix determined by the right-hand side of (7.78) we see that the mean curvature is also equal to κ. The Willmore functional is given for the surface in S^3 by

$$\mathcal{W}(M) = \int_F (1 + \kappa^2(t)) \, d\phi \, dt$$

where F is a fundamental region for the covering $x: \mathbb{R}^2 \to M$ and $ds = 2dt$ denotes arc length of the curve $\mathfrak{p}$ on S^2. Thus we have

$$\mathcal{W}(M) = \pi \int_0^L (1 + \kappa^2(s)) \, ds.$$

It follows that $\mathfrak{r}$ is an extremal surface for the functional $\mathcal{W}$ if and only if $\mathfrak{p}$ is an extremal curve for $\int (1 + \kappa^2) \, ds$.

However, it was shown by Langer and Singer (1984) that there are infinitely many simple closed curves on S^2 that are critical points for $\int (1 + \kappa^2) \, ds$. It follows that there are infinitely many embedded Hopf tori that are critical points for $\mathcal{W}$. The stereographic projection of these tori are then embedded Willmore tori in $\mathbb{R}^3$.

Pinkall then shows that, *with one exception, the Willmore tori obtained in this way cannot possibly be obtained by stereographic*

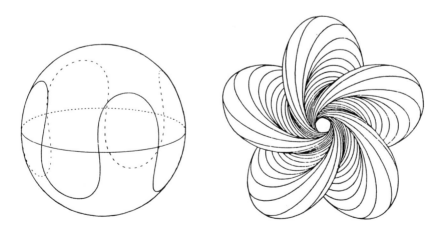

Fig. 7.12 Stereographic projection of embedded Hopf torus.

projection of minimal surfaces in S^3. The exceptional case is none other than the Clifford torus which projects to the '$\sqrt{2}$ torus' or its conformal equivalent. An example of a Willmore torus obtained in this way is given in Fig. 7.12, which shows the closed curve on S^2 and the corresponding torus. It can be shown, however, that with the exception of the Clifford torus, all Willmore tori that arise in this way cannot give an absolute minimum value to the Willmore functional.

A recent paper from Ferus and Pedit (1990) obtains, following a method due to Hsiang and Lawson (1971), a 1-parameter family of non-congruent minimal tori IH in S^4 (and hence Willmore tori in S^3) invariant under the action of S^1. They show that these include *all* minimal tori in S^4 invariant under circle action and hence *all* such Willmore tori in S^3. The case previously studied by Pinkall, namely Willmore tori invariant under the Hopf action on S^3, is a particular case of this more general theory. However the values of $\int H^2 \, dA$ for the resulting Willmore tori in $\mathbb{R}^3$ obtained by stereographic projection seem to be much higher than the value $2\pi^2$ given by the exceptional case of the Clifford torus.

A very recent and rather deep paper by Ferus *et al.* (1990) provides a general background for the differential-geometric classification of submanifolds. We remind the reader that if Σ is a compact Riemann surface and $f: \Sigma \to S^4$ is a conformal minimal immersion, then we say that f is *super-minimal* if at each point its curvature ellipse is a circle. Analytically this condition is equivalent to the relation $\langle f_{zz}, f_{zz} \rangle = 0$ where a subscript stands for differentiation with respect to a local holomorphic coordinate z on Σ. In this extensive paper the authors give a rather complete description of all *non-super-minimal* tori in S^4. This gives rise to a classification of all non-super-minimal Willmore tori in $\mathbb{R}^3$ of so-called polynomial type. However it is not known whether every non-super-minimal Willmore torus is of polynomial type. This is closely related to the problem: Can a non-super-minimal Willmore torus have umbilics?. At Oberwolfach in 1991 K. Voss showed that there are Willmore surfaces with rotational symmetry containing umbilics but these surfaces do not close.

7.10 OTHER RECENT CONTRIBUTIONS TO THE SUBJECT

Simon (1986) proved the existence of a minimizing immersion $f: T^2 \to \mathbb{R}^3$ with the property $\mathcal{W}(f) = \mathcal{W}(T^2)$. This implies that for all immersed tori we must have $\mathcal{W}(T^2) > 4\pi$.

Kusner (1989), working independently of the German school, published a paper in which he estimated the infimum of $\mathcal{W} = \int H^2 \, dA$ for each regular homotopy class of immersions into $\mathbb{R}^3$. In particular for

$\mathbb{R}P^2$ he provides an example of a Willmore surface with $\int H^2 \, dA = 12\pi$, a result already known to Pinkall. Kusner considers a path of immersed surfaces as a regular homotopy, and the path component, that is, the regular homotopy class of the immersion $M \to \mathbb{R}^3$ is denoted by $[M]$. His main result is the following.

Theorem (7.10.1) *The infimum $\mathcal{W}_{[M]}$ for $\mathcal{W}$ over any regular homotopy class $[M]$ of compact immersed surfaces $M \to \mathbb{R}^3$ satisfies $\mathcal{W}_{[M]} < 20\pi$. The infimum of $\mathcal{W}$ among compact surfaces of a given topological type $M \to \mathbb{R}^3$ is strictly less than $(<)$;*

8π *if M is orientable;*

12π *if M is non-orientable and $\chi(M)$ is even;*

16π *if M is non-orientable and $\chi(M)$ is odd.*

In 1986 I lectured on conformal geometry at a mathematics conference at the University of Southampton, England. An amplified version of this lecture appeared as Pedit and Willmore (1988). In this paper we generalized the definition of the Willmore functional to apply to the isometric immersion of a Riemannian manifold (M, g) into another Riemannian manifold (M', g'), and we calculated the Euler equation for the critical values of this functional. The results obtained agree with those of previous authors in the particular cases considered by them.

A very recent paper by Voss (1991) in the form of a pre-print considers an immersion $X: M \to M'$ of an m-dimensional Riemannian manifold isometrically into a Riemannian manifold M' of dimension $m + 1$. As particular cases he considers minimal hypersurfaces, Willmore surfaces, and elastic curves considered previously by Langer and Singer.

Finally we refer to a surprising result by Wente (1986). Hopf had previously shown that there cannot exist *embedded* in $\mathbb{R}^3$ a torus of constant mean curvature. Whether or not such a torus exists if one allows self-intersections remained for some years an open question. I am informed by Peter Dombrowski that if you read very carefully what Hopf has to say on the problem, one sees that Hopf considered the existence of such a torus on the whole unlikely but his feeling was insufficient to be deemed a definite conjecture on the matter. However, this in no way diminishes Wente's remarkable contribution, showing that such a torus exists.

It had previously been proved by Garsia and Ruedy (1962) that any compact Riemann surface can be conformally immersed in $\mathbb{R}^3$. In order to find such an immersion which, in some sense, is the optimum, we can ask for an immersion which minimizes the Willmore functional. This gives rise to the concept of 'constrained Willmore surface',

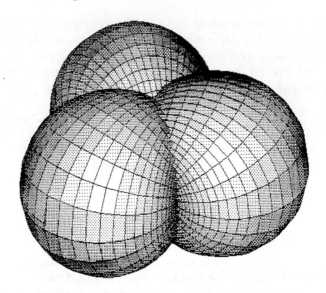

Fig. 7.13 Abresch torus — a constrained Willmore surface.

corresponding to the critical points of the Willmore functional when restricted to the space of all *conformal* immersions $f(M, g) \to \mathbb{R}^3$. This gives rise to a larger class of surfaces than the usual Willmore surface. It has been observed by J. Langer that any compact surface in $\mathbb{R}^3$ with constant mean curvature is a constrained Willmore surface. In particular Wente's constant mean curvature torus is such an example — see Fig. 7.13 due to U. Abresch.

The reader will see that this subject is a very flourishing research topic in modern Riemannian Geometry.

Further light is shed on the problem by a private communication from Dr. Alexander Bobenko of the Technical University, Berlin, who kindly sent me Figs 7.14 and 7.15 as examples of Willmore surfaces with a line of umbilics. Figure 7.14 is an example of a rotationary invariant Willmore surface. Parts of the surface lying above and below a certain plane are minimal surfaces in hyperbolic space, realized as the upper and lower half-space with the Poincaré metric. The line of umbilics is formed by the intersection of these surfaces with the plane. Figure 7.15 represents a Willmore surface which is a surface of revolution but not a minimal surface in hyperbolic space.

These pictures were obtained by Mattias Heil and Alexander Bobenko. Further details will be found in a forthcoming paper by M. Babich and A. Bobenko, 'Willmore tori with umbilic lines and minimal surfaces in hyperbolic space', pre-print SFB 288, Berlin 1992.

7.10 Other recent contributions to the subject

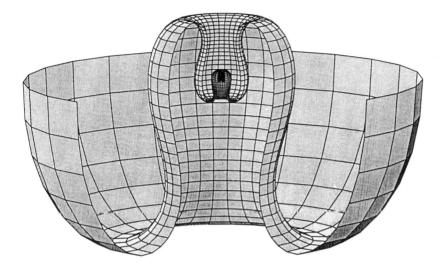

Fig. 7.14 Willmore surface with line of umbilics.

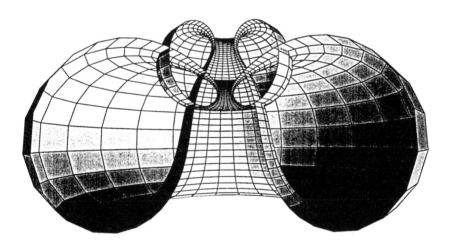

Fig. 7.15 Willmore surface of revolution, not minimal in hyperbolic space.

A recent paper by Palmer (1991) investigates the second variation of the Willmore functional by relating it to the second variation of area of the conformal Gauss map in the case when the surface is free from umbilics. A sufficient condition for instability is derived and examples are provided to show applicability. A second paper Palmer (1993) extends these ideas to surfaces with umbilics. A fourth order conformally invariant elliptic operator is derived which is the Jacobi operator for Willmore surfaces. This leads to a Morse-Smale type index formula analogous to the one for minimal submanifolds.

A brief survey on Willmore surfaces is given by Willmore (1992).

7.11 EXERCISES

1. Read (and understand!) Ferus *et al.*
2. Read (and understand!) Bryant (1984) and Bryant (1988).

Appendix: Partitions of unity

The essential idea underlying this concept is to provide a means of extending, globally over a manifold M in a consistent manner, differentiable objects which are defined locally in some neighbourhood of each point of M. Moreover, this construction is available for C^∞-manifolds but not for analytic manifolds. Since we find the topic complicated we proceed in steps.

Step 1

Consider the function $f: \mathbb{R} \to \mathbb{R}$ given by

$$f(x) = \begin{cases} e^{-1/x^2}, & x > 0, \\ 0, & x \leq 0. \end{cases}$$

Clearly this function is smooth, except possibly at $x = 0$. However, since $\lim_{x \to +0} x^{-1} e^{-1/x^2} = 0$, it is easily seen that f is differentiable at $x = 0$. Indeed $f'(0) = 0$. In a similar way it follows that $f''(0) = 0, \ldots, f^{(n)}(0) = 0$ and f is of class C^∞ on $\mathbb{R}$.

Step 2

Consider the interval $[a, b]$ where $0 \leq a < b$. We wish to construct a function $g(x)$ which is of class C^∞ on $\mathbb{R}$, is strictly positive on (a, b), and is zero on $(-\infty, a] \cup [b, \infty)$. Clearly this is achieved by defining $g(x)$ as

$$g(x) = f(x - a) f(b - x).$$

Step 3

Suppose now that we wish to construct a C^∞-function h on $\mathbb{R}$ which takes the value 0 for $x \leq a$, takes the value 1 for $x \geq b$ and satisfies $0 < h(x) < 1$ for $a < x < b$. Write

$$A = \int_{-\infty}^{\infty} g(x) \, dx,$$

and define h by

$$h(x) = \frac{1}{A} \int_{-\infty}^{x} g(t) \, dt.$$

Then h has the required properties.

Step 4

We obtain a result analogous to that of step 3. More specifically we obtain the following theorem.

Theorem (A.1) Let $B_\varepsilon(a)$ be an open ball in $\mathbb{R}^n$ of centre a and radius ε. Then there exists a C^∞-function $p(x)$ on $\mathbb{R}^n$ which is positive on $B_\varepsilon(a)$, has the value 1 on $\bar{B}_{\varepsilon/2}(a)$ and the value 0 outside $B_\varepsilon(a)$.

Using the function $f(t)$ defined in step 1, we let

$$\bar{g}(x) = \frac{f(\varepsilon - \|x\|)}{f(\varepsilon - \|x\|) + f(\|x\| - \tfrac{1}{2}\varepsilon)},$$

where $x = (x^1, \ldots, x^n)$, $\|x\| = \sqrt{(\Sigma_{i=1}^n((x^i)^2)}$.

Clearly $\bar{g}(x)$ is a C^∞-function, since the denominator cannot take the value 0. When $\|x\| \geqslant \varepsilon$ the numerator is zero, otherwise it is positive. When $0 \leqslant \|x\| \leqslant \varepsilon/2$ the value of $\bar{g}(x)$ is identically 1. Thus $\bar{g}(x)$ is C^∞, vanishes outside $B_\varepsilon(0)$, and is positive on its interior. If we define the function $p(x)$ by

$$p(x) = \bar{g}(x - a)$$

we see that $p(x)$ has the required properties.

We make a slight digression at this point, as the previous argument leads to the following theorem which is interesting in its own right.

Theorem (A.2) Let $F \subset \mathbb{R}^n$ be a closed set and let $K \subset \mathbb{R}^n$ be compact such that $F \cap K = \varnothing$. Then there is a C^∞-function $\sigma(x)$ whose domain is the whole of $\mathbb{R}^n$ and whose range of values is the closed interval $[0,1]$ such that $\sigma(x) \equiv 1$ on K and $\sigma(x) \equiv 0$ on F.

To prove the theorem let $B_i(a_i)$, $i = 1, \ldots, k$ be a finite collection of n-balls in $\mathbb{R}^n \setminus F$ so that $\cup_{i=1}^k B_i(a_i) \supset K$. That such a collection exists follows from the fact that $K \cap F = \varnothing$ and because K is compact. We apply the results of Theorem (A.1) to each $B_\varepsilon(a_i)$. Thus, for each i, let $p_i(x)$ have the property that it is positive on $B_\varepsilon(a_i)$, identically 1 $B_{\varepsilon/2}(a_i)$, and is zero outside $B_\varepsilon(a_i)$. Define a new function $\sigma(x)$ by

$$\sigma(x) = 1 - \prod_{i=1}^k (1 - p_i).$$

We claim that $\sigma(x)$ has the required properties. This follows because for each $x \in K$ there is at least one p_i with value 1 so $\sigma(x) = 1$, and hence σ is identically 1 on K. Moreover, outside $\cup_{i=1}^k B_\varepsilon(a_i)$ each p_i vanishes so $\sigma(x) = 0$, and since F lies outside this union, $\sigma \equiv 0$ on F. This completes the proof of the theorem.

Appendix

Step 5

Here we make four definitions and quote a standard theorem from topology.

Definition (A.3) *A covering $\{A_\alpha\}$ of a manifold M by subsets (not necessarily open) is said to be locally finite if each $m \in M$ has a neighbourhood U which intersects only a finite number of the sets A_α.*

Definition (A.4) *Let $\{A_\alpha\}, \{B_\beta\}$ be coverings of M. Then $\{B_\beta\}$ is called a refinement of $\{A_\alpha\}$ if each $B_\beta \subset A_\alpha$ for some α.*

Definition (A.5) *A space is σ-compact if it is the union of a countable number of compact sets.*

We note that any manifold is σ-compact because it is locally compact and because it admits a countable basis of open sets $O_1, O_2, \ldots$ such that each $\bar{O}_i$ is compact.

Definition (A.6) *A space which has the property that every open covering has a locally finite refinement is called paracompact.*

Now we appeal to the following standard result in general topology.

Theorem (A.7) *A locally compact Hausdorff space with a countable basis is paracompact.*

It follows as an immediate corollary that a manifold is paracompact. However, following the arguments used by Boothby (1986), we prove the following analogous result which is more suitable for subsequent applications.

Step 6

Theorem (A.8) *Let $\{A_\alpha\}$ be a covering of the manifold M by open sets. Then there exists a countable, locally finite refinement $\{U_i, \phi_i\}$ of coordinate neighbourhoods with $\phi_i(U_i) = B_3^n(0)$ for $i = 1, 2, \ldots$. Moreover, if $V_i = \phi_i^{-1}(B_1^n(0)) \subset U_i$ then $\{V_i\}$ is a covering of M.*

The reader is warned that the proof of this theorem is rather subtle and is best read with the aid of pencil and paper. One thing is immediately clear, namely that the choice of radii 1 and 3 is arbitrary.

We start with the countable basis of open sets $\{O_i\}$ with $\bar{O}_i$ compact, and define inductively a sequence of compact sets $K_1, K_2, \ldots$ as follows. We define $K_1 = \bar{O}_1$; assume now that $K_1, K_2, \ldots, K_i$ are defined, and let r be the first integer for which $K_i \subset \cup_{j=1}^r O_j$. Then we define K_{i+1} by

$$K_{i+1} = \bar{O}_1 \cup \bar{O}_2 \cup \cdots \cup \bar{O}_r = \overline{O_1 \cup O_2 \cup \cdots \cup O_r}.$$

Denote the interior of K_{i+1} by $\mathring{K}_{i+1}$ and note that it contains K_i. Then, for each $i \geq 2$ consider the open set

$$(\mathring{K}_{i+2} - K_{i-1}) \cap A_\alpha.$$

Around each point m in this set choose a coordinate neighbourhood $U_{m,\alpha}, \phi_{m,\alpha}$ lying inside the set and such that $\phi_{m,\alpha}(m) = 0$ and $\phi_{m,\alpha}(U_{m,\alpha}) = B_3^n(0)$. We define $V_{m,\alpha} = \phi_{m,\alpha}^{-1}(B_1^n(0))$ and observe that $V_{m,\alpha} \subset (\mathring{K}_{i+2} - K_{i-1})$. Now allow m, α to vary; we see that a finite number of the $V_{m,\alpha}$ covers $K_{i+1} - \mathring{K}_i$ which is itself closed, and we denote these by $V_{i,k}$, the suffix k serving as a label for this finite set. For each integral value i, the suffix k takes only a finite number of values, and this ensures that the collection $V_{i,k}$ is denumerable. We renumber these as $V_1, V_2, \ldots$ and denote by $(U_1, \phi_1), (U_2, \phi_2), \ldots$ the corresponding coordinate neighbourhoods containing them. We claim that these neighbourhoods satisfy the requirements of the theorem. For each $m \in M$ there is an index i such that $m \in \mathring{K}_{i-1}$ but by the method of construction only a finite number of the neighbourhoods U_j, V_j meet $\mathring{K}_{i-1}$. Therefore $\{U_i\}$ and $\{V_i\}$ are locally finite coverings refining $\{A_\alpha\}$. This completes the proof of the theorem.

We call the refinement U_i, V_i, ϕ_i of Theorem (A.8) a *regular covering by spherical coordinate neighbourhoods subordinate to the open covering* $\{A_\alpha\}$.

Step 7

We have now reached the stage of defining a partition of unity. We recall that the *support* of a function f on a manifold M is the closure of the set where f does not vanish. We write

$$\mathrm{supp}(f) = \overline{\{x \in M | f(x) \neq 0\}}.$$

Definition (A.9) A C^∞-*partition of unity on M is a collection of C^∞-functions $\{f_\gamma\}$ defined on M such that:*

(i) $f_\gamma \geq 0$ *on M;*
(ii) $\{\mathrm{supp}(f_\gamma)\}$ *form a locally finite covering of M;*
(iii) $\Sigma_\gamma f_\gamma(x) = 1$ *for every $x \in M$.*

Definition (A.10) A partition of unity is said to be *subordinate* to an open covering $\{A_\alpha\}$ of M if for each γ there is an A_α for which $\mathrm{supp}(f_\gamma) \subset A_\alpha$.

Step 8

We now state and prove the main theorem for the existence of partitions of unity.

Theorem (A.11) We can associate to each regular covering $\{U_i, V_i, \phi_i\}$ of M a partition of unity $\{f_i\}$ such that $f_i > 0$ on $V_i = \phi^{-1}(B_1(0))$ and $\text{supp} f_i \subset \phi^{-1}(\overline{B_2(0)})$. In particular every open covering $\{A_\alpha\}$ admits a partition of unity subordinate to it.

To prove the theorem, we see from Theorem (A.1) that for each i, we can construct a non-negative function $\bar{p}(x)$ on $\mathbb{R}^n$ which takes the value 1 on $B_1^n(0)$ and zero outside $B_2^n(0)$. The corresponding function p_i defined by

$$p_i = \begin{cases} \bar{p} \circ \phi_i & \text{on } U_i \\ 0 & \text{on } M \setminus U_i \end{cases}$$

is a C^∞-function defined on M. Moreover it has support in $\phi_i^{-1}(\overline{B_2^n(0)})$, has value 1 on $\bar{V}_i$, and is never negative. Since $\{V_i\}$ is locally finite covering, it follows that

$$f_i = \frac{p_i}{\sum_j p_j}, \qquad i = 1, 2, \ldots$$

are functions which have the required property. This completes the proof of Theorem (A.11).

The reader will be aware of the rather arduous task to reach and prove Theorem (A.11). Of course, if the manifold M is compact there is no need to use partitions of unity. However, in general, it is a powerful tool for extending globally over a manifold a differential geometric object which is defined locally in a neighbourhood of each of its points. For example, one of its earliest applications is to prove that every C^∞-manifold admits a C^∞-metric.

Bibliography

Abe, N. (1982). On generalized total curvatures and conformal mappings. *Hiroshima Math. J.* **12**, 203-7.

Ahlfors, L. and Sario, L. (1960). *Riemann Surfaces.* Princeton University Press.

Allamigeon, A. (1965). Propriétés globales des espaces de Riemann harmoniques. *Ann. Inst. Fourier* **15**, 91-132.

Allendoerfer, C. B. and Weil, A. (1943). The Gauss-Bonnet theorem for Riemannian polyhedra. *Trans. Amer. Math. Soc.* **53**, 101-29.

Avez, A. (1963). Applications de la formule de Gauss-Bonnet-Chern aux variétés à quatre dimensions. *C. R. Acad. Sci. Paris* **256**, 5488-90.

Berger, M. (1961). Sur quelques variétés d'Einstein compactes. *Annali di Math. Pura e Appl.* **53**, 89-96.

Berger, M., Gauduchon, P., and Mazet, E. (1971). *Le spectre d'une variété riemannienne.* Lecture Notes in Mathematics, 194, Springer, Berlin.

Besse, A. L. (1978). *Manifolds all of whose geodesics are closed.* Ergebnisse der Mathematik und ihrer Grenzgebiete, 93, Springer, Berlin.

Besse, A. L. (1987). *Einstein Manifolds*, Springer, Berlin.

Bishop, R. L. and Crittenden, R. J. (1964). *Geometry of Manifolds*, Academic Press, New York.

Bishop, R. L. and Goldberg, S. I. (1964). Some implications of the generalized Gauss-Bonnet theorem. *Trans. Amer. Math. Soc.* **112**, 508-35.

Blaschke, W. (1929). *Vorlesungen über Differentialgeometrie*, III. Springer, Berlin.

Boothby, W. (1986). *An Introduction to Differentiable Manifolds.* Academic Press, London.

Borel, A. and Serre, J. P. (1951). Détermination des p-puissances reduites de Steenrod dans la cohomologie des groupes classiques. Applications. *C. R. Acad. Sci. Paris* **233**, 680-2.

Bryant, R. (1984). A duality theorem for Willmore surfaces. *J. Diff. Geom.* **20**, 23-53.

Bryant, R. (1988). Surfaces in Conformal Geometry. *Amer. Math. Soc. Proc. Symp. Pure Maths.* **48**, 227-40.

Burstall, F. E. and Rawnsley, J. H. (1990). *Twistor Theory for Riemannian Symmetric Spaces.* Lecture Notes in Mathematics 1424, Springer, Berlin.

Carpenter, P., Gray, A., and Willmore, T. J. (1982). The curvature of Einstein symmetric spaces. *Quart. J. Math. Oxford*, (2), **33**, 45-64.

Chen, B.-Y. (1973a). An invariant of conformal mappings. *Proc. Amer. Math. Soc.* **40**, 563-4.

Chen, B.-Y. (1973b). On a variational problem on hypersurfaces. *J. Lond. Math. Soc.* **2**, 321-5.

Chen, B.-Y. (1973c). *Geometry of Submanifolds*. Dekker, New York.
Chern, S. S. (1944). A simple intrinsic proof of the Gauss–Bonnet formula for closed Riemannian manifolds. *Annals of Maths* **45**, 747–52.
Chern, S. S. (1955). An elementary proof of the existence of isothermal parameters on a surface. *Proc. Amer. Math. Soc.* **6**, 771–82.
Chern, S. S. (1956). On the curvature and characteristic classes of a Riemannian manifold. *Abh. Math. Sem. Univ. Hamburg* **20**, 117–26.
Chern, S. S. (1967). *Studies in Global Geometry and Analysis*. Mathematical Association of America, 16–58.
Chevalley, C. (1946). *Theory of Lie Groups*, I. Princeton University Press.
Courant, R. and Hilbert, D. (1962). *Methods of Mathematical Physics*, Vol. 2, Interscience, New York.
Damek, E. and Ricci, F. (1991). A class of non-symmetric harmonic Riemannian spaces. Rapporto Interno N. 12, Dipartimento di Matematica, Politecnico di Torino.
D'Atri, J. E. and Nickerson, H. K. (1969). Divergence-preserving geodesic symmetries. *J. Diff. Geom.* **3**, 467–476.
De Rham, G. (1952). Sur la réducibilité d'un espace de Riemann. *Comment. Math. Helvetici*, **26**, 341.
Deturk, D. M. and Kazdan, J. L. (1981). Some regularity theorems in Riemannian geometry. *Ann. Scient. Ec. Norm. Sup.* **14**, 249–60.
Do Carmo, M.(1976). Differential geometry of curves and surfaces. Prentice-Hall, Englewood Cliffs, NJ.
Dombrowski, P. (1979). 150 years after Gauss ... *Astérisque* **62**, société mathématique de France, Paris.
Duff, G. F. D. and Spencer, D. C. (1952). Harmonic tensors on Riemannian manifolds with boundary. *Annals of Maths* **56**, 128–56.
Eisenhart, L. P. (1926). *Riemannian Geometry*, Princeton University Press.
Fary, I. (1949). Sur la courbure totale d'une courbe gauche faisant un noeud. *Bull. Soc. Math. France*, **77**, 128–38.
Fenchel, W. (1929). Über die Krümmung und Windüng geschlossener Raumkurven. *Math. Ann.*, **101**, 238–52.
Ferus, D. and Pedit, F. (1990). S^1-equivariant minimal tori in S^4, and S^1-equivariant Willmore tori in S^3. *Math. Zeits.* **204**, 269–82.
Ferus, D., Pedit, F., Pinkall, U., and Sterling, I. (1992). Minimal Tori in S^4. *J. reine angew. Math.*, **429**, 1–47.
Garsia, A. and Ruedy, R. (1962). On the conformal types of algebraic surfaces of Euclidean space. *Comment. Math. Helv.* **7**, 49–60.
Gray, A. (1969). Vector cross-products on manifolds. *Trans. Amer. Math. Soc.* **141**, 465–504.
Gray, A. (1973). The volume of a small geodesic ball of a Riemannian manifold. *Michigan Math. J.* **20**, 329–44.
Gray, A. (1976). Curvature identities for Hermitian and almost-Hermitian manifolds. *Tohoku Math. J.* **28**, 601–12.
Gray, A. (1978a). Chern numbers and curvature. *Amer. J. Maths.* **100**, 463–76.
Gray, A. (1978b). Einstein-like manifolds which are not Einstein. *Geometrica Dedicata*, **7**, 259–80.

Gray, A. and Vankecke, L. (1979). *Almost-Hermitian manifolds with constant holomorphic sectional curvature.* Casopis pro pestovani matematiky, roc. 104, Praha.

Gray, A. and Willmore, T. J. (1982). Mean-value theorems for Riemannian manifolds, *Proc. Roy. Soc. Edin.* **92A**, 343–64.

Griffiths, P. and Harris, J. (1978). *Principles of Algebraic Geometry.* Wiley-Interscience, New York.

Gromov, M. (1983). Filling Riemannian manifolds. *J. Diff. Geom.* **18**, 1–147.

Gunning, R. C. and Rossi, H. (1965). *Analytic functions of several complex variables.* Prentice-Hall, Englewood Cliffs, NJ.

Hamilton, R. S. (1982). Three-manifolds with positive Ricci curvature. *J. Diff. Geom.* **17**, 255–306.

Hawley, N. S. (1953). Constant holomorphic sectional curvature, *Canada Math. J.* **5**, 53–6.

Hopf, H. (1925). Über die curvatura integra geschlossener Hyperflachen, *Math. Ann.* **95**, 340–67.

Hopf, H. and Rinow, W. (1931). Über der Begriff der vollständigen differential geometrischen. *Comment. Math. Helvetici* **3**, 209–25.

Hsiang, W. Y. and Lawson, H. B. (1971). Minimal submanifolds of low cohomogenity. *J. Diff. Geom.* **5**, 1–38.

Hurewicz, W. (1958). *Lectures on Ordinary Differential Equations.* MIT Press, Cambridge, Mass.

Igusa, J. (1954). On the structure of a certain class of Kahler manifolds. *Amer. J. Math.* **76**, 669–78.

Karcher, H., Pinkall, U., and Sterling, I. (1988). New minimal surfaces in S^3. *J. Diff. Geom.* **28**, 169–85.

Kervaire, M. (1960). A manifold which does not admit any differentiable structure. *Comment. Math. Helvetica* **34**, 257–70.

Kobayashi, S. and Nomizu, K. (1963). *Foundations of Differential Geometry*, Vols 1, 2. Wiley-Interscience, New York.

Kusner, R. (1989). Comparison surfaces for the Willmore problem. *Pacific Journal of Maths.* **138**, 317–45.

Langer, J. and Singer, D. (1984). Curves in the hyperbolic plane and mean curvature of tori in R^3 and S^3. *Bull. Lond. Math. Soc.* **16**, 531–4.

Lawson, H. B. (1970). Complete minimal surfaces in S^3. *Annals of Maths* **92**, 335–74.

Li, P. and Yau, S. T. (1982). A new conformal invariant and its application to the Willmore conjecture and first eigenvalues of compact surfaces. *Invent. Math.* **69**, 269–91.

Lichnerowicz, A., (1958). Géométrie des Groupes de Transformations. Dunod, Paris.

Lichnerowicz, A. (1963). *C. R. Acad. Sci. Paris*, **256**, 3548–50.

Lichnerowicz, A. and Walker, A. G. (1945). Sur les espaces riemanniens harmoniques de type hyperbolique normal. *C. R. Acad. Sci.* **221**, 394–6.

Milnor, J. M. (1950). On the total curvature of knots. *Annals of Maths* **52**, 248–57.

Milnor, J. M. (1956). On manifolds homeomorphic to the 7-sphere. *Ann. Math.* **64**, 394–405.

Newlander, A. and Nirenberg, L. (1957). Complex analytic coordinates in almost-complex manifolds. *Annals of Maths* **65**, 391–404.

Palmer, B. (1991). The Conformal Gauss Map and the Stability of Willmore Surfaces. *Ann. Global Anal. Geom.* **9**, 305–17.

Palmer, B. (1993). Second Variation Formulas for Willmore Surfaces. *The Problem of Plateau*, ed. T. M. Rassias. World Scientific Press, Singapore 221–8.

Parrots, S. (1987). *Relativistic Electrodynamics and Differential Geometry*. Springer, Berlin.

Pedit, F. and Willmore, T. J. (1988). Conformal geometry. *Sem. Fis. Univ. Modenna*, **XXXVI**, 237–45.

Pinkall, U. (1985). Hopf tori in S^3. *Invent. Math.* **81**, 379–86.

Pinkall, U. (1986). Inequalities of Willmore type for submanifolds. *Math. Zeits.* **193**, 241–6.

Pinkall, U. and Sterling, I. (86-29). Willmore surfaces. Max Planck Institute für Mathematik. preprint.

Pinkall, U. and Sterling, I. (1987). Willmore surfaces. *Math. Intelligencer*, **9**, 38–43.

Roberts, P. H. and Ursell, H. D. (1960). Random walk on a sphere and on a Riemannian manifold. *Phil. Trans. Roy. Soc. London*, Ser. A, **252**, 317–56.

Rodrigues, O. (1814–16). Paris, École Polytechnique Corresp. III, 162–82.

Ruse, H. S., Walker, A. G., and Willmore, T. J. (1961). *Harmonic Spaces*. Edizioni Cremonese, Roma.

Sakai, T. (1971). On the eigenvalues of the Laplacian and curvature of Riemannian manifolds. *Tohoku Math. J.* **23**, 589–603.

Saloman, S. (1984). Harmonic 4-spaces. *Math. Ann.* **269**, 168–78.

Salamon, S. (1989). *Riemannian geometry and holonomy groups*. Pitman Research Notes in Mathematics Series, **201**. Longman Scientific and Technical, Harlow, Essex.

Segigawa, K. and Vanhecke, L. (1986). Volume-preserving geodesic symmetries on 4-dimensional 2-stein spaces. *Kodai Math. J.* **9**, 215–24.

Shiohama, K. and Takagi, A. (1970). A characterization of a standard torus in E^3. *J. Diff. Geom.* **4**, 477–85.

Simon, L. (1986). Existence of Willmore surfaces. Miniconference, Canberra on geometry and partial differential equations. *Proc. Cent. Math. Anal., Australian National University*, **10**, 187–216.

Singer, I. M. and Thorpe, J. A. (1969). The curvature of 4-dimensional Einstein spaces. In *Global Analysis*, Papers in honour of K. Kodaira. Princeton University Press, pp. 355–65.

Steenrod, N. (1951). *The Topology of Fibre Bundles*. Princeton University Press.

Szabo, Z. (1990). The Lichnerowicz conjecture on harmonic manifolds. *J. Diff. Geom.* **31**, 1–28.

Thomsen, G. (1923). Über Konforme Geometrie, I: Grundlagen der Konformen Flächentheorie. *Abh. Math. Sem. Hamburg*, 31–56.

Totaro, B. (1990). Space curves with non-zero torsion. *International Journal of Mathematics* **1**, 109–17.

Vanhecke, L. (1981). A note on harmonic spaces. *Bull. Lond. Math. Soc.* **13**, 545-6.
Voss, K. (1991). Variation of curvature integrals. Preprint.
Walker, A. G. (1947). Note on pseudo-harmonic functions, II. *J. Lond. Math. Soc.* **22**, 101-4.
Wang, H. C. (1952). Two-point homogeneous spaces. *Ann. Math.* **55**, 177-91.
Warner, F. W. (1983). *Foundations of Differentiable Manifolds and Lie Groups.* Springer, New York.
Weiner, J. L. (1978). On a problem of Chen, Willmore et alia. *Indiana University Math. J.* **27**, 19-35.
Wente, H. (1986). Counter-example to a conjecture of H. Hopf. *Pacific Journal of Mathematics* **121**, 193-243.
Weyl, H. (1918). Reine Infinitesimalgeometrie. *Math. Zeits.* **2**, 384-411.
White, J. H. (1973). A global invariant of conformal mappings in space. *Proc. Amer. Math. Soc.* **38**, 162-4.
Whitehead, J. H. C. (1932). Convex regions in the geometry of paths. *Quart. J. Math. Oxford* **3**, 33-42.
Willmore, T. J. (1950). Mean-value theorems in harmonic Riemannian spaces. *J. Lond. Math. Soc.* **25**, 54-7.
Willmore, T. J. (1959). *An Introduction to Differential Geometry.* Oxford University Press.
Willmore, T. J. (1965). Note on embedded surfaces. Ann. Stiint. Univ. "Al. I. Cuza", Iasi. Ia. Mat., 493-6.
Willmore, T. J. (1971). Mean curvature of Riemannian immersions. *J. Lond. Math. Soc.* (2) **3**, 307-10.
Willmore, T. J. (1982). *Total Curvature in Riemannian Geometry.* Ellis Horwood, Chichester.
Willmore, T. J. (1992). A survey on Willmore surfaces. In *Geometry and Topology of Submanifolds*, (eds F. Dillen and L. Verstraelen). World Scientific, 11-16.
Willmore, T. J. and Hitchen, N. (eds) (1984). *Global Riemannian Geometry.* Ellis Horwood, Chichester.
Wolf, J. A. (1967). *Spaces of Constant Curvature.* McGraw-Hill, New York.
Yano, K. (1970). *Integral Formulas in Riemannian Geometry.* Dekker, New York.
Yano, K. and Bochner, S. (1953). *Curvature and Betti Numbers.* Annals of Math. Studies, No 32, Princeton University Press.
Yano, K. and Ishihara, S. (1973). *Tangent and Cotangent Bundles.* Dekker, New York.
Zalcman, L. (1973). Mean-values and differential equations. *Israel Journ. Math.* **14**, 339-52.

Index

adapted frame 125
almost-complex 159
anti-self-dual 197
arc length 65

Bianchi identities 42, 53
Bochner's lemma 98
Bryant, R. L. 287

canonical involution 221
Cartan's lemma 33
catenoid 135
Cayley numbers 163
change of basis 25
chart 3
Chern form 174, 177
Chow's theorem 169
Christoffel symbols 57
Clifford torus 300
Codazzi equations 123, 128, 186
Codazzi tensor 123
coderivative 87
commutative metrics 253
completeness 72
complex structure 152
complexification 153
conformal connexions 110
conformal metrics 105
conformal volume 291
connexion 39, 49
contact element 18
contraction 25
contravariant tensor 22
cotangent bundle 14
covariant differentiation 38, 45
covariant tensor 22
curvature form 50
curvature tensor 42, 47

D'Atri space 257
decomposable tensor 23
differentiable manifold 2, 3
differentiable map 5
differential form 28
differential ideal 32
differential of a map 10

discriminant function 206
distance function 206
distribution 18
dual map 10

Einstein spaces 104
equation of Gauss 118
equation of Weingarten 119
Euclidean Laplacian 247
Euler-Poincaré characteristic 146
exponential map 67
exterior derivation 28
exterior product 23

Fenchel's theorem 130, 271
frame bundle 189
Frenet-Serret 129
Frobenius theorem 19, 33

Gauss equations 122, 127
Gauss lemma 69
Gauss map 135
Gauss-Bonnet global 146-51
Gauss-Bonnet local 142-6
Gauss-Kronecker curvature 139
Gaussian curvature 134
geodesic 65
geodesic curvature 131
geodesic polars 210
germs 6

harmonic form 90
harmonic metric 223
helicoid 135
Hermitian metric 155
holomorphic sectional curvature 174
Hopf cylinder 297
Hopf fibration 169

imbedding 14
immersion 14
implicit function theorem 14
integrability 162, 192, 195
integrable distribution 19

Index

Interior multiplication 26
isotropic immersions 187
inverse function theorem 14
involutive distribution 19

Jacobi field 213–17
Jacobi identity 18

k-harmonic 267
k-stein 257
Kähler metrics 166, 170
Koszul connexion 43

Laplacian 82, 90
Ledger's formula 232
Lichnerowicz's formula 239
Lie bracket 17
Lie derivative 82

Maxwell's equations 93
mean curvature 134
mean curvature vector 119
metric 56
Milnor's theorem 131
minimal immersion ($\mathbf{H} = 0$) 151

Nijenhuis tensor 162
non-singular map 10
normal bundle 117
normal connexion 119
normal coordinates 72, 202

orientation 13, 34

parallel transport 46
partition of unity 37, 305
Pinkall's inequality 284
Pizetti's formula 247
Poincare lemma 92
Principal curvatures 141
principal directions 141
product manifolds 4, 56
projective spaces 37, 168, 175
pull-back 31

quaternions 163

recurrent spaces 113
Ricci equations 124, 128
Ricci form 173
Ricci tensor 64

scalar curvature 64
Schur's theorem 62
second fundamental form 118
sectional curvature 60
Singer–Thorpe basis 241
skew-symmetric tensors 23
star operator 84
submanifold 14
super-Einstein 251
super-minimal 300
symmetric tensors 23

tangent bundle 14
tangent vectors 5, 6
tensor field 27
tensor product 21
tensorial form 54
theorema egregium 134
torsion form 50
torsion tensor 41, 47
torsional derivative 55
Totaro's theorem 131
twistor 180, 184

umbilic 136

vector field 16
visual volume 291
volume density function 217
volume form 79

W-surface 136
Walker derivation 164
wedge product 23
Wente's torus 301
Weyl tensor 103, 196–9
Willmore conjecture 276
Willmore dual 288
Willmore surface 280–4